MW01027268

AGENT-BASED MODELING AND NETWORK DYNAMICS

Agent-Based Modeling and Network Dynamics

Akira Namatame
National Defense Academy, Japan

Shu-Heng Chen
National Chengchi University, Taiwan

Great Clarendon Street, Oxford, OX2 6DP,
United Kingdom

Oxford University Press is a department of the University of Oxford.
It furthers the University's objective of excellence in research, scholarship,
and education by publishing worldwide. Oxford is a registered trade mark of
Oxford University Press in the UK and in certain other countries

First Edition published in 2016

Reprinted 2016, 2017
Impression: 3

Published in the United States of America by Oxford University Press
198 Madison Avenue, New York, NY 10016, United States of America

British Library Cataloguing in Publication Data
Data available

Library of Congress Control Number: 2015944353

ISBN 978–0–19–870828–5

Printed and bound by
CPI Group (UK) Ltd, Croydon, CR0 4YY

Abstract

Social environments influence much of what people do. To be more precise, individual behavior is often influenced by social relations with others. Social influence refers to the behavioral change of individuals affected by other people. Individual behavior is often governed by formal and informal social networks. Networks are created by individuals forming and maintaining relationships that, in turn, influence individual behavior. Hence, considering individual behavior within the context of an explicit network structure gives us a model that helps to explain how people connect with their peers, and why they become increasingly similar to their peers over time.

Agent-based modeling has grown out of many disparate scientific fields, including computer science, physics, biology, economics, and other social sciences. In conjunction with the increasing public interest in complex networks, there has been a coalescence of scientific fields interested in agent-based approaches and network science. Each of these fields brings important ideas to the discussion, and a comprehensive understanding seems to require a synthesis of perspectives from all of them. These are ideas that have traditionally been dispersed across multiple disciplines. Thus, understanding highly connected individual behavior requires a set of ideas for reasoning about network structures, strategic behavior, and feedback effects across large populations.

This text attempts to bridge such multidisciplinary approaches in the ultimate hope that it might lead to a better understanding of complex social and economic problems. By constructing agent-based models, we develop methods and tools useful for exploring the deeper mechanisms underlying socioeconomic dynamics. We begin by extrapolating theories of structure and behavior, two bodies of theory that will serve as the foundation for agent-based network models. With this set of ideas in mind, we introduce some of the main topics to be considered in the book, and the means by which these topics articulate the mechanisms of socioeconomic dynamics.

This book also serves as an introduction for a general reader to socioeconomic systems. Its breadth and readability make it especially useful as a starting point for those who wish to dig deeper into the socioeconomic systems from any academic background and level. It is richly peppered with facts and simulation results aimed at situating the models discussed within actual socioeconomic systems. Hardware developments will soon make it possible to construct large-scale models, with agents numbering in the millions and hundreds of millions. It will be argued that the main impediment to creating empirically relevant agents on this scale is our current lack of understanding regarding the realistic behavior of agents. This epistemic bottleneck obfuscates efforts to write the rules dictating agent behavior, which is the primary challenge faced by researchers studying such models. Network theory is the study of network structures, whereas agent-based model theory studies models of individual behavior in settings where the outcome depends on the behavior of others. The agent approach to networks is at the forefront of such models.

The authors are deeply indebted to several people. The contributors were our friends and close colleagues, along with colleagues known to us only by name, and the authors of the books and papers to which we refer in the text. We thank those colleagues who provided stimulating academic discussion and debate provoked by the various readings and ideas, including Professors Bikas K. Chakrabarti, Yuji Aruka, Dirk Helbing, Thomas Lux, and Frank Schweitzer. We also thank our wonderful students at the National Defense Academy and National Chengchi University, who provided a quiet retreat from the pressures of teaching and administration during the later editing stages.

Finally, we want to thank all of you who take the time to read this book. We will be grateful if you gain as much from reading it and thinking about it as we have from writing it.

Contents

1
Introduction

1.1 Agent-Based Modeling: A Brief Historical Review

1.1.1 From Equations to Networks

The book is about the integration of agent-based modeling and network science. Agent-based modeling in social sciences has a long history, while in the early days it was recognized by different terms in different disciplines, for example, *individual-based modeling* in ecology (Grimm and Railsback, 2005), *actor-based modeling* in sociology (Macy and Willer, 2002), and *interaction-based modeling* in social economics (Durlauf and Young, 2004). In many disciplines, before the advent of agent-based modeling, the dominant approach in modeling social phenomena was *equation-based modeling*, frequently characterized by differential equations or systems of differential equations.

The famous examples are the Lokta-Volterra equation in ecology, the Kermack-McKendric equation, more popularly known as the SIR (standing for susceptible-infected-recovered) model in epidemics, and the Bass equation in marketing (Bass, 1969). Those equations describe a dynamic or evolutionary process in a space. The Lokta-Volterra equation describes the evolution of the population of different species in an ecological system, such as in the Amazon River, in the Sub-Saharan Desert, or in the Chernobyl nuclear disaster area. On the other hand, the Kermack-McKendric equation describes the diffusion of the epidemics or pandemics from one single individual to a large part of society. The Bass equation is the application of the Kermack-McKendric equation to the diffusion of new products or technology. While these processes can be meaningfully pictured in a spatial context, the spatial concern is often assumed away or abstracted away by the much simpler equation-based modeling. Therefore, it leaves a general question for social scientists: can a space-free thinking be good enough? A possibly negative answer to this question is the reason that we see the burgeoning and the development of the agent-based modeling methodology in social sciences.

What distinguishes agent-based modeling from the conventional equation-based modeling lies in the spatial factor or the geographical specificity. However, the significance of the spatial concern is not limited to just spatial mobility or transportation (Liedtke, 2006). The spatial specificity turns out to be another attribute of agents, which not only introduces a heterogeneity of agents in a natural way, but also helps define and

Agent-Based Modeling and Network Dynamics. First Edition. Akira Namatame and Shu-Heng Chen.

operate various interaction rules. Hence, the ultimate goal of including the spatial factor in agent-based modeling is to enable us to have an explicit expression of the feasible interactions among heterogeneous agents.

Hence, through various spatial models, interaction and heterogeneity (including interaction platforms and prototypes) as the essences of agent-based models are neatly encapsulated. Once we have come to this realization, it is not hard to see that spatial modeling is too limited for capturing various interactions interesting social scientists; after all, many social interactions are beyond the common sense of the Euclidean geometry. Social scientists need a more extended notion of distance, neighborhood, and locality than the Euclidean geometry can offer. Therefore, to enhance the expressive power of our experiences with interactions, a more general geometry (map) for *social locations*, rather than just the *physical locations*, of various agents is indispensable. However, compared to a map of a campus, a city, or a country, maps regarding humans, from a genome map of humans to a genealogic tree of a family, are harder to draw and require lots of creative thinking. Yet the map reflecting a general social structure, a specific group of people, or even just an individual's surrounding is more demanding.

Graphs and networks are currently the tools typically used to make these otherwise very difficult maps partially visualizable, as if we have a social network represented by a mathematic graph or its various hierarchical and multi-level extensions. The basic constitutes of the graph, i.e., nodes and links, can be endowed with more flexible interpretations so that the constructed graph or network can powerfully represent various geometries of social relations (Chen, 2006).

1.1.2 Networks as Checkerboards: Cellular Automata

When did we see the first application of agent-based modeling in social sciences? This question can be debatable, but few will doubt that the *checkerboard model* proposed by Sakoda (1971) and Schelling (1971) is one of the origins.[1] It is then clear to see that spatial specificity, as an essential element of agent-based modeling, was already there. Alternatively put, right from the beginning of the literature on agent-based modeling in social sciences and humanities, the network element was already there, and the first network employed in this literature is the checkerboard or the lattice model.

In fact, the checkerboard model was much earlier known as cellular automata in mathematics and computer science (see more in Section 2.4). In 1975, Albin first introduced cellular automata to economics and social sciences (Albin, 1975). It then became clear that what can be considered as the earliest integration of agent-based modeling with networks is this cellular automata tradition (the CA tradition) (Chen, 2012). As will be reviewed in this book, the CA tradition has accompanied the early development of agent-based models throughout the last two decades. Specifically, it has been applied

[1] Both of these articles are published in the inception issue of the *Journal of Mathematical Sociology* in 1971. After 40 years, the journal has a special issue as a celebration of the four-decade-long research on the micro–macro link in sociology (Raub, Buskens, and Van Assen, 2011).

to generate a new class of models, known as *spatial games*, which gives one of the illustrations of the significance of spatial structure or social structure in social behavior (Nowak, 2006).

1.1.3 From Networks to Agents: Autonomous Agents

Although the CA model gave the initial impetus for the early development of agent-based models, this spatial or network concern had only limited effect on further development of them. One of the reasons is that the lattice (checkerboard) structure is not an appealing interaction platform for many other applications, and the second reason is that in late 1980s, various tools for constructing artificial agents become available, which inevitably placed more research focuses on *agents* rather than on *networks*.

Among these tools, computational intelligence is the one which has the greatest influence (Chen, 2008). Computational intelligence is a consortium of fuzzy logic, artificial neural networks, evolutionary computation, and swarm intelligence. These tools are linguistically, neuroscientifically, and biologically oriented; they together give a powerful engine to develop a stronger notion of artificial adaptive agents, known as autonomous agents. In addition to this, learning models from psychology and decision science, such as reinforcement learning, decision trees, and fast-and-frugal heuristics, also contribute quite significantly to this development (Chen, 2013). Hence, what we have seen from the late 1980s and throughout the 1990s is the efforts focusing on the design of artificial adaptive agents, who are able to learn, adapt, and interact with others. This certainly levels up the functionality of agents or automata in lattice models and formally distinguishes agents in the agent-based models from atoms in the statistical-mechanical models. Nevertheless, interactions in this new stream of agent-based model became space-free or network-free.

The models of artificial adaptive agents or autonomous agents care less of how interaction happens but more on how it ends. For example, social learning or social influence is one of the most frequently-seen social interactions; in evolutionary computation it is normally implemented in the vein of the selection principle or the survival-of-the-fittest principle. Unlike replicate dynamics in which imitation or reproduction is done on the population base, in evolutionary computation it is implemented on the individual base (the agent base) through various selection schemes, such as the roulette wheel selection or the tournament selection. Interactions operated with these schemes only depend exclusively on a kind of random bumping without reference to any social structure. Therefore, implicitly, it requires either a fully-connected network or a random network, depending on whether the link is weighted or binary.

1.1.4 From Agents to Networks: Network Science

Coming to the twenty-first century, while a large group of agent-based models is still space-free or network-free, the fast development of network science has contributed to "bringing back" the network structure to the agent-based models. The modern network science begins in 1998 when the small-world network (the Watts-Stogatz network) was proposed (Watts and Strogatz, 1998); in the following year, 1999, the scale-free network (the Barabási-Albert network) was proposed (Barabasi and Albert, 1999). The essence of these two contributions and many follow-up studies suggest some generation procedures of networks and formalize the research questions on the formation of networks or the dynamics and evolution of networks. It then becomes clear that this issue can be accessible for agent-based models.

In fact, the two proposed network generation procedures, either the Watts-Stogatz one or the Barabási-Albert one, can both be given an agent-based interpretation. For social scientists, the only question is whether the behavioral rule (the linkage or the connection rule) assumed for the agents (the nodes) are socially or economically sensible. The answer to this question distinguishes the social and economic networks from other physical or mathematical networks. Economists, for example, will be more interested in knowing under what circumstance randomly rewiring or randomly connecting can be a boundedly rational behavior. Clearly, if agents can choose the connection based on their free-will, they would tend to consider each such connection as a welfare-enhancing opportunity. If so, random connection in general is not economically sensible. On the other hand, the preferential attachment in the Barabási-Albert model may be satisfied with a sound economic behavioral principle, such as the herding effect or information cascade; nonetheless, without a concrete context, no network formation rules can be sensible justified.

1.1.5 Networks and Games

The game-theoretic approach for network formation attempts to provide a general framework to understand the emergence of various network topologies from a viewpoint of deliberating economic calculations (Jackson and Wolinsky, 1996). However, its development relies on very stringent assumptions, specifically, perfect information on costs of links and values of the networks, which largely restricts its usefulness and relevance, apart from providing minimum theoretical light.

On the other hand, if we treat network formation rules individually and leave them to evolve over time, then it is immediately clear that agent-based models can make themselves available as an alternative network-generating mechanism. Therefore, in the 2000s, the integration of agent-based modeling and network science has a new dimension. In addition to making the network topology an additional attribute of agents and as part of the interaction rules, as seen in the checkerboard or lattice model, there is room for developing agent-based models of networks. As said in the very beginning, this book is about the integration and it will cover both of these dimensions, to be further detailed in Section 1.2.

While the purpose of this book is provide an integration of agent-based modeling and the network science, in many parts of the book, we try to bring together networks and games. This is probably one of the most exciting progresses in agent-based modeling over the last three decades. Certainly, it does not mean that without agent-based modeling these two research areas cannot be tied together, but, as we shall see in this book, agent-based modeling does allow us to explore the richness of the space synergizing networks and games. On the one hand, agent-based modeling advances our understanding of the effect of network topologies on game-like strategic behavior, and, on the other hand, it advances our understanding of the formation of various network topologies through the lens of the game-theoretic approach.

Let us illustrate this point with prosocial behavior, a long-standing research subject. Why do people choose to cooperate, to be kind, even under the presence of the high temptation to defect?[2] Despite various possible competitive explanations offered before the early 1980s, *social structure* in the manifestation of spatial configurations and network topologies has never been considered as an option. In this book, we shall see how a substantial progress in *spatial reciprocity* or *network reciprocity* has been made since the early 1990s (Albin, 1992; Nowak and May, 1992).

In this line of research, two major types of games are involved: one is the prisoner's dilemma game and the other is the public goods game. The two games are used because the former has sometimes been understood as a pairwise game only, and the latter is introduced as an effort to extend the two-person pairwise network games into truly N-person network games.[3]

The ignorance of N-person prisoner's dilemma games makes the development of the literature on the network games seemingly less coherent; after all, the switch from the two-person prisoner's dilemma game to the N-person public goods game is not smooth and inevitably leave a *hole*. Nevertheless, the progress made over the last two to three decades is not just to show that network topologies can be cooperation-enhanced, but also to pinpoint what make them matters. A pile of literature indicates that *clustering* is the key. The route from the regular network (lattice), to small-world network, to scale-free network, then further to degree distribution, to degree–degree correlation, and then to discovery of clustering in various forms, from triangles to cliques, makes this scientific navigation almost like a scenic drive specifically for those young talents.

1.1.6 Networks and Markets

Earlier we mentioned that the integration of agent-based modeling and network science had a new face after the late 1990s, because the formation algorithms were proposed by

[2] For example, in the preface of the book *Evil Genes*, Jim Phelps wrote,

> "...[Humans] are remarkably good. This is not just a wishful statement. Sophisticated brain science ... shows how deeply our tendency to trust and cooperate with others is rooted in brain anatomy and function–even when such behaviors place us at direct personal risk." (Oakley (2007), p. 20).

[3] This can be quite evidently seen in many similar passages as in the following quotation.

> "However, the prisoner's dilemma game is *different from* the public good games in the sense that the former describes pairwise interactions but the latter are played by groups of individuals. (Rong, Yang, and Wang (2010), p.1; Italics added.)"

What has been overlooked by a substantial amount of the literature is the existence of the N-person prisoner's dilemma game, in which the public good game can be shown as a special case of the N-person prisoner's dilemma game. For a formal definition of the N-person prisoner's dilemma game, the interested reader is referred to Coleman (2014, Section 8.7). The payoff matrix defined there is more general than the usual public good game. In economics, the N-person prisoner's dilemma game has been used to model oligopolistic competition (Chen and Ni, 2000).

physicists and, later on, extended by agent-based social scientists (Skyrms and Pemantle, 2000; Eguiluz et al., 2005). While the latter studies are initially motivated by the social relations characterized by games, empirical studies accumulated earlier provide another impetus for reflecting on the network formation processes (Granovetter, 1973; Kranton and Minehart, 2001; Montgomery, 1991). Instead of addressing game-theoretical questions, the empirically-inspired studies attempt to address network formation in each domain of real life, such as networks in markets, from goods-and-service markets and labor markets to financial markets. Agent-based models of these empirical networks are developed in the 2000s in parallel to the agent-based models of network games. They together consolidate both the theoretical and empirical foundations of the integration of agent-based modeling and network science.

1.2 Agent Network Dynamics

On the one hand, our thoughts, feelings, and actions depend to a considerable degree on social context. That is, our social environment influences much of what we do, think, and feel. On the other hand, our behavior contributes to the social environment of which we are a part. Social interdependence is defined by the interactions among individuals in a society, and it can be understood to exist when one individual's behavior depends on another's. Social interdependence is a fundamental way by which individuals are connected. Social interdependence concerns not only individual strategies and interactive patterns that facilitate or inhibit cooperation, but also the formation of social structures, the emergence of solidarity, mutual assistance networks, power structures, and trust networks.

Most individual interactions with the rest of the world happen at the local, rather than the global, level. People tend not to care as much about the entire population's decisions as they do about the decisions made by their friends and colleagues. At work, for example, individuals are likely to choose technology that is compatible with the technology used by those with whom they collaborate directly, regardless of whether it is universally popular. When such phenomena are analyzed, the underlying social network can be considered at two different levels: at one level, the network is a relatively amorphous collection of individuals, and particular effects are perceived as forming part of an aggregate; at another level, however, the network has a fine structure and an individual's behavior is partly determined by that individual's neighbors in the network.

What we can observe is a growing awareness and adoption of a new innovation that is visible as an aggregate across an entire society. What are the underlying mechanisms that lead to such success? Standard refrains are often invoked in such situations: the *rich get richer*, *winners take all*, a small advantage is magnified in a *critical mass*, and new ideas *go viral*. Nevertheless, the rich do not always get richer, and initial advantages do not always lead to success. To understand how this diffusion process works, and how it is manifest in the interconnected actions of massive numbers of individuals, we need to study the network dynamics of connected behavior. Simple rules concerning individual behavior can result in complex properties at the aggregate level. So-called emergent behavior at the macroscopic level influences individual behavior and interactions at the microscopic level.

We are interested in answering the following questions: How do agents, who interact and adapt to each other, produce the aggregate outcomes of interest? How do we identify the micro behavior that contributes to consistency in the behavior of interest at the macroscopic level? These questions raise other questions about how socioeconomic systems work and how they can be made to work better. To answer these questions, we may need to precisely identify the micro behavior that produces the macroscopic behavior of interest. People constantly interact with each other in different ways for different purposes. Somehow, these individual interactions exhibit a degree of coherence at the macroscopic level, suggesting that aggregates have structure and regularity. The individuals involved may have a limited view of the whole system, but their activities are coordinated to a large degree, often exhibiting so-called emergent properties—properties to the system that do not exist in its individual components. Emergent properties are not only the result of individual behavior; they are also the result of the interactions among them.

This approach to social networks formalizes the description of networks of relations in a society. In social networks, individuals can be represented as nodes, and relations or information flows between individuals can be represented as links. These links represent different types of relations between individuals, including the exchange of information, the transfer of knowledge, collaboration, and mutual trust. Social networks depend on social relationships, which, in turn, shape the relations between individuals. Networks are constituted individuals who form and maintain mutual relationships that, in turn, influence individual behavior. In this way, individual behavior is subject to an explicit network structure that conforms to a model that can help us understand how people are connected to their peers, and why they come to resemble them over time.

Individual behavior is often governed by these formal and informal social networks. To be more precise, individual behavior is often influenced by the social network of which the individual is a part. For instance, when we form an opinion about a political issue, we are likely to be influenced by the opinions of our friends, family, and colleagues. Moreover, we tend to influence the political opinions held by many of them as well. Social influence refers to the behavioral change in individuals insofar as others affect it in a network. Social influence is an intuitive and recognized phenomenon in social networks. The degree of social influence depends on several factors, such as the strength of relationships between people in the network, the network distance between users, temporal effects, and the characteristics of the network and the individuals in it. Much sociological research has been devoted to showing how various forms of social influence shape individual action (Marsden and Friedkin, 1993). One fundamental point is that, in a network setting, we should evaluate individual actions not in isolation, but with the expectation that the network will react to what people do. This means that there is considerable subtlety to the interplay of causes and effects. Moreover, the influence of the network is present regardless of whether its members are aware of the network. In a large and tightly interconnected group, the individuals comprising it will often respond in complex ways that are apparent only at the level of the entire population, even though these effects may come from implicit networks that are not directly apparent to the individual.

In many cases, the topic of interest depends on certain topological properties of networks. The majority of recent studies involve the following two key lines of research: the first involves determining the values of important topological properties in a network that evolves over time, and the second involves evaluating the effects of these properties on network functions. Important processes take place within a social network, such as the diffusion of opinions and innovations, and these processes are mostly influenced by the topological properties of the social network. In turn, the opinions and practices of individual agents exhibit a clear impact on the network's topology, as when, for example, conflicting opinions lead to the breakup of a social interaction.

The specific pattern of connections defines the network's topology. For many applications, it is unnecessary to model the exact topology of a given real-world network. Social networks are not always rigid structures imposed on us. Often, we have considerable control over our personal social relations. Our friends may influence us in forming our political opinions, for instance, but we are also to a large extent free to choose our friends. Moreover, it is likely that these same opinions are a factor in choosing our friends. Thus, social networks and individual behavior within social networks develop interdependently–they *coevolve*, in other words. Hence, our political opinions coevolve according to our social network and our individual characteristics, a process that holds for many other types of opinions and behavior. With respect to social networks, a lot is known about measuring their structural properties. Lately, such work has concentrated on the rules of network formation and on the relation between the process of network creation and the structure of the resulting network. However, little is known about the dynamics of the processes occurring in networks, and to what degree these processes depend on the network properties. Two questions are important here: What structure can be expected to emerge in a social network, and how will behavior be distributed in that network? These are central questions for social science and network science. Future research is directed at understanding how the structure of a social network determines the dynamics of various types of social processes occurring in the network.

The mutual relationship between the individual and the network results in the self-organization of social systems. Social relations depend on and, in turn, influence the relationships between individuals and groups within society. Of chief interest is the manner by which the structure of a social network determines the dynamics of various types of emergent properties occurring within that network. For emergence at the social level, patterns of social interactions are critical. Much effort has been devoted to understanding how the behavior of interacting agents is affected by the topology of these interactions. Various models have been proposed to study how locally engendered rules lead to the emergence of global phenomena. These topics are the predominant concern of statistical physicists, and they have converged with the social sciences to form the discipline of socio-physics (Ball, 2004).

Most networked systems seen in real life have associated dynamics, and the structural properties of such networks must be connected to their dynamic behavior. In general, each node in a network undergoes a dynamic process when coupled with other nodes.

The efficiency of a communication network depends on communication paths, and these are dictated by the underlying network topology. In this way, the network structure determines to a large extent the possibility of a coherent response. Thus, the topology of a network remains static despite dynamic changes to the states of the nodes comprising it. In this respect, one question of particular significance is whether there is any correlation between the network's resistance to small perturbations of dynamic variables and the specific arrangement of the network's connections. A network is said to be stable when perturbations decay rapidly, such that they are unable to spread to the rest of the network. Stability of this sort is necessary for networked systems to survive in the noisy environment that characterizes the real world.

Here, the topology of the network itself is regarded as a dynamical system. It changes in time according to specific, and often local, rules. Investigations in this area have revealed that certain evolutionary rules give rise to peculiar network topologies with special properties. The interplay between the topology and the network dynamics raises many questions. It is clear that in most real-world networks, the evolution of the topology is invariably linked to the state of the network, and vice versa. For example, consider a traffic network. The topology of such a network influences its dynamic state, such as the flow and density of each node and link. However, if traffic congestion is a common occurrence for some links, it is probable that new links will be built to mitigate the traffic on the congested links. A high load for a given node or link can also cause component failures that have the potential to sever links or remove nodes from the network. On a longer timescale, a high load is an incentive for the installation of additional connections to alleviate the load. Consequently, a feedback loop between the state and topology of the network is formed, and this feedback loop can lead to a complicated mutual interaction between a time-varying network topology and the network dynamics. Certain topological properties strongly impact the network dynamics, and topological network properties should be derived from the dynamics according to which the networks are created. Until recently, these two lines of network research were generally pursued independently. Even where strong interaction and cross-fertilization are evident, a given model would either describe the dynamics *of* a certain network or the dynamics *on* a certain network (Gil and Zanette, 2006).

Agent network dynamics is another approach to the study of complex dynamic processes. Agent networks are represented by graphs, where nodes represent agents, and edges represent the relationships between them. Links can be directed or undirected, and they can be weighted or unweighted. Agent networks can be used to create systems that support the exchange of information in real communities, for instance. Agent systems are joined according to the network topology. Thus, the topology of the network remains static while the states of agents change dynamically. Important processes studied within this framework include contact processes such as the spread of epidemics, the diffusion of innovations, and opinion-formation dynamics. Studies of this sort reinforce the notion that certain topological properties have a strong impact on the dynamics *of* and *on* agent networks.

Much effort has been devoted to understanding how a model describing the behavior of interacting agents is affected by the network of interactions. More recently, the focus has shifted to take into account the fact that many networks, and in particular social networks, are dynamic in nature: links appear and disappear continually over time, and across timescales. Moreover, such modifications to the network's topology do not occur independently from the agents' states, but rather as a feedback effect: the topology determines the evolution of the agents' opinions, and this, in turn, determines how the topology is modified. That is, the network becomes adaptive. Under this framework, the coevolution of an adaptive network of interacting agents is investigated to show how the final state of the system depends on this coevolution. The direction of the research is currently aimed at understanding how the structure of agent networks determines network dynamics of various types. The most promising models are those in which the network structure determines the dynamic processes, and in which the dynamic processes correspondingly shape the network structure. The first line of research is concerned with the dynamics *of* networks. Here, the topology of the network itself is regarded as a dynamical system. It changes over time according to specific, often local, rules. Investigations in this area have revealed that certain evolutionary rules give rise to network topologies with special properties. The second line of research, by contrast, concerns the dynamics *on* networks (Gross and Blasius, 2008).

In agent networks, the network topology usually emerges as a result of the interactions between agents. This approach is useful for determining whether the observable phenomena are explained by the difference in the network topologies or by the difference in the dynamics of agents. Furthermore, agent-based models are used to explore the evolution of networks. The hypothesis used here is that networked systems have a topology that is characterized by the interactions of agents, and that, in addition, the topology imposes some sort of influence on them. Fluctuations in the behavior of agents invoked by the topology will alter the topology. Such a multi-level relationship to the interaction between the agents and the global topology is called a micro–macro loop.

Agent interactions also occur in structured networks. However, there is a relative lag to the development of tools for modeling, understanding, and predicting dynamic-agent interactions and behavior in complex networks. Even recent progress in complex network modeling has not yet offered the capability of modeling dynamic processes among agents who interact in networks. Computational modeling of dynamic agent interactions in structured networks is important for understanding the sometimes counterintuitive dynamics of such loosely coupled adaptive agents. Despite the multiplicity of concepts used to describe these mechanisms, each mechanism affects the agent system at both the microscopic and the macroscopic level. At the microscopic level, these mechanisms help individual agents mitigate complexity and uncertainty. At the macroscopic level, they facilitate coordination and cooperation among agents, and consequently affect the success of the agent system as a whole. From an applied point of view, it is desirable to collect an inventory of the microscopic dynamics at the individual level and to determine their corresponding effects at the macroscopic network level. Such an inventory would provide researchers with specific guidelines concerning the phenomena observed in socioeconomic systems.

1.3 Outline of the Book

The foundations of this book are interdisciplinary. We turn to sociology and economics, network science, and agent-based modeling. These disciplines have made important contributions and provided the mechanisms used to understand several socio-economic problems. From sociology and economics, we draw on a broad set of theoretical frameworks for studying the structure and dynamics of socioeconomic systems. From network science, we derive the language for elucidating the complexity of network structures and systems with interacting agents. From agent-based modeling, we glean models for the strategic behavior of individuals who interact with each other and operate as members of larger aggregates. This book is unique insofar as it is a focused effort to bring together these traditions in the framework of agent-based modeling.

1.3.1 Foundation Chapters

Chapter 2 gives a historical review of the development of network-based agent-based models. From lattices to graphs, we separate the development into two stages. We should be able to see that the issues dealt with in these two stages can be completely the same, and they are distinguished simply by our network thinking or our awareness of networks. This chapter can be viewed as the foundation chapter of the whole book. It covers two major parts, namely, the network-based discrete-choice models and the network-based agent-based models. If the interaction of agents is the essence of agent-based models, then the network-based discrete-choice models just inform us how these interactions among agents are defined and operated by the underlying networks. The network-based agent-based models are the agent-based models that explicitly specify the networks for interactions. Since networks define interactions and affect decision-making, they and network-based discrete-choice models are the two sides of the same coin. In a chronological order, we demonstrate the network-based agent-based models via a number of pioneering works. The purpose of these demonstrations is to enable us to see how network topologies can affect the operation of various economic and social systems, including residential segregation, pro-social behavior, oligopolistic competition, market sentiment manipulation, sharing of public resources, market mechanism, marketing, and macroeconomic stability. Cellular automata as the theoretical underpinnings of the complex dynamics of these models are also introduced. This theoretical background is a backbone for the undecidability and unpredictability properties of many other models to be introduced in other chapters.

Based on the general understanding of the network-based agent-based models, Chapter 3 presents the collective behavior of adaptive agents—that is, agents who coadapt to other agents. When agents coadapt to each other, the aggregate behavior is extremely difficult to predict. What makes this interactive situation both interesting and difficult to evaluate is that we must evaluate the aggregate outcome, and not merely the way agents behave within the constraints of their own environments. The following questions are confronted. How do interacting agents who coadapt to each other produce certain aggregate outcomes, sometimes, also called the reverse engineering problem?

How do we identify the microscopic behavior of agents who contribute to the regularities observed at the macroscopic level?

In the network-based agent-based models, network topologies are exogenously given. This leaves us an issue on how these network topologies are determined in the first place. In other words, what is the network formation process of the economic and social networks? Chapter 4 address this fundamental question by introducing agent-based models of social networks. This chapter considers agent-based models as formation algorithms for networks. Such formation algorithms are distinguished from sociological, socio-physical, and game-theoretical formation algorithms.

The chapter begins with a review of cooperative and non-cooperative game theory of network formation, followed by a series of human subject experiments carried out in light of game theory. The limitations of these models, in particular their empirical relevance, are discussed, which naturally motivates the use of agent-based modeling as an alternative approach to network formations. As we shall see in this chapter, the agent-based models of social networks can be considered as an extension of the spatial games and network games reviewed in Chapter 2. The difference is that games are no longer played in a given network topology; instead, they become part of the determinants of network formation. Different games with different strategic and adaptive behavior will lead to different network topologies. This chapter reviews the work done using the friendship game, the prisoner's dilemma game, the stag hunt game, the public goods game, and the trust game (the investment game). They together provide formal treatments to our intuitive understanding on how human relation is shaped by various competitive and cooperative opportunities. The chapter ends up with a brief remark on the multiplex games and the evolved social networks.

If we take Chapters 2, 3, and 4 as the foundations of the exogeneity and the endogeneity of network topologies, then the following three chapters focus on the primary interactions that either can be affected by the exogenous network topologies or can endogenously shape the evolution of network topologies. These three focusing interaction processes are diffusion dynamics (Chapter 5), cascade dynamics (Chapter 6), and influence dynamics (Chapter 7).

1.3.2 Primary Dynamics on and of Networks

Chapter 5 addresses agent-based diffusion dynamics. Powerfully influential information, ideas, and opinions often spread quickly throughout society. Diffusion is a daily social phenomenon of propagation within any society. In this chapter, we explore the diffusion process. This line of inquiry is often formulated as epidemic diffusion. Whether certain information succeeds in penetrating society depends primarily on the probability that it will be passed from the individual who knows it to other interactive partners. Epidemic information-transmission dynamics are known as threshold phenomena. That is, a probability threshold exists below which information will spread merely within a limited group and above which it will penetrate the whole society. Diffusion processes also take place within social networks, and they are influenced by the topological properties of networks. This chapter investigates the particular network conditions that determine whether information will penetrate the whole society.

Chapter 6 is concerned with agent-based cascade dynamics. Much social phenomena differs from diffusion-like epidemics. Epidemic models are limited in that they treat the relationships between all pairs of agents as equal. They do not account for the potential effects of individual heterogeneity on decisions to behave in certain ways or to adopt ideas. In this chapter, we consider a diffusion model for deliberate behavior. For each agent there is a threshold for action, determined in part by the number of neighboring agents who approve of the action. Such a threshold-based model can also be applied to the study of diffusion in terms of innovation, cascading failures, and chains of failures. Cascade dynamics are modeled as coordinated processes where each agent decides to adopt a particular action based on the number or ratio of neighboring agents who already have adopted it. The chain process of failures is discussed as another example of cascade dynamics. The failure of one of the components can lead to a global failure, as in a chain of bankruptcies.

Chapter 7 pertains to agent-based influence dynamics. The interaction between individuals does not merely involve sharing information; its function is to construct a shared reality consisting of agreed-upon opinions. In consensus formation, individuals must influence one another to arrive at a common interpretation of their shared reality. One of the main questions surrounding this idea concerns the possibility of a global consensus, defined as the coincidence of internal states without the need for a central supervisory authority. Social influence is a recognized phenomenon in social networks. The degree of social influence depends on many factors, such as the strength of the relationships between people, the type of individuals in the network, the network distance between users, the effects of time, and the characteristics of the network. We focus on the computational aspects of analyzing social influence and describe the measurements and algorithms used to interpret it. More specifically, we discuss the qualitative and quantitative measures used to determine the level of influence in an agent network. Using agent-based modeling and network analysis, we shed light on these issues and investigate the kinds of connections that tend to foster mutual influence in egalitarian societies.

1.3.3 Applications

The last part of the book, consisting of three chapters, is mainly on the applications of the agent-based modeling of economic and social networks to practical and challenging issues. They include the three prominent applications, the practical network formation in different domains (Chapter 8), network risks (Chapter 9), and economic growth (Chapter 10).

Chapter 8 is concerned with economic and social networks in reality. This chapter focuses on the empirical aspect of networks, i.e., various economic and social networks built using real data. In this chapter, we review three major market-oriented networks, namely, the buyer–seller network, the labor contact network, and the world trade web.

Chapter 9 discusses agent-based models of networked risks. Networked systems provide many benefits and opportunities. However, there are also negative aspects to networking. Systemic instability is said to exist in a system that will eventually lose control, despite each component functioning properly. This chapter explores how the complexity of networked systems contributes to the emergence of systemic instability. An agent-based model can explain how the behavior of individual agents contributes to our understanding of potential vulnerabilities and the paths through which risk can propagate across agent networks. Various methodological issues related to managing systemic risk and cascading failures are also discussed.

Chapter 10 is concerned with agent-based modeling of economic growth. This chapter presents agent-based models as a means for studying the interactions between the financial system and the real economy. Such a model can reproduce a wide array of empirical regularities to fundamental economic activities as stylized facts. An agent-based model will contribute to enriching our understanding of potential vulnerabilities and the avenues by which economic crises propagate in economic and financial systems. Combining agent-based modeling with network analysis can shed light on our understanding of how the failure of a firm propagates through dynamic credit networks and leads to a cascade of bankruptcies.

2

Network Awareness in Agent-based Models

2.1 Network Awareness

The *network awareness* is a recent scientific phenomenon. The neologism *network science* is just a decade old. However, the significance of networks in various human behaviors and activities has a history as long as human history. Almost all solutions to real-world problems involve the network element. The whole life of any individual, from cradle to grave, is determined by his/her family ties, neighbors, classmates, friends, and colleagues. This personal social net provides him/her with various opportunities and information, which affects his/her choices, actions, and achievements in a substantial way.

From an individual level to a social level, networks remain important and, sometimes, become even more important. The end of the period of the warring states and the great unification of China by Qin Shihuang (260–210 BC) in year 221 BC was due to the *horizontal alliance strategy* proposed by Qin's important strategist, Zhang Yi, which helped Qin break down the "network" of the other six states. Networks can promote war and peace, and threaten or consolidate national security. Hence, regardless of our awareness of networks, their influences have been already there; however, if we are aware of their existence and influences, then we can be proactive to change the network topologies, as Zhang Yi had done for Qin and China.

In fact, China has long already been aware of networks, but not just networks of humans, but networks of humans and their objects. This is known as *Feng Shui* (Rossbach, 1983; Webster, 2012). In Feng Shi, the focus is the layout of all articles or objects in a space in which agents are situated. In our familiar terminology, these articles by their attributes have to be placed properly so as to facilitate their "interactions" with agents. If so, the welfare, fortune, business, health, and human relations of agents can be enhanced. Nowadays, Feng Shui still plays a pivotal role in Chinese architecture (Tam, Tso, and Lam, 1999).

Different degrees of network awareness are also reflected in agent-based modeling. At a lower degree, researchers are aware of the existence of networks and their impacts, but are not much aware of how these networks are formed. At this stage, networks are

Agent-Based Modeling and Network Dynamics. First Edition. Akira Namatame and Shu-Heng Chen.

taken as exogenous settings, as they are fixed or given (see this chapter 2). At a higher degree, researchers go further to find the origins of these networks. At this stage, humans' decisions on networking are taken into account, and networks are endogenously generated. Generally, these networks may not have steady state and can evolve over time (see Chapter 4).

The network awareness provides a simple taxonomy of the literature overarching agent-based models and economic and social networks. The literature pertaining to the former case (the exogenous treatment) is called *network-based agent-based models* (covered in this chapter), and the literature pertaining to the latter case (the endogenous treatment) is called *agent-based models of social networks* (see Chapter 4). It is interesting to see that these two streams of the literature do not develop in a parallel manner, maybe partially due to the network awareness. It starts from the research questions on how networks or, more specifically, network topologies can affect the operation of the economy or the society. The questions addressed include the effect of network topologies on financial security, effectiveness of macroeconomic regulation policy, marketing promotion, market efficiency, coordination of social dilemma, and employment opportunities.

In principle, the network topology can be taken as an additional exogenous variable and placed into all kinds of agent-based models that we have studied. For example, it can be placed into agent-based financial markets to address the network effect on the price efficiency. This doing is very natural because the interaction of agents is the core element of agent-based models, which automatically imply the existence of an underpinning network to support interactions. Notwithstanding with this feature, we totally concur on what Bargigli and Tedeschi (2014) have said, "...many contributions in the field of AB modeling, as well as in other fields of economics, do not *explicitly* make use of network concepts and measures. (Bargigli and Tedeschi (2014), p. 2; Italics, added.)"

2.2 First-Generation Models: Lattices

While many agent-based economic models do not *explicitly* take networks into account, one should not ignore the fact that agent-based models are originated from the biophysical computational models, known as *cellular automata* (Hegselmann and Flache, 1998; Chen, 2012). Some well-known cellular automaton models, such as John Conway's *game of life* (Adamatzky, 2010) or Stephen Wolfram's *elementary cellular automata* (Wolfram, 1983, 1994), have been used by social scientists as spatial networks for their models. Two earliest examples in the literature are James Sakoda's *social interaction models* and Thomas Schelling's *segregation models* (Sakoda, 1971; Schelling, 1971). Their models are known as the *checkerboard model* for social scientists before the names of cellular automata were propagated, probably through Albin (1975) who first introduced the neologism to economics. Therefore, it is fair to say that networks have been already explicitly used in the agent-based social scientific models right from the beginning of the stream of the literature, much earlier than the advent of the literature on *spatial games*, specifically, spatial prisoner's dilemma games (Nowak and Sigmund, 2000).

We may characterize the first generation of the network-based agent-based models as the models built upon lattices or checkerboards. The network is closely related to the physical distance or geographical proximity, i.e., the *neighborhood* in the conventional sense. In a regular two-dimensional lattice, with a radius of one, i.e., within a one-step range, an agent can normally reach four neighbors (*von Neumann Neighborhood*) or eight neighbors (*Moore Neighborhood*), depending on how a step is defined. The essence of *social economics* is that agents' preferences and decisions are not independent but are generally influenced by other agents' preferences and decisions (Durlauf and Young, 2004). When the lattices (networks) are introduced into the agent-based model, they help us to specify a clear indication on this interdependence relationship, in particular when the influence is subject to locality. A specific class of decision rules, known as the *neighborhood-based decision rules*, can then be developed and operated in the agent-based models.

2.2.1 Schelling's Segregation Models

The neighborhood-based decision rule assumes that an agents' decision (preference) will under the influence of their neighbors. The Schelling segregation model is a good illustration of this rule (Schelling, 1971). The major decision rule in the Schelling model is a migration rule or a residential choice rule. Agents have to decide whether they should stay put or migrate to somewhere else. In Schelling's segregation model, this decision is based on who are the neighbors or, more specifically, their ethnicity. Schelling considered two ethnic groups in his model, and the rule is simple, as follows. If agent i in his community (neighborhood) $\mathbf{N}_\mathrm{i}$ is "minority," then agent i will decide to migrate; otherwise, he will decide to stay. Notice that here the definition of majority or minority is not necessary based on the usual division figure, 50%, but is determined by agents' psychological perceptions; indeed, the tolerance capacity for the fraction of the unlike people around us is, to some extent, psychological.

The network-based decision rule is then parameterized by a threshold, the migration threshold or tolerance capacity, denoted by θ. If the migration threshold or the tolerance capacity is set to 30%, then the agent will migrate if more than 30% of his neighbors belong to the other ethnicity

$$a_i(t) = \begin{cases} 1 \text{ (migrate)}, & \text{if } f_{\mathbf{N}_\mathrm{i}}(t) > \theta_i, \\ 0 \text{ (stay put)}, & \text{otherwise}, \end{cases} \tag{2.1}$$

where

$$f_{\mathbf{N}_\mathrm{i}}(t) = \frac{|\{j : j \in \mathbf{N}_\mathrm{i}(t), j(x) \neq i(x)\}|}{|\mathbf{N}_\mathrm{i}(t)|}. \tag{2.2}$$

In eqn. (2.2), $i(x)$ gives the ethnical group of agent i, and "$j(x) \neq i(x)$" means that agents i and j have different ethnical groups. Hence, $f_{\mathbf{N}_\mathrm{i}}(t)$ is the fraction of agent i's neighbors who do not share the same ethnical group as agent i.

Indexing θ by i implies that agents may not be homogeneous in their thresholds; some may have lower tolerance capacity, and some may have higher tolerance capacity. Hence, introducing this heterogeneity can cause a wave of migration or a domino effect of migration. Depending on the distribution of the threshold values, one may identify a *tipping point* or *critical point*, at which a morphogenetic change in demographics can occur (Schelling, 1972).

The Schelling model, as a pioneering agent-based model, has been extensively studied in the literature. The main result is that the emergence of segregation is a quite robust result. As Banos (2012) has stated, "Schelling's model of segregation is so robust that it is only weakly affected by the underlying urban structure: whatever the size and shape of city, this undesired emerging phenomenon will happen. (Banos (2012), p. 394)". Hence, the first primary lesson from the network-based agent-based model is the *emergent structure of segregation*.

The robustness of the segregation result provides a classic illustration for the robustness check for agent-based models: a robustness check of a single simple model can be so demanding and so extensive that it still has not been accomplished after two dozen studies have been devoted to it.[1] Rogers and McKane (2011) has proposed a generalized analytical Schelling model to encompass a number of variants of the Schelling model. Both Rogers and McKane (2011) and Banos (2012) can give readers a list of extensive studies along the line initiated by Schelling.

Not all variants of the Schelling model are the first generation of network-based agent-based models. As we shall see, the second-generation network-based models (see Section 2.3) have been quickly extended to "advance" the Schelling model. Among those variants belonging to the first generation, a research line on the sensitivity of the segregation result deserves our attention (Sander, Schreiber, and Doherty, 2000; Wasserman and Yohe, 2001; Laurie and Jaggi, 2003; Fossett and Dietrich, 2009). Basically, this research line deals with a question that is well anticipated. The question can be presented as follows. The standard Schelling model, relying on either the von Neumann neighborhood or Moore neighborhood, only uses a radius of one to define the neighborhood. What happens if this radius changes, say, from $R = 1$ to $R = 5$?

The base tone of these studies is that *the degree of segregation increases with the vision.* A particularly illustrating example is demonstrated in Laurie and Jaggi (2003), Figures 2 and 3. These two figures compare the result of the case of $R = 1, \theta = 0.5$ with that of the case of $R = 5, \theta = 0.5$. When the radius of the neighborhood increases from one to five, we can see that the size of the clusters formed in the latter are much larger and evident than those formed in the former. Something similar is also found in Sander, Schreiber, and Doherty (2000) and Wasserman and Yohe (2001). However, Laurie and Jaggi (2003) have carried out a thorough examination of the domain of the parameters. In addition to the parameter domain, called the *unstable regime*, where a society

[1] We use the Schelling model to make this point explicit for a detouring reason. Robustness check is a virtue, but it is equally important to know that robustness check may essentially know no bound. When a referee asks an author to revise his paper by checking its robustness, a reasonable bound should be delineated. We, being editors of journals for years, have the feeling that this awareness should be widely shared by the community.

invariably segregates and segregation increases as vision, R, increases, their simulations also identify the *stable regime*, where the integrated societies are stable and segregation decreases as vision increases. A narrow intermediate regime where a complex behavior is observed is also founded. This complex pattern is also found in Fossett and Dietrich (2009), "We *do find* important effects of the size or scale of neighborhoods involved in agent vision: model-generated segregation outcomes vary in complex ways with agent vision. . . . Thus, what is important is the scale of agent vision—that is, the number of neighbors they "see"—not the particular spatial arrangement of those neighbors. (Fossett and Dietrich (2009), 149; Italics added)"

This research, called the *visionary variant* of the Schelling model, is less well noticed than other variants. The ignorance of this research line, not just limited to the few studies cited here but the entire research idea, may lead us to claim the robustness of the segregation result in a premature way. When the segregation result is too robust, one may tend to attribute the general failure of the integration policy to the prediction of "Schelling theory." Roughly speaking, the "shadow" of the negative result of the "Schelling theory" may generate a pessimistic psychology toward the effectiveness of any policy toward integration. It should not be, but it may be cautiously read in a form of fatalism.[2]

Under this circumstance, the result provided by Laurie and Jaggi (2003) gives us a positive qualification to the negative side of the Schelling model and brings in a more positive thinking of the policy, particularly, education of preference.[3] "The central policy implication of the study is an optimistic note: contrary to popular belief, rather modest decreases in xenophobia and/or preferences for one's own kind, when coupled with increased vision, can lead to stable and integrated neighbourhoods. (Laurie and Jaggi (2003), p. 2703)." We do not mean to involve ourselves in a debate on whether education of preference can be sufficient for shaping an integration policy. The point that we would like to make is that it is premature to claim the robustness of the segregation result and use it to explain or forecast the effectiveness of an integration policy.

The agent-based model as a powerful tool for thought experiments can help us to enhance our expressiveness of the concern for the issue under examination. The reason that it is premature to make a claim on the robustness of segregation is simply because many concerns related to the issue have not been incorporated into the model. Vision is just an example, and preference is another example (Pancs and Vriend, 2007). However, all of these models only consider homogeneous agents. Hence, R and θ are homogeneous

[2] For example, Pancs and Vriend (2007) state, "This policy has failed in that integration simply did not happen. The spatial proximity model provides a possible answer why this could have been expected. (Pancs and Vriend (2007), p. 4)." Later on, they add,

> "The stark conclusion is that a wide class of preferences for integration results in extreme segregation. Any integration policy must be based on a good understanding of the mechanisms underlying this result. The analysis suggests that education of preferences is not sufficient to achieve integration." (Pancs and Vriend, 2007, p. 5)

[3] As well shown in Laurie and Jaggi (2003), Figure 7, when $\theta = 0.7$, an integrated society eventually emerges if R is large enough. More interestingly, this integration also happens for the case of $\theta = 0.65$ in a nonlinear way: initially the society segregates with an increase in R, but up to a point ($R = 4$), the society increasingly integrates with an increase in R.

in all agents. For a next step of research, we echo the remark made by Banos (2012), "Carefully introducing heterogeneity in agent-based models, both at the level of agents' attributes and behaviors but also in the way we define agents' environment is therefore a key issue the community should handle more vigorously, in order to improve our understanding of the 'real-world dynamics of the city'. (Banos (2012), p. 405)"

As a concrete example, one can allow for agents' heterogeneity in the size of neighborhoods, characterized by different radius. This relaxation can be motivated by the possibility that the perception of neighborhood is psychological and subjective. It therefore differs from agent to agent. To incorporate this possibility into the model, we can conveniently assume that agents are heterogeneous in their *vision*. Naturally, an agent with a better vision can "see" a larger radius of agents as his neighbors, whereas an agent with poor vision can only see the neighbors immediately around, for example, the von Neumann neighborhood with a radius of one only. To the best of our knowledge, this heterogeneous-agent extension of the Schelling model has not been attempted in the literature.

2.2.2 Spatia Games

The second application of the lattice model to economics and social sciences is the *spatial games*, which is the predecessor of network games (to be reviewed in Section 2.3.2 and Section 4.4.1). In the spatial game, each agent i is placed in a lattice and plays a game with each of his neighbors, j ($j \in \mathbf{N}_i$). Part of the origin of the spatial game can also be traced back to Schelling's multi-person games.[4] In Schelling's original setting of the multi-person game, there is neither lattice nor network. Albin (1992) transforms Schelling's multi-person game into a two-person game with multi-persons (neighbors).[5] Albin considered that these two-person variants "do identify properties of significance in MPD and several other multi-person games described by Schelling (1978). (Albin (1992), p. 193)" This is so because in the spatial game all agents simultaneously choose whether to cooperate or to defect, regardless of the specific neighbor whom they encounter.

Let $\mathbf{A}$ be the action or strategy space, and $\mathbf{A} = \{1, 0\}$, where "1" denotes the action "cooperation" and "0" denotes the action "defection". In addition, let $\mathbf{X}(t)$ be the set of the choice that each agent makes at time t: $\mathbf{X}(t) = \{a_i(t)\}$, $a_i(t) \in \mathbf{A}$. A strategy of playing the game can be formulated as follows. Let $A_i(t)$ be the history of the actions taken by agent i up to the time t, i.e., $A_i(t) \equiv \{a_i(t), a_i(t-1), \ldots a_i(1)\}$. Let $A_i^k(t)$ be the subset of $A_i(t)$, which only keeps the most recent k periods, $A_i(t) \equiv \{a_i(t), a_i(t-1), \ldots, a_i(t-k+1)\}$. A strategy for agent i can then be defined as the following mapping:

$$f_i : \{A_j^k(t)\}_{j \in \{i\} \cup \mathbf{N}_i} \rightarrow \mathbf{A}. \tag{2.3}$$

[4] See the final chapter (Chapter 7) of Schelling (1978) (also see Dixit (2006)). In this section, Schelling uses a number of multi-person games to illustrate that individual rational choices do not necessarily lead to the most efficient results for the society as a whole.

[5] We shall see these two different versions of games again in Section 4.9.

Hence, for example, if agent i has four neighbors (the von Neumann neighborhood), and assuming $k = 1$, then including agent i, there are a total of 2^5 possible combinations of actions (scenarios) that a decision can be contingent upon. For each scenario, the decision can be either "1" or "0"; hence, there are a total of 2^{2^5} ($=2^{32}$) possible strategies, which is a number far beyond the reasonable set of alternatives that human normally can handle, not to mention the more general case that induces a total of $2^{2^{5\times k}}$ strategies, and the even more daunting case when the neighborhood changes from the von Neumann neighborhood to the Moore neighborhood, a total of $2^{2^{9\times k}}$ strategies.

Albin (1992) considered a different format of strategies, which is in spirit close to the format used by (Schelling, 1978). This format only looks at a local aggregate, namely, the number of cooperative neighbors in the preceding period,

$$N_i^a(t-1) = \#\{j : a_j(t-1) = 1, j \in \{i\} \cup \mathbf{N}_i\}.$$

Obviously, $N_i^a(t-1) = 0, 1, \ldots, N_i$, where N_i is the cardinality of $\mathbf{N}_i$. With this aggregate information, agent i no longer cares about the action taken by each of his neighbors; hence, the number of possible scenarios can be reduced from the $2^{2^{N_i+1}}$ to only $N_i + 1$, and the number of possible strategies has also been reduced to the 2^{N_i+1}, accordingly. It is easy to code this kind of strategy by a binary string with $N_i + 1$ bits. From the left to the right, each bit indicates the action to take when the number of the cooperative agents corresponds to the order of position of that bit. Hence, for example,

$$f_i = 000001111$$

indicates

$$f_i = \begin{cases} 0, & \text{if } N_i^a(t-1) < 5, \\ 1, & \text{if } N_i^a(t-1) \geq 6. \end{cases} \tag{2.4}$$

As for another example, say,

$$f_i = 010101010$$

indicates

$$f_i = \begin{cases} 0, & \text{if } N_i^a(t-1) \text{ is an even number,} \\ 1, & \text{if } N_i^a(t-1) \text{ is an odd number.} \end{cases} \tag{2.5}$$

Some of these strategies make good sense. For example, the threshold-based rule, such as eqn. (2.4), says that the agent i will take the cooperative action only if his number of cooperative neighbors has come to a minimum point. Furthermore, $f_i = 000000000$ and $f_i = 111111111$ refers to the strategy "always defect" and "always cooperate," respectively.

Once we have this structure, it is not hard to see that the neighborhood-based decision rules f_i, as formulated by Albin, have actually connected the spatial games to *cellular automata*.[6] von Nuemann's original intention was to understand the self-replication capability, a fundamental capability of biological entities or entities with life. The self-replicating automaton that von Neumann constructed was so complex that it had 29 states (cells can be in one of 29 states). Conway has simplified the von Neumann machine into two-state two-dimensional cellular automata. The two states are "alive" or "dead." Hence, at each point in time, each cell can be either alive or dead. These two states correspond to the two actions chosen by agents in the spatial game. At each point in time, their state can be either "cooperate" or "defect." The strategy employed by agents corresponds to the state-transition rule operated on each cell in cellular automata.

The one-dimensional two-state cellular automata have been thoroughly studied by Wolfram, which is called the *elementary cellular automata*. The state transition rule operated on each cell depends on the cell itself and its neighbor both on the left and on the right. Hence, there are a total of 2^3 configurations (scenarios) and 2^{2^3} (256) transition rules. Wolfram systematically studied these 256 rules and generated their time–space patterns. Based on the patterns generated, the 256 rules are taxonomized into four classes, which represent a hierarchy of complexity. Briefly speaking, in terms of system dynamics, these four classes are fixed points, limit cycles (periodic cycles), chaos (pseudo randomness, strange attractors), and on the edge of chaos (complex patterns).

For the first class after some periods we can have, for example, agents who either always defect (0) or always cooperate (1). For the second class, after some periods we can have, for example, agents alternate between defection and cooperation in a periodic way; some agents defect at the odd periods and cooperate at the even periods, whereas some agents just do the opposite. More generally, each agent can have a cooperation-defection cycle with specific periods; while the periods can be heterogeneous from agent to agent, it is fixed for each agent. For the third class, the behavior of each agent becomes pseudo random. Completely different from the previous two classes, where each agent can precisely predict the behavior (action) of each of his neighbors when t is sufficiently large, the third class in general defies this predictability. For the fourth class, the emergent time–space pattern is not entirely pseudo random all the time. A pattern may emerge and remain for an unknown duration and then disappear before the time–space pattern of the system transits into another chaotic or irregular configuration. Situated in this dynamic, each agent may sometimes find that it is possible to trace (anticipate) the behavior (action) of their neighbors, for example, they either all cooperate or all defect or cooperate and defect with some discernible patterns (cycles), but then in some other time this tractability is lost and neighbors' behavior become unpredictable for the agent.

In Wolfram's elementary cellular automata, out of the 256 possible state-transition rules (strategies), it turns out that roughly 85% of the automata (222 automata) behave in a uniform and uninteresting way (Classes I and II), 5% (10 automata) in a seemingly random and chaotic way (Class III), and 10% (22 automata) in a way that seems ordered

[6] For those readers who are less familiar with cellular automata, see Albin (1975) and Albin (1998).

and complex. However, the above results are obtained by assuming that all agents (cells) follow the same rule ($f_i = f, \forall i$). In the case of spatial games, agents generally follow different rules. Furthermore, the rules followed by them may not be time-invariant when their learning and adaptive behavior are also taken in to account. These differentiations may lead to a different distribution of the limit behavior of spatial games; nonetheless, the four-class taxonomy remains valid. Hence, the main contribution of Albin's work is to use the automata theory (computational theory) to shed light on the significance of the network topologies to the results of spatial games.

While in this section we use Albin (1992) to motivate the spatial game through Schelling's multi-person game, it is worth noting that the earliest model on the spatial game is given by Axelrod (1984), and the motivation is different from the one given by Albin (1992).[7] In Axelrod (1984) (his Chapter 8), the spatial game is introduced as an imposed social structure, and Axelrod wants to know the consequence of social structure (network topology) on the pro-social behavior. He has interestingly shown that the imposed social structure can protect clusters of cooperators from the invasion of defectors.

This is done by first assuming that all agents in a regular lattice use the tit-for-tat strategy and all agents initially are cooperators. Obviously, this initial state is a steady state. Then, one tries to perturb the steady state by introducing an intruder (a mutant) at the center of the lattice, who plays the strategy of always defecting instead of using the tit-for-tat strategy. Given the simple imitation dynamics, i.e., each agent will copy the strategy of his most successful neighbor, Axelrod shows that, under the von Neumann neighborhood, the always defecting strategy will spread from the center to entire space in snowflake-like patterns of defectors bypassing islands of cooperators.

The essential message of Axelrod's spatial game is to show that by imposing social structure (network topology) one can accommodate both the existence of the cooperators and defectors. Therefore, the social structure (network topology) is sufficient enough to support the emergence of cooperative behavior. The upshot of this research line initiated by Axelrod (1984) and extended by others has been well described by Nowak and Sigmund (2000).

> "The main message so far is that neighborhood structure seems to offer a promising way out of the Prisoner's Dilemma toward the emergence of cooperation. There are many alternative explanations of the prevalence of cooperation, but, arguably, none require less sophistication on the part of the individual agents than those with spatial structure. The latter need no foresight, no memory, and no family structure. *Viscosity suffices* (Nowak and Sigmund (2000), p. 145; Italics added)"

This snowflake pattern generated in Axelrod (1984) is, however, entirely predictable (computationally reducible). In other words, the cellular automaton composed of agents, who switch between "tit-for-tat" and "always defect" by imitating the most successful

[7] However, Axelrod (1984) is often ignored in many studies, which attribute the earliest work on spatial games to Matsuo (1985).

neighbor, has dynamics of Class I, conditional on the initial configuration of all being cooperative except one.

Axelrod did not use the automata theory to further analyze the complex patterns that may emerge from his formulation of spatial games, like Albin (1992) does in Schelling's formulation of multi-person games. However, almost at the same time that the paper of Albin was published, a formal research line in the spirit of Axelrod also began, and contributed substantially to what nowadays is known as spatial games.[8] The pioneering work known as the *Nowak-May model* has shaped the study of the spatial game in the frame of cellular automata (Nowak and May, 1992, 1993). This is done by systematically parameterizing the game.

Let $\pi_i(a_i, a_j)$ be the payoff of a two-person game that agent i plays with his opponent agent j, when agent i takes the action a_i ($a_i \in \mathbf{A}_i$), and agent j takes the action a_j ($a_j \in \mathbf{A}_j$). In the prisoner's dilemma game, $\mathbf{A}_i = \mathbf{A}_j = \{1(C), 0(D)\}$. Since the game is symmetric, $\pi_j(a_j, a_i) = \pi_i(a_i, a_j)$, we shall omit the subscript appearing in the payoff function and let Π be the payoff matrix:

$$\Pi = \begin{bmatrix} \pi_{11} & \pi_{10} \\ \pi_{01} & \pi_{00} \end{bmatrix}, \tag{2.6}$$

where $\pi_{11} = \pi(1,1)$, $\pi_{10} = \pi(1,0)$, and so on. As a prisoner's dilemma game, the following inequality should be satisfied: $\pi_{01} > \pi_{11} > \pi_{00} > \pi_{10}$. What Nowak and May (1992) did is to "normalize" the game by setting $\pi_{11} = 1$ and $\pi_{00} = \pi_{10} = 0$, and leave only one parameter be free, i.e., π_{01}. Under this normalization, the game can be characterized by a single parameter, namely,

$$\text{the temptation to defect} \equiv \frac{\pi_{01}}{\pi_{11}} = \pi_{01}.$$

Based on the value of π_{01}, one can delineate the boundary of the cluster of cooperators and the dynamics of its advancement.

$$\begin{vmatrix} 1&1&1&0&0&0 \\ 1&1&1&0&0&0 \\ 1&1&1&0&0&0 \\ 1&1&1&0&0&0 \\ 1&1&1&0&0&0 \\ 1&1&1&0&0&0 \end{vmatrix} \longrightarrow \begin{vmatrix} 1&1&1&\mathbf{1}&0&0 \\ 1&1&1&\mathbf{1}&0&0 \\ 1&1&1&\mathbf{1}&0&0 \\ 1&1&1&\mathbf{1}&0&0 \\ 1&1&1&\mathbf{1}&0&0 \\ 1&1&1&\mathbf{1}&0&0 \end{vmatrix} \longrightarrow \begin{vmatrix} 1&1&1&1&\mathbf{1}&0 \\ 1&1&1&1&\mathbf{1}&0 \\ 1&1&1&1&\mathbf{1}&0 \\ 1&1&1&1&\mathbf{1}&0 \\ 1&1&1&1&\mathbf{1}&0 \\ 1&1&1&1&\mathbf{1}&0 \end{vmatrix} \tag{2.7}$$

The mapping (2.7) is an illustration borrowed from Nowak and Sigmund (2000). Consider that each agent is given a Moore neighborhood. Then as shown in (2.7),

[8] It is a misfortune that Albin's work has been largely ignored in the mainstream literature on spatial games. For example, it has not been mentioned even in the survey on spatial games (Nowak and Sigmund, 2000).

each cooperative agent in the boundary is surrounded by five cooperative neighbors and three defective neighbors; his one-round payoff is

$$5\pi_{11} + 3\pi_{10}.$$

On the other hand, each defective agent in the boundary is surrounded by three cooperative neighbors and five defective neighbors and his one-round payoff is

$$3\pi_{01} + 5\pi_{00}.$$

Hence, as long as

$$5\pi_{11} + 3\pi_{10} > 3\pi_{01} + 5\pi_{00},$$

the boundary of cooperation will expand, i.e., advance to the right, as shown in (2.7). Similarly, for the case of (2.8), if

$$3\pi_{11} + 5\pi_{10} > 2\pi_{01} + 6\pi_{00},$$

the block of the cooperators in the middle also expands outward. Therefore, depending on π_{01}, various possible movements can occur, which together show that the Nowak-May models can also have different classes of cellular automata. The complex temporal-spatial pattern may emerge with the coexistence of the cooperators and defectors.

$$\begin{vmatrix} 0&0&0&0&0&0 \\ 0&0&0&0&0&0 \\ 0&0&1&1&0&0 \\ 0&0&1&1&0&0 \\ 0&0&0&0&0&0 \\ 0&0&0&0&0&0 \end{vmatrix} \rightarrow \begin{vmatrix} 0&0&0&0&0&0 \\ 0&\mathbf{1}&\mathbf{1}&\mathbf{1}&\mathbf{1}&0 \\ 0&\mathbf{1}&1&1&\mathbf{1}&0 \\ 0&\mathbf{1}&1&1&\mathbf{1}&0 \\ 0&\mathbf{1}&\mathbf{1}&\mathbf{1}&\mathbf{1}&0 \\ 0&0&0&0&0&0 \end{vmatrix} \rightarrow \begin{vmatrix} \mathbf{1}&\mathbf{1}&\mathbf{1}&\mathbf{1}&\mathbf{1}&\mathbf{1} \\ \mathbf{1}&1&1&1&1&\mathbf{1} \\ \mathbf{1}&1&1&1&1&\mathbf{1} \\ \mathbf{1}&1&1&1&1&\mathbf{1} \\ \mathbf{1}&1&1&1&1&\mathbf{1} \\ \mathbf{1}&\mathbf{1}&\mathbf{1}&\mathbf{1}&\mathbf{1}&\mathbf{1} \end{vmatrix} \quad (2.8)$$

2.2.3 Oligopolistic Competition

One year after the publication of Albin (1992), Keenan and O'Brien (1993) also applied a one-dimensional lattice in a torus layout to model firms' local competitive behavior. In their model, a large number of firms, such as gasoline stations, retailers, or supermarkets, evenly distributed on a one-dimensional lattice or ring (Figure 2.1). Each small circle is situated by one firm. The price decision made by firm 'i' depends on the price set by its local rival firms, the left neighbor (firm $i-1$) and the right neighbor (firm $i+1$), as in the following:

$$P_i(t+1) = f(P_{i-1}(t), P_{i+1}(t)). \quad (2.9)$$

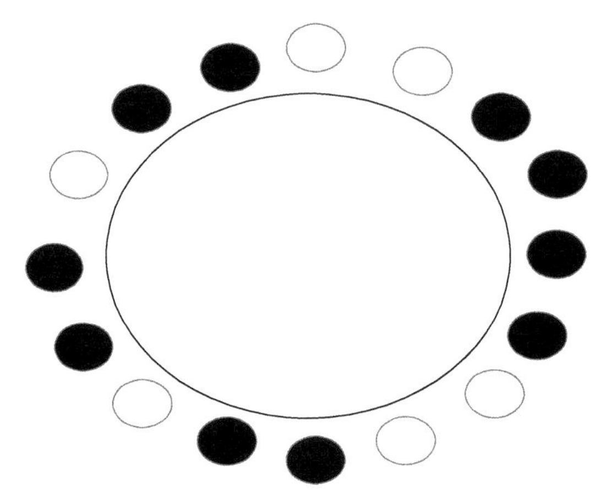

Figure 2.1 *Ring of Firms*

The above figure shows that a group of firms distribute evenly around a circle. Customers are also evenly distributed along the circle. A locally oligopolistic competition for the firms exists. The firms may charge a high (cartel) price, denoted by the black color, or may charge a low (competitive) price, denoted by the white color.

By replacing $P_{i-1}(t)$ and $P_{i+1}(t)$ recursively using eqn. (2.9), we can have

$$\begin{aligned} P_i(t+1) &= f(f(P_{i-2}(t-1), P_i(t-1)), f(P_i(t-1), P_{i+2}(t-1))) \\ &= f^2(P_{i-2}(t-1), P_i(t-1), P_{i+2}(t-1)), \end{aligned} \tag{2.10}$$

or simply

$$P_i(t+2) = f^2(P_{i-2}(t), P_i(t), P_{i+2}(t)). \tag{2.11}$$

Furthermore, instead of considering the price to be continuous, we assume that the price will be discretized at a much coarser level into only either a high price, denoted by 1, or a low price, denoted by 0. Even so, the pricing rule does not immediately render itself to fit cellular automata, but if we separate the firms in the one-dimensional lattice (ring) by odd numbers and even numbers and simply focus on the dynamics of either the odd-numbered firms or even-numbered firms, then by eqn. (2.11), we can actually have two cellular automata running in parallel, one corresponding to the odd number firms and one corresponding to the even-number firms.

Hence, cellular automata can be used to simulate the geographical distribution of prices over time, after a specific pricing rule is given. One price rule studied in Keenan and O'Brien (1993) is *Edgeworth process*.[9] This process, as shown in eqn. (2.12), basically says that firm i will charge the high price (the cartel price) if either both of its

[9] For the details of the Edgeworth process, see Keenan and O'Brien (1993).

neighboring firms $i-2$ and $i+2$ (not the immediate neighbors, $i-1$ and $i+1$) also charge the cartel price or both of them charge the low price (competitive price); otherwise, firm i will charge the low price.

$$P_i(t+2) = \begin{cases} 1 \text{ (Black)}, & \text{if } P_{i-2}(t) = P_{i+2}(t) = 1 \text{ or } 0, \\ 0 \text{ (Green)}, & \text{otherwise.} \end{cases} \tag{2.12}$$

Eqn. (2.12) is essentially equivalent to CA Rule 90 in Wolfram's elementary cellular automata. Denoting the high price by the black color and the low price by the gray color, we also demonstrate this pricing rule in terms of a cellular automaton and the triggered dynamics of the geographical price distribution in Figure 2.2. From Figure 2.2, one can see that cartels persist over time, as shown by the black triangles. Therefore, similar in vein to the analysis of Albin (1992), Keenan and O'Brien (1993) also used the spatial configuration to show that without consideration of the strategic behavior of firms, the resulting market dynamics are capable of demonstrating considerable collusive activity and cooperative and competitive behavior continue to coexist throughout time.

Keenan and O'Brien (1993) actually studied a number of other rules as well, including the case of three-state cellular automata so that the price can be low, middle, and high, but they also considered the case where different cells actually follow different pricing rules. For example, they considered a subset of firms (the odd number firms) following CA Rule 18 and another subset of firms (the even-number firms) following CA Rule 126. There are more intriguing patterns of the geographical distribution of prices being generated.

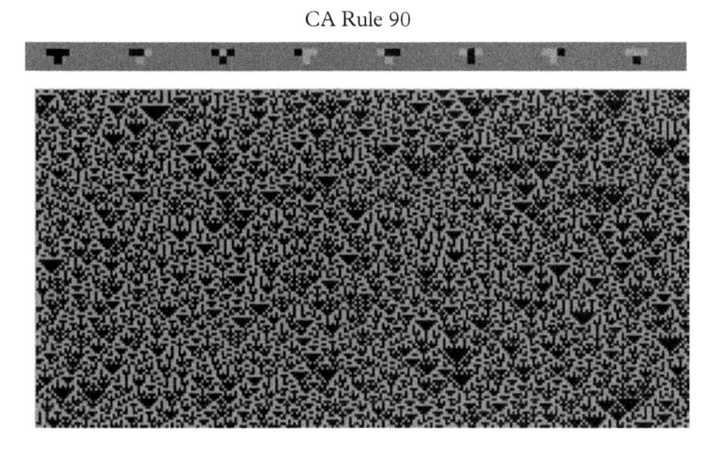

Figure 2.2 *Geographically Oligopolistic Competition*

The figure above shows the evolution of the one-dimensional cellular automation where alternate firms are allowed to change prices at alternate times according to the Edgeworth process (2.12). Firms charging the low (competitive) prices are represented as gray rectangles; firms charging the high (cartel) price are represented as black rectangles.

The upshot of the Keenan and O'Brien (1993) research is to enable us to see that the reaction function of each firm to different opponents' strategies, including prices, advertisement, R&D, etc., can be understood as network-based (neighborhood-based) decision rules. These decision rules are either derived from simple heuristics or more deliberating calculation. Regardless of their formation process, once they are formed as network-based (neighborhood-based) decision rules, cellular automata can be applied as a theoretical underpinning of these models, specifically for the understanding of the observed Class III and Class IV phenomena. Of course, the kind of cellular automata must also be more general than the "standard" cellular automata, as formulated by Conway and Wolfram.

2.2.4 Sentiment Indexes

The neighborhood-based (network-based) decision rules naturally can be applied to the opinion formation or to the formation of sentiment index (see, also, Chapters 5 to 7 of this book). To see this, one can begin with one-dimensional elementary cellular automata and consider that there are two possible valences, one positive, say, being optimistic, and one negative, say being pessimistic. Each agent's valence at time $t + 1$ will depend on the configuration of his optimistic and pessimistic neighbors. Hence, let us consider a simple majority rule: the agent has three neighbors (including him) and the valence of his expectation (sentiment) in the next period will follow the majority of expectations (sentiments) in this period. If so, the CA Rule 128, as shown at the top of Figure 2.3,

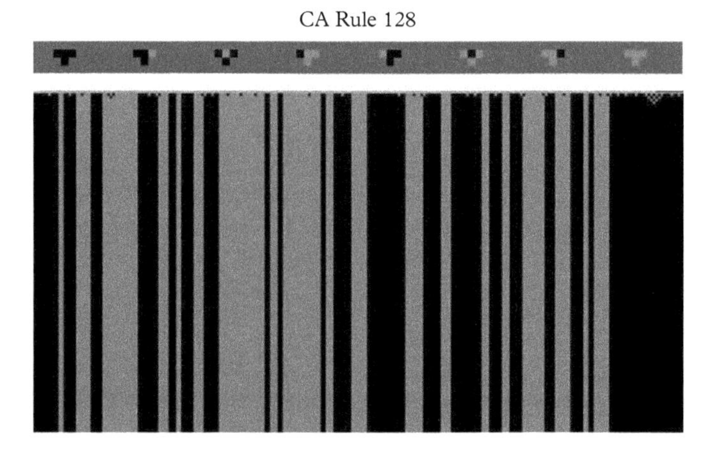

Figure 2.3 *Opinion Dynamics: The Majority Rule*

In the above figure, we start with an initially highly mixture of opinions: 50% of the agents are optimistic (colored by gray), and 50% of the agents are pessimistic (colored by black). By the behavioral rule CA Rule 128, we can see that very quickly the society have been clustered into a region of optimism and pessimism. The proportion of agents who are optimistic and pessimistic may not change much as compared to the initial distribution; their locations are, however, much more structured.

should represent this behavior rule well, and the dynamics of the opinion formation is shown in Figure 2.3.

We now consider a slightly different behavioral rule. In this case, as in the previous case, agents still follow the majority rule; however, when there is a unanimous optimism or pessimism appearing in the previous period, agents will switch to the opposite sentiment. This behavioral rule is motivated by a mixture of momentum rule (majority rule) and contrarian rule. This behavioral rule can be well described by the CA Rule 105, as shown at the top of Figure 2.4, and the opinion dynamics as appearing in the spatiotemporal space become rather erratic.

The behavioral rules above can be generalized in many different ways. In a two-dimensional cellular automata, Chen (1997) considered both local and global influences. Basically, in addition to the above simple majority rule, Chen (1997) further assumed that there is a census institute, for example, a governmental agency, who constantly samples agents' sentiment and publishes a global outlook. Each agent i will also take into account the global outlook, while forming their sentiment. This is essentially to add a *hub agent*, who is able to transmit the information of agents from corner to corner. For some reasons, the hub agent may have the motivation to make agents believe that there is a global promising prospect ahead and, through the expectations-driven morale (see more details in Chapters 5 to 7), generate a self-fulfilling process. With this motivation, bad news may be manipulated and mitigated to some degree. Given this framework, Chen (1997) asks the question: will the hub agent simply tell truth and reveal what he knows about the global outlook? In other words, is the "cooked-up" sentiment index reliable? Would the officially-made sentiment index be manipulated? Chen (1997)

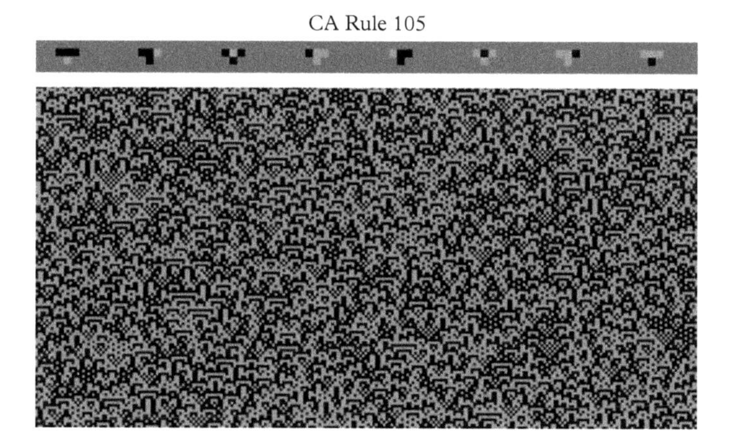

Figure 2.4 *Opinion Dynamics: The Mixture of the Majority and the Contrarian Rule*

In the above figure, we start with an initially highly mixture of opinions: 50% of the agents are optimistic (colored by gray), and 50% of the agents are pessimistic (colored by black). By the behavioral rule CA Rule 105, we can see that the opinion dynamics, as appearing in the spatiotemporal space, are rather erratic.

analyzed this issue by assuming that when forming their own expectations, agents will assign weights to both local signal and global signal, and these weights are constantly adjusted through Bayesian learning in a form of Kalman filtering. The learning mechanism prevents agents from being "brain washed" by the hub agent. The essential lesson learned from this study is that even though agents are equipped with some learning abilities, *honest*, "always telling the truth," is still not the best policy for the hub agent; instead, because of the network effect, a degree of manipulation may be favorable for the hub agent, and that degree is also state-dependent.

In general, this state-dependent truth revealing policy can vary with different network topologies. In other words, the truth revealing policy may impact the society differently when the underlying network topology changes, but this channel has been largely ignored in the literature. As Chen (1997) argues, whether government should intervene in the market or save the market can be understood as a form of truth revealing policy. The hands free policy is essentially equivalent to the truth revealing policy, and any degree of intervention, price manipulation, or price support can be considered as a kind of distortion to the true market sentiment. When and how much intervention should be involved has long been debated; however, the effect of the social network topologies on intervention in a form of opinion formation process has not been given sufficient thoughts. Would the intervention (policy) effectiveness or ineffectiveness remain the same when social media networks have dominated the society to a larger degree than 20 years ago? This is issue has yet to be addressed.

2.2.5 Good Society

We have long been questioning the role of government or the role of central (top-down) intervention and regulation. Is it possible to leave citizens themselves to coordinate and solve a resource allocation problem purely from individual actions, not even making an attempt to form an alliance or union? Can the purely individual actions alone bring in a change for the society? Can the good society emerge under an extremely minimal degree of coordination? In this section we shall show, again, network topology matter for this issue, and we shall demonstrate this point through the well-known El Farol bar problem. As we shall see, the El Farol bar problem as a theoretical environment for the study of the coordination problem and as a complex adaptive system can be used to demonstrate how the coordination problem can sometimes be solvable and sometimes be unsolvable.

El Farol Bar Problem

In the original El Farol bar problem (Arthur, 1994), N people decide independently, without collusion or prior communication, whether to go to a bar. Going is enjoyable only if the bar is not crowded, otherwise the agents would prefer to stay home. The bar is crowded if more than B people show up, whereas it is not crowded, and thus enjoyable, if attendees are B or fewer. If we denote the agent's decision "to go" by "1" and "not to go" by "0", and the actual number of attendees by n ($n \leq N$), then the agent's payoff function has the general form:

$$U(x,n) = \begin{cases} u_1, & \text{if } x = 0 \text{ and } n \geq B, \\ u_2, & \text{if } x = 0 \text{ and } n < B, \\ u_3, & \text{if } x = 1 \text{ and } n > B, \\ u_4, & \text{if } x = 1 \text{ and } n \leq B. \end{cases} \tag{2.13}$$

The payoffs have either the order $u_4 = u_1 > u_2 = u_3$ or the order $u_4 > u_1 = u_2 > u_3$.[10]

The El Farol Bar (hereafter, EFB) problem or the congestion problem has been regarded as a classic model for the study of the allocation or the use of public resources, particularly when central or top-down coordination is not available. Through the modeling and simulation of the EFB problem, one hopes to have a general understanding of when and how a good coordination can emerge from the bottom up. While over the last two decades various agent-based models have been proposed to address the EFB problem, most of them have been concerned with the overuse (congestion) or the underuse (idleness) of public resources. Efficiency certainly is one important concern of the coordination problem, but it is not the only one. The one concern missing in the literature is *equity* or *fairness*; after all, public resources belong to all members of a community, not just a few of them.

Chen and Gostoli (2015) reformulated the original EFB problem; hence, both *efficiency* and *equity* have been taken into account in evaluating the emerging bottom-up coordination. In their study, it is found that the *social network* plays an important role as an efficient means of coordination, i.e., an outcome with neither idleness nor congestion. However, many of these efficient outcomes are not *equitable*, so that the public resource is not equally shared by all community members. A typical example is the emergence of two clusters of agents; one always goes to the bar, and one never goes to the bar—a familiar phenomenon known as social segregation or exclusion. They, nonetheless, found that if some agents can be endowed with a kind of *social preference*, then both efficient and equitable outcomes become likely. In fact, they found that, as long as a proportion of the community members has social preference, then the convergence to the equilibrium which is both efficient and equitable is guaranteed. They refer to this outcome as the *"good society" equilibrium*.

Two network typologies considered by them are shown in Figure 2.5: the *circular network (circular neighborhood)*, where each agent is connected to the two agents to his left and the two agents to his right, and the *von Neumann network (von Neumann neighborhood)*, with the agents occupying a cell in a bidimensional grid covering the surface of a torus. Hence, in each of the two networks, the agent is connected to four neighbors, denoted by $N1$, $N2$, $N3$, and $N4$. The reason why they started with the circular network and the von Neumann network rather than the typical social networks is because

[10] For some discussions of these two inequalities, the interested reader is referred to Chen and Gostoli (2015).

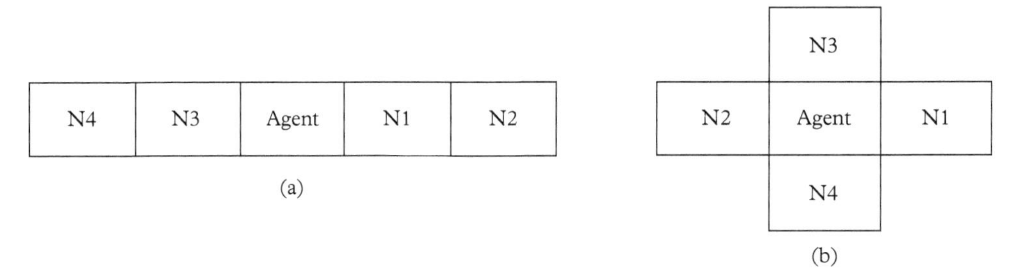

Figure 2.5 *Circular (a) and von Neumann (b) neighborhoods*

these chosen networks have a topology that is very similar to those of the elementary cellular automata studied by Wolfram (2002). It therefore endows us with a basic theoretical underpinning which may make it easier to recast the classical EFB problem into a familiar environment of cellular automata, as a step toward more complex situations.

Behavioral Rules

Contrary to the prototypical EFB problem, each agent is assigned, at the beginning of the simulation, only one strategy z, randomly chosen from the whole strategy space. Their representation of the strategy is based on the binary string as normally used in cellular automata. The idea is that each agent will first look at what his four neighbors did in the previous period and only then will decide what he will do in the current period, i.e., mapping from the neighbors' previous decisions to his current decision. Denote the action "going to the bar" by 1 and "staying at home" by 0. Then there are 2^4 possible states, each corresponding to one combination of the decision "1" or "0" made by the four neighbors. Each strategy is composed of 16 rules specifying the action a the agent has to take in the current period, one rule for each state. Each strategy can then be represented by a 16-bit long string. If we fix the numbering order of the 16 (2^4) states, then the corresponding 16-bit representation for the strategy exemplified there is simply, for example, "0010001110101110", i.e., an array of the decisions corresponding to each of the sixteen states, respectively. All together, there are 2^{16} possible strategies in the strategy space.

Performance of a Behavioral Rule

They defined the variable $a_i(t)$ as the action taken by agent i in period t: it takes the value 1 if the agent goes to the bar and the value 0 otherwise. Moreover, they defined the variable $s_i(t)$ as the outcome of agent i's decision in period t: it takes the value 1 if the agent took the *right* decision (that is, if he went to the bar and the bar was not crowded or if he stayed at home and the bar was too crowded) and it takes the value 0 if the agent took the *wrong* decision (that is, if he went to the bar and the bar was too crowded or if he stayed at home and the bar was not crowded). The agents are endowed with a memory of length m. This means that they store in two vectors, $\mathbf{a}$ and $\mathbf{s}$ of length m, the last m values of a and s, respectively. So, at the end of any given period t, agent i's vectors $\mathbf{a}_i$ and $\mathbf{s}_i$, are composed, respectively, of $a_i(t), a_i(t-1), \ldots, a_i(t+1-m)$ and of $s_i(t), s_i(t-1), \ldots, s_i(t+1-m)$.

Agent i's *attendance frequency* over the most recent m periods, d_i, is defined by:

$$d_i = \frac{1}{m} \sum_{j=t=1-m}^{t} a_i(j). \tag{2.14}$$

The attendance frequency's value can go from 1, if the agent always went to the bar, to 0, if the agent never went to the bar, in the last m periods. Moreover, agent i's *decision accuracy rate*, f_i, is given by (2.15):

$$f_i = \frac{1}{m} \sum_{j=t+1-m}^{t} s_i(j). \tag{2.15}$$

The decision accuracy rate can go from 1, if the agent always made the right decision, to 0, if the agent always made the wrong decision, in the last m periods. We define the duration of agent i's current strategy (the number of periods the agent is using his current strategy) as r_i. In order for the average attendance and the decision accuracy associated with any strategy to be computed, it has to be adopted for a number of periods equal to the agents' memory size m: so, we can think of m as the *trial period* of a strategy. This value is set to ten for all the agents in all their simulations.

Agents in their model may have an inequity-averse preference, which is characterized by a parameter called the *minimum attendance threshold*, denoted by α_i, that is, a fair share of the access to the public resources or a fair attendance frequency expected by the agent. It can take any value from 0, if the agents do not care about their attendance frequency, to 0.6. They did not consider a higher value than 0.6 because these agents with equity concern do not claim to go with an attendance frequency higher than the threshold $B/N(= 0.6)$.

Simonian Boundedly Rational Agents

The agents' strategies evolve through both social learning (imitation) and individual learning (mutation). So, the social network plays a role both in the agents' decision process, allowing the agents to gather information regarding their neighbors' choices, and in the agents' learning process, allowing the agents to imitate their neighbors' strategies. In any given period, an agent i imitates the strategy of one of his neighbors if the following six conditions are met:

(a) $f_i < 1$ and/or $d_i < \alpha_i$

(b) $r_i \geq m_i$

and the agent has at least one neighbor j for which the following conditions are verified:

(c) $f_j > f_i$
(d) $d_j \geq \alpha_i$
(e) $r_j \geq m_j$
(f) $z_j \neq z_i$.

Condition (a) is quite obvious. It simply states that the agent will have the tendency to imitate if he is not satisfied with his current situation (strategy). There are two possibilities that may cause this dissatisfaction. First, there are errors in his decision ($f_i < 1$) so there is room for an improvement, and, second, he is not satisfied with his attendance frequency ($d_i < \alpha_i$). Notice that, by this later qualification, the agent may still look for change even though all his decisions are accurate ($f_i = 1$). Condition (b) shows that the agent will not change his strategy frequently and will consider doing so only if the strategy has been tested long enough, i.e., after or upon the completion of the trial period with a given duration of m_i. When imitating neighbors, agent i will only consider those strategies which not only lead to more accurate outcomes, but also lead to a satisfactory attendance frequency (Condition (c) and (d)). The above promising strategy should be based on long testing, with a duration of m_j periods, rather than sheer luck (Condition (e)). Finally, agent i will not imitate the same strategy which he is currently using (Condition (f)).

If the first two conditions are met but at least one of the last four is not, or, alternatively put, if the agent has not yet reached the optimal strategy and in the current period he cannot imitate any of his neighbors, then the agent, with a probability p ($p << 1$), will mutate a randomly chosen rule on its strategy while with probability $1 - p$ he will keep using his present strategy. While the imitation process ensures that the most successful strategies are spread in the population, the mutation process ensures that new, eventually better, strategies are introduced over time. Once the agent has adopted a new strategy (either through imitation or mutation) he will reset his memory to zero and will start keeping track of the new strategy's fitness. The agent *stops* both the imitation and the mutation processes if the following two conditions are met:

1. $f_i = 1$
2. $d_i \geq \alpha_i$.

When these two conditions are verified for all the agents, the system reaches equilibrium: no further change in the agents' behavior takes place after this point as the agents always make the right decision and go to the bar with a satisfying attendance frequency.

Under the influence of Simon's notion of bounded rationality, behavioral economists characterize each decision process with three main stays, namely, a search rule, a stopping rule, and a decision rule (Gigerenzer and Gaissmaier, 2011). The proposed learning process above can basically be read with this Simonian framework. Conditions (b)–(f) give the search rules, including when to start searching, (a) gives the stopping criteria, and the rest give the decision rules. Notice that the inequity-averse preference, together with the forecasting accuracy, in this model plays exactly the role of the stopping criteria. Hence, agents in their model are bounded rational in the sense of the Simonian satisfying agents, rather than the expected-utility maximizing agents.

Network Topologies and Coordination Patterns

Chen and Gostoli (2015) began with the simulation of the EFB system with both the circular neighborhood and the von Neumann neighborhood. In this series of simulations, the agents, as in the original model, do not care about their attendance frequency. In this case, the same learning mechanism applies but with the minimum attendance threshold set to 0 ($\alpha_i = 0, \forall i$). Accordingly, the agents decide whether or not to imitate their neighbors only on the basis of the strategies' accuracy rates. In this way, they were able to assess how the outcomes were affected by the introduction of social networks and, in particular, the effect of different network structures on the kinds of equilibria reached by the system, the equilibrium distribution, and, in particular, the emerging likelihood of a "good society."

Each network topology is run 1000 times. The results show that each simulation of both settings always reaches perfect coordination, that is, the state where the bar attendance is always equal to the threshold and, consequently, the agents never make the wrong choice. While the EBF in both networks eventually converges to the same aggregate outcome (a 60% attendance rate all the time), from the mesoscopic viewpoint, they differ from run to run. To effectively characterize these equilibria at the mesoscopic level, Chen and Gostoli (2015) focused on the attendance frequency of agents when the perfect coordination is formed, d_i^*. In this way, their equilibrium can be represented by the heterogeneity in this attendance frequency over all agents. More precisely, the perfect coordinating equilibrium of the EBF problem is given by the set that shows the observed attending frequencies, b_j^*, and the share of the agents with b_j^*, π_j^*:

$$\Xi \equiv \{(b_j^*, \pi_j^*)\}_{j=1}^{c} \equiv \{(b_1^*, \pi_1^*), (b_2^*, \pi_2^*), \ldots (b_c^*, \pi_c^*)\}, \tag{2.16}$$

where $b_1^* > b_2^* > \ldots > b_c^*$.

In (2.16), c refers to the number of clusters, and π_j^* is the size of the corresponding cluster. Taking the bimodal perfect coordination equilibrium as an example, we have two clusters of agents, one that always goes ($b_1^* = 1$) and one that never goes ($b_2^* = 0$); 60% of agents are of the first kind, and 40% agents are of the second kind. Hence, this equilibrium is characterized by

$$\Xi_{Bi} \equiv \{(1, 0.6), (0, 0.4)\}. \tag{2.17}$$

Alternatively, the "good society" is an equilibrium characterized as

$$\Xi_G \equiv \{(0.6, 1)\}. \tag{2.18}$$

For convenience, we shall call these equilibria, based on the number of emerging clusters, $1C$ equilibrium, $2C$ equilibrium, etc. Hence the "good society" equilibrium, Ξ_G, is a $1C$ equilibrium, and the segregated equilibrium, Ξ_{Bi}, is a $2C$ equilibrium. Then one way to present their simulation result is to show the histogram of each of these C equilibria over our 1000 runs. Here, we use *equilibria* because, except for the 1C equilibrium, we can have multiple equilibria for each C ($C \geq 2$). For example, for the 2C equilibria, in addition to Ξ_{Bi} as shown in (2.17), the other observed $2C$ equilibrium is:

$$\Xi_2 \equiv \{(1, 0.2), (0.5, 0.8)\}. \tag{2.19}$$

Similarly, for $C = 3$, we can have

$$\Xi_{3\text{-}1} \equiv \{(0.7, 0.1), (0.6, 0.8), (0.5, 0.1)\} \tag{2.20}$$

or

$$\Xi_{3\text{-}2} \equiv \{(1, 0.4), (0.5, 0.4), (0, 0.2)\}. \tag{2.21}$$

Furthermore, even for two equilibria having the same $\{b_j^*\}$, their $\{\pi_j^*\}$ can still be different. For example, one alternative for $\Xi_{3\text{-}1}$ is

$$\Xi_{3\text{-}3} \equiv \{(0.7, 0.3), (0.6, 0.4), (0.5, 0.3)\}. \tag{2.22}$$

Figure 2.6 shows the histogram of the C equilibria from $C = 1, 2, \ldots, 8$ for both the circular network (CN) and the von Neumann network (vNN).[11]

From Figure 2.6 we can see that, while the literature on the EBF problem had identified only one kind of equilibrium, that is, Ξ_{Bi} in (2.17), the introduction of social networks associated with the use of local information leads to the emergence of many different kinds of equilibria. While the 2C equilibria remain the most likely outcome in both networks, with the von Neumann network the system has a non-negligible probability (18%) of reaching the $1C$ equilibrium (the "good society" equilibrium), Ξ_G. The fact that the system has relatively good chances to reach the perfectly equitable equilibrium is a quite interesting result considering that, in this version, agents have no minimum attendance thresholds; yet, it is the second most likely equilibrium, with a probability up to almost one third of the probability of the $2C$ equilibria. Different network structures are, however, characterized by different equilibria distributions: for example, the probability of reaching Ξ_G declines to only 2% in the circular network.

A finer look at the results further shows that, within the equilibria characterized by the emergence of two clusters ($2C$), the great majority (over 90%) are represented by the Ξ_{Bi}. The rest (less than 10%) are represented by Ξ_2 (see eqn. 2.19). The great majority of the $3C$ equilibria are represented by an equilibrium where some agents never go to the bar, some always go, and the rest go with an attendance frequency of 0.5, i.e., $\Xi_{3\text{-}2}$

[11] The equilibrium with more than eight clusters of agents has not been found in any of their simulations.

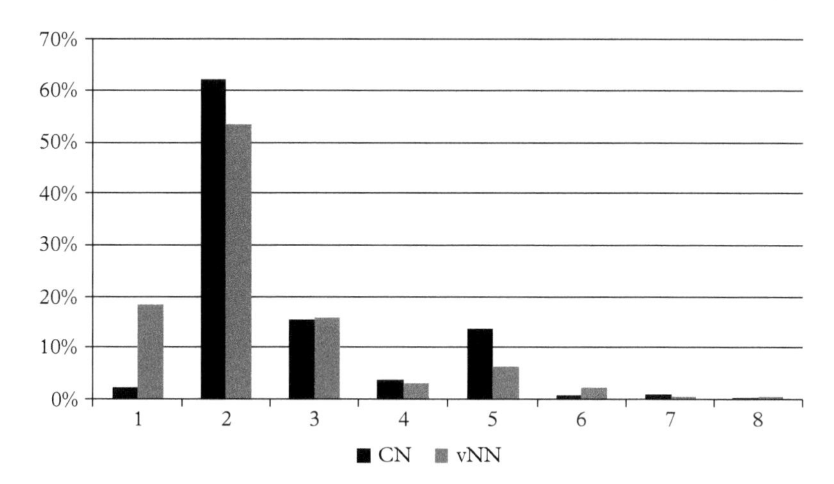

Figure 2.6 *Histogram of the C Equilibria with the circular (CN) and the von Neumann (vNN) neighborhoods*

in (2.21). Another relatively frequent outcome is the emergence of the $5C$ equilibria. Within this case, the great majority is represented by a configuration where, besides the three clusters mentioned for the $3C$ case, two more groups of clusters, going to the bar respectively with a frequency of 0.4 and 0.6, emerge, such as

$$\Xi_5 \equiv \{(1, 0.58), (0.6, 0.01), (0.5, 0.02), (0.4, 0.01), (0, 0.38)\}. \tag{2.23}$$

2.2.6 Summary

The network-based agent-based models allow us to query the significance of networks on the emergent properties at two separate levels. First, the formation of the decision rules is subject to the networks, the so-called network-based decision rules. Second, the network-based decision rules followed by the agents can then grow various complex economic and social patterns. From what we have presented in this chapter, we can see that network topologies associated with the network-based decision rules can have at least two effects on the agent-based models. First, at the mesoscopic level, they determine how economic activities and human relations are distributed spatially, an issue typically studied in regional sciences. In this chapter, we have seen three such examples, namely, the spatial distribution of ethnicity (Section 2.2.1), human relations (Section 2.2.2), and pricing (Section 2.2.3). Second, at the macro level, we show the effect of the spatial structure on information dissemination and aggregation and on coordination of economic activities (Sections 2.2.4 to 2.2.5).

While the notion of the network used in the first-generation model is only in a spatial sense, it helps us to see the immediate connection between the automata theory and the network theory and, based on that connection, we can taxonomize the emerged patterns into four classes, some of which are computationally irreducible and their properties cannot be known ex ante but can only be known through "experiencing" (simulation). As is well said, "[L]ife cannot be predicted; it must be lived." (Lamm and Unger (2011), p. 78).

2.3 Second-Generation Models: Graphs

Models built upon cellular automata characterize the first generation of network-based agent-based models. In these models, the network is synonymous with the *spatially defined "neighbors."* When communication is primitive, man's mobility is severely limited, the assumption that the economic and social network is geographically defined is plausible. However, in the second machine age (Brynjolfsson and McAfee, 2014) or the era of digital revolution, man's social mobility is basically free of the spatial restriction. It is, therefore, well anticipated that the first-generation models can be largely extended if the notion of the network can be extended. This comes to the second-generation network-based agent-based models, which appears almost a decade later than the first-generation model.

Graphs, instead of the regular lattice, become the underpinning frame for the second-generation models. Graph theory is an old subject in mathematics (Biggs, Lloyd, and Wilson, 1986) and it was used by sociologists in the 1950s to give a conceptual model of social networks (Barnes, 1954). Nonetheless, the foundations of social networks, specifically the non-trivial network formation algorithms and the characterizations of social networks, were not available until the middle and late 1990s. Therefore, it is not surprise to see that the burgeoning of the second-generation models only appeared at the turn to the twenty-first century.

It is, however, worth noticing that before the advent of the second-generation models, some elementary graphs had already been used in social sciences, while not explicitly. For example, the fully connected network is implicitly assumed in many economic contexts, such as competitive markets. In fact, the law of one price, to some extent, does rest upon a fully connected (bipartite) graph in the following sense: buyers know the prices of all sellers, and sellers also know the prices of each other. In addition to the fully-connected network, the random graph is also used in social sciences as a model to capture the randomly local encountering. This can be seen in a large class of economic models. Many agent-based models relying on social learning mechanisms, such as evolutionary computation, allow agents to randomly meet each other. The random graph, therefore, becomes a companion to the entropy-maximization agents (the zero-intelligent agents) and they together become a twin device for model simplicity. Finally, the ring is another graph which is occasionally used to characterize the geographical restriction. For example, the ring network was used to study bilateral trade in a non-tatonnement process (Albin and Foley, 1992).

The second-generation models began with the formulation of *small-world networks* (Watts and Strogatz, 1998). This network topology has two nice features. First, it can be regarded as a class of networks encompassing both the random network and the ring network. In between the two, one can have a network which is neither totally random nor entirely regular. Second, before the introduction of its formation algorithm, the *small-world phenomenon* had already been noticed and studied by sociologists for decades, specifically by Milgram and Travers who obtained the famous experimental findings on the *six degrees of separation* (Milgram, 1967; Travers and Milgram, 1969). Hence, it is a network topology not only theoretically sound but also empirically relevant.

At the turn of this century, two network-based agent-based models were proposed. Both began to extend the existing research of the economic effects of network topology from the original regular grids to small-world networks. Among the two, Wilhite (2001) studies the effect of the small-world network on the efficiency of the decentralized market in the form of bilateral trades, whereas Abramson and Kuperman (2001) examine the effect of the small-world network on spatial games. The agent-based models of networks reviewed in this section enable us to examine the effect of network topologies on market mechanism (Section 2.3.1), on prosocial behavior (Section 2.3.2), on consumer behavior and marketing (Section 2.3.3), and on macroeconomic stability (Section 2.3.4). Of course, this is not an exhaustive list, but it is comprehensive enough to give a general flavor of many other works that are inevitably left out due to space constraint.

2.3.1 Bilateral Trade

While one focus of economics is trade and market, their spatial and social embeddedness have been largely ignored in the general introductory textbook. Partially under the influence of sociology, economists have a surging interest in the effect of social network on trade (see, also, Section 8.1). This influence is not limited to empirical studies, but is also seen in theoretical studies. Wilhite (2001) is the first published work that addresses the economic role of network topology in the Walrasian general equilibrium framework, specifically the trading processes and the economic outcome due to different network topologies.

Wilhite (2001) studied the role of network topologies in bilateral trade. To achieve this, he actually considered four different kinds of network topologies. The choice of network topology is not purely based on the literature in network science, but a unique economic consideration regarding the extent of a market. The four network topologies: a fully connected network, a locally disconnected network (Figure 2.7), a locally connected network (Figure 2.8), and small-world network (Figure 2.9). Strictly speaking, the small-world network topology considered by Wilhite (2001) is not the Watts-Strogatz kind (Watts and Strogatz, 1998), but just a random additional link connecting two different locally disparate groups. The two ends of the link must not be involved in any existing inter-group link, as shown in Figure 2.9. Therefore, from Figures 2.7 to 2.9, Wilhite (2001) actually studied the change in trading process when there are more and more inter-group links, being added; they correspond to a completely open global market, isolated local markets, and local but connected markets with different degrees of connection. This setting actually allows us to have an economic geographic underpinning of the bilateral trade model.

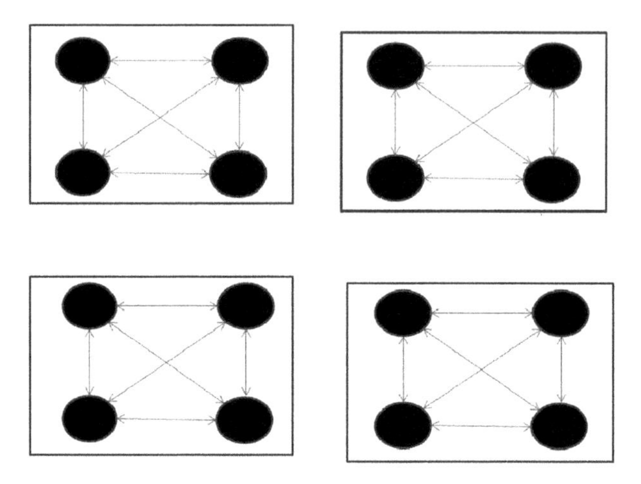

Figure 2.7 *Local Disconnected Network*

The above figure demonstrates four groups of agents, each composed of a number of individuals (in this illustration, four individuals). The bilateral trade is possible for agents belonging to the same group, as shown by the links, but inter-group trade is infeasible.

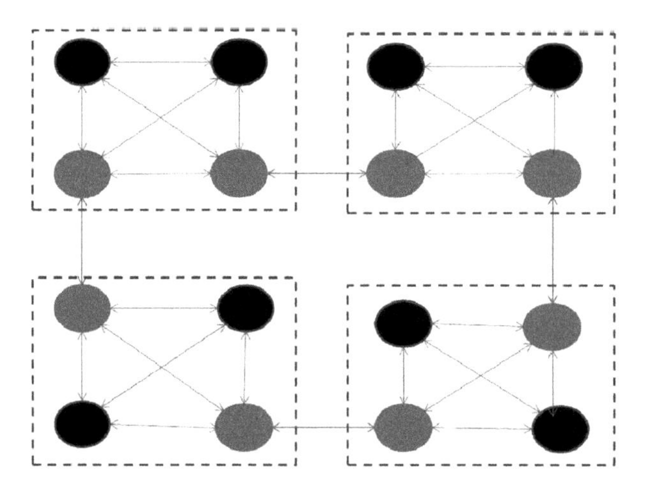

Figure 2.8 *Local Connected Network*

The above figure demonstrates four groups of agents, each composed of a number of individuals (in this illustration, four individuals). The bilateral trade is possible for agents belonging to the same group, as shown by the links. Inter-group trade is feasible only for those who are connected (the gray color agents).

In Wilhite (2001), the network serves as a precondition for chance discovery. Agents are assumed to know the marginal rates of substitution (reservation prices) of all agents belonging to the same group, and can only trade with the agents within the same group. Agents in the Wilhite model are not strategic, maybe for the consideration of

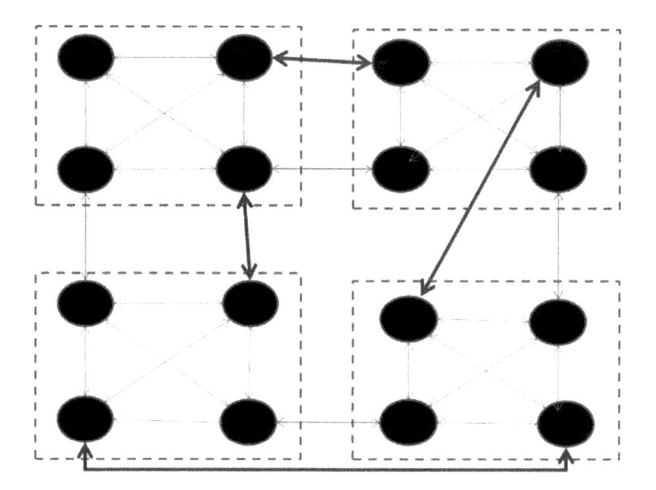

Figure 2.9 *Local Connected Network with Additional Interior Connections*

The above figure demonstrates four groups of agents, each composed of a number of individuals (in this illustration, four individuals). The bilateral trade is possible for agents belonging to the same group, as shown by the links. Inter-group trade is feasible only for those who are connected.

model simplicity. They are not only myopic but also not greedy (having no bargaining strategy) at all. They are myopic in the sense that all pairs of agents can only accept welfare-enhancement deals. The price is simply set to be the market-clearing price of a two-person economy. This is equivalent to assume that there is a well-established culture or norm to decide the fair price. Under this circumstance, there is no need for the agents to consider any arbitrage behavior. Forecasting the forthcoming prices, buying low, and selling high so as to maximize their utilities in the long run are not considered in this model. Furthermore, in Wilhite (2001), there is an issue of information transparency. If all agents are globally connected, then the marginal rates of substitution of all agents become publicly known, and each agent only needs to figure out what would be the most favorable deal in the market and try to make a deal with the potential partner.

The most interesting result of the Wilhite model is the effect of network topologies on the speed of convergence to the competitive equilibrium (Walrasian equilibrium). The superiority of the small-world network relative to the fully-connected network is found by Wilhite (2001). Wilhite (2001) actually balanced the convergence speed and the search intensity and found that, while the small-world network may converge to the Walrasian equilibrium price more slowly, it is less costly if we also take search intensity into account. In fact, Wilhite (2001) can be regarded as one of the earliest studies showing the significance of network topologies in economic performance. In particular, Wilhite (2001) showed the economic efficiency of the small-world network.

In addition to the convergence speed, Wilhite (2001) also identified the important nodes (middle men), such as the centrality of the agents, and found some positive connections between centrality and wealth. The appearance of the middle men not only helps wealth creation (in terms of utility) in the society, but also contributes to the inequality of wealth distribution (again in terms of utility of final holdings).

2.3.2 Games in Social Networks

Given the development of Albin (1992) and Nowak and May (1992), and the advent of network science in the late 1990s, it is well anticipated that the lattice used in the early spatial games can be more generally replaced with social networks. This extension allows us to use the characterizations of network topologies to shed light on the properties of games, and motivates a series of new questions to address. Notice that the first-generation models allow us to see how the spatial structure can lead to computational irreducibility when certain kinds of behavioral rules are applied. A more restrictive issue focusing on the social dilemma games is how the introduced spatial structure can enhance (or inhibit) prosocial behavior, a kind of cooperative behavior that Nowak called *spatial reciprocity* (Nowak, 2006). The first question in general becomes harder when cellular automata (lattices) are replaced by social networks, but the subsequent studies focusing on the issue on prosocial behavior becomes even richer. This latter enrichment is probably the most profound achievement of the second-generation models. In this section, due to the space constraint, we only give a light flavor of the second-generation models of spatial games.

The first contribution to the second-generation model is Abramson and Kuperman (2001), who embedded the two-person prisoner's dilemma game with a *small-world network* (Watts and Strogatz, 1998). The setting of their model of the prisoner's dilemma game is standard. They began with a payoff matrix, which is also used in spatial games, for example, see eqn. (2.6), i.e.,

$$\Pi = \begin{bmatrix} \pi_{11} & \pi_{10} \\ \pi_{01} & \pi_{00} \end{bmatrix} = \begin{bmatrix} 1 & 0 \\ \pi_{01} & 0 \end{bmatrix}. \tag{2.24}$$

As in eqn. (2.6), π_{01} is normalized as the parameter for the *temptation to defect*.

The small-world network starts from a ring as shown in Figure 2.10 (the left panel), and each agent is connected to n agents (n is an even number). Half of the n neighbors are situated at the left-hand side of the agent, and another half are situated at the right-hand side of the agent. As shown in Figure 2.10, we have n equal to two, and hence one on each side. Then there is a rewiring probability for each of the links, p_r. By rewiring, we mean changing one end of an edge of nodes, and keeping the other end the same. An example of $p_r = 0.3$ is shown in Figure 2.10 (the right panel). To make the total number of links constant, not being perturbed by rewiring, some additional constraints have to be imposed. Given the set of neighbors of agent i, N_i, agent i plays a pairwise prisoner's dilemma game for each of his neighbors. In each round of the game, agent i chooses one

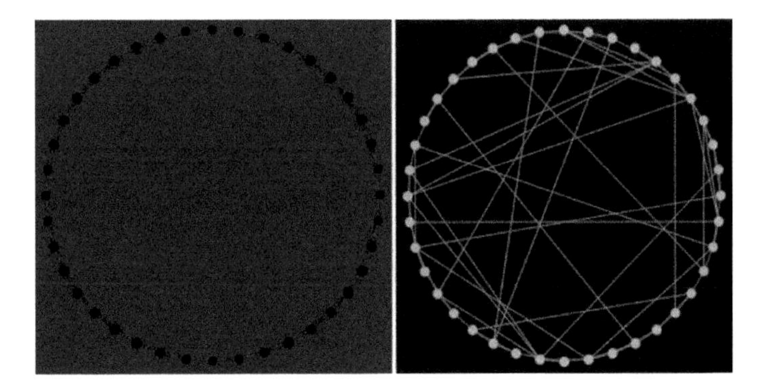

Figure 2.10 *Small World Network*

The above figure demonstrates a ring network (left) and a small-world network (right). In both networks, there are 40 agents (N = 40), and the small-world network on the right is generated from the left using a rewiring probability of 0.3.

of the two actions, either cooperating or defecting, and that action is applied to all his paired neighbors. Based on the actions chosen by him, $a_i(t)$, and by his neighbors, $a_j(t)$ ($j \in N_i$), his payoff of the round is summed up as follows.

$$\Pi_t(a_i(t)) = \sum_{j \in \mathbf{N}_i} \pi_{i,j}(a_i(t), a_j(t)). \tag{2.25}$$

At the end of the each round of the game, the agent imitates his most successful (wealthiest) agent.

$$a_i(t+1) = a_{j^*}(t) \tag{2.26}$$

where

$$j^* = \arg\max_{j \in \mathbf{N}_i} \Pi_t(a_j(t)). \tag{2.27}$$

In addition to imitation, a small probability of mutation is also introduced.

The model is then characterized by three parameters, namely, the temptation to defect (π_{01}), the rewiring probability (p_r), and the number of neighbors, n.[12] Abramson and Kuperman (2001) then studied how the changes in the values of these three parameters can affect the behavior of the game. Among the three parameters, the last two, i.e., n and p_r, are related to the network topology; hence, the effect of the network topology on prosocial behavior is also addressed in Abramson and Kuperman (2001).

[12] The total number of agents, N, is also an parameter, but this parameter has not been fixed at 1000 in Abramson and Kuperman (2001).

In this very initial step, instead of looking at the "fractal geometry" or any chaotic dynamics of cooperative behavior with their simply-behaved agents, they focus on the mesoscopic structure of prosocial behavior, namely, the fraction of cooperating agents. Through their simulation, they are able to show that when the p_r is small but positive, such as around a niche of 0.1, and n is small, say $n = 4$, there is a noticeable uprising fraction of defecting agents (a noticeable decline of the fraction of cooperating agents). "We have found surprising collective behaviors corresponding to the small-world systems, put on evidence by the enhancement of defection in situations where cooperation is the norm. (Abramson and Kuperman (2001), p.1)"

Using a game slight different from the prisoner's dilemma game, known as the snowdrift game[13], Hauert and Doebeli (2004) shows that cooperation is inhibited whenever evolution takes place in a spatially structured population, like the one studied in Nowak and May (1992). This result renders the role of spatial structure as game specific and not necessarily beneficial in promoting cooperative behavior.

Motivated by this result, the research direction is driven to search whether there is a network topology which can promote cooperative behavior and is not so game specific. A great attention was then drawn to the scale-free network (Barabasi, 1999). This is mainly because the scale-free network has great empirical relevance as opposed to many other network topologies; specifically, its degree distribution is more interesting and realistic when applied to real data. Santos and Pacheco (2005) is the first article which finds the significance of the scale-free network in cooperation enhancement. They found that cooperation even dominates over defection in heterogeneous, scale-free networks where the distribution density of local connections follows a power law. Specifically, contrary to Hauert and Doebeli (2004), cooperation becomes the dominating trait not only in the prisoner's dilemma game, but also in the snowdrift game, and that holds for all values of the parameters of both games.

For this interesting finding related to the scale-free network, Gómez-Gardeñes et al. (2007) provide a careful analysis. They compared the dynamical organization of cooperation on the scale-free networks and Erdös-Rényi random networks. The former has a power-law degree distribution, whereas the latter has a Poisson degree distribution. They found that while in homogeneous random networks pure cooperators are grouped into several clusters, in heterogeneous scale-free networks they always form a single cluster containing the most connected individuals (hubs). Hence, they elucidated the strong impact the degree distribution has on the social structure of prosocial agents.

Newman and Park (2003) argued that social networks differ from most other types of networks, including technological and biological networks, in two important ways. First, they have non-trivial clustering or network transitivity, and second, they show positive degree correlations, also called assortative mixing. Intuitively, degree correlation answers the question: do the high-degree vertices in a network associate preferentially with other high-degree vertices; hence, a hub is very likely to be connected to another hub? Pusch, Weber, and Porto (2008) find that both assortative degree–degree correlations

[13] In the prisoner's dilemma game, $\pi_{10} > \pi_{11} > \pi_{00} > \pi_{10}$, whereas in the snowdrift game the last inequality is reversed: $\pi_{10} > \pi_{00}$.

and enhanced clustering substantially enhance cooperation for a very high temptation to defect. Therefore, in addition to degree distribution, as suggested by Pusch, Weber, and Porto (2008), it appears reasonable to consider degree–degree correlations and enhanced clustering as additional factors in the evolution of sustainable cooperation.

Assenza, Gómez-Gardeñes, and Latora (2008) also have a similar finding on the clustering effect on cooperation enhancement. In addition, the transition to the zero level of cooperation becomes sharper as the clustering of the network increases. The sudden drop of the cooperation in highly clustered populations is because high clustering homogenizes the process of invasion across all degree classes by defectors and hence even decreases the chances of survival of low densities of cooperators in the network.

2.3.3 Network Consumption and Marketing

In the area of consumer behavior and marketing, the effects of networks have been known for a long time. The familiar *bandwagon effect* indicates the desire to be in style or to have a good simply because almost everyone else has it. This property, known as *network externalities*, has been well analyzed in Katz and Shapiro (1985). *Metcalfe's law*, formulated by Metcalfe in regard to Ethernet in 1973, explains many of the network effects of communication technologies and networks such as the Internet and the World Wide Web (Gilder, 1993). By this law, the value of a network equals approximately the square of the number of users of the system. The price which somebody is willing to pay to gain access to a network is based solely on the number of other people who are currently using it. Fax machines and the Internet are prime examples. The more people using the service, the more others will be willing to buy in. Hence, network topology can matter because it can introduce critical points to market demand, which in turn cause the demand curve to no longer be continuous as conventional economics assumes. This phenomenon is known as the *avalanche effect*.

In this section, we shall use Chen, Sun, and Wang (2006) as an illustration of effects of network topologies on consumption behavior. Their agent-based model is simple but contains many core ideas which are still drawing researchers' attention. Some other related studies will be mentioned in Section 2.4.

Neighborhood-Based Discrete Choice Model

For a market network model, Chen, Sun, and Wang (2006) started with modeling agents' decisions to buy, as in the following optimization problem:

$$\max_{a_i \in \{0,1\}} V_i = \max_{a_i \in \{0,1\}} a_i \left(H_i + \sum_{j \in N_i} w_{ij} a_j - P \right), \tag{2.28}$$

where a_i is an indicator for a binary decision.

$$a_i = \begin{cases} 1, & \textit{if agent } i \textit{ decides to buy}, \\ 0, & \textit{otherwise}. \end{cases} \tag{2.29}$$

What is inside the bracket is the classical linear willingness to pay function. H_i represents agent i's idiosyncratic utility of consuming the commodity. Written in terms of monetary value, it functions as the *reservation price* of agent i, i.e., his idiosyncratic willingness to pay for the commodity. P is the price of the commodity. The difference between H_i and P can be taken as the familiar *consumer's surplus* when agent i decides to buy ($a_i = 1$).

Now, agent i is situated in a network, and is directly connected to a collection, $\mathbf{N}_i$, of some other agents, called agent j, where $j \in \mathbf{N}_i$. The network effect is captured by the parameter w_{ij}, which characterizes the influence on agent i of the decision of agent j. Clearly the larger the absolute value of w_{ij}, the larger the influence of agent j on agent i.[14] If $w_{ij} = 0$, for all $j \in \mathbf{N}_i$, then there is no network effect on agent i; essentially, agent i makes his decision *independently*.

Consider the literature on *attention control*. If agent i's attention to the external world has an upper limit, say W_i, then

$$\sum_{j \in \mathbf{N}_i} w_{ij} = W_i. \tag{2.30}$$

Furthermore, if agent i is indifferent to all his neighbors j, then W is uniformly distributed over $\mathbf{N}_i$, i.e.,

$$w_{ij} = \frac{W_i}{N_i}, \quad \forall \ j \in \mathbf{N}_i, \tag{2.31}$$

where N_i is the number of agents (vertices) connected to agent i. Under this setting, as the size of his neighborhood becomes larger, the influence of each individual neighbor on agent i becomes smaller.

The *external influence* of the network is demonstrated by the feature that agents are *heterogeneous* in preference, H_i.[15] It should be clear now that the optimization problem (2.28) leads to a very simple solution, namely,

$$a_i = \begin{cases} 1, & \textit{if } H_i + \sum_{j \in \mathbf{N}_i} w_{ij} a_j - P \geq 0, \\ 0, & \textit{otherwise}. \end{cases} \tag{2.32}$$

To fully represent the network dynamics, it will be useful to extend eqn. (2.28) as an individual decision to collective decisions, as in (2.33).

$$\mathbf{A(t+1)} = g(\mathbf{H} + \mathbf{W} \cdot \mathbf{D} \cdot \mathbf{A(t)} - \mathbf{P}), \tag{2.33}$$

[14] w_{ij} can in general be both positive and negative, while in their application, Chen, Sun, and Wang (2006) only consider the positive externality.

[15] Obviously, if all agents share the same H, there will be no need to study the network influence in this specific context, since "influence" means causality: someone will have to buy first, and someone will follow. If the agents are homogeneous in terms of preference, then from eqn. (2.28), their decision will be perfectly homogeneous, including their timing, so that the network influence does not exist.

where

$$\mathbf{H} = \begin{bmatrix} H_1 \\ H_2 \\ \cdot \\ \cdot \\ \cdot \\ H_n \end{bmatrix}, \mathbf{W} = \begin{bmatrix} W_1 & 0 & \dots & 0 \\ 0 & W_2 & \dots & 0 \\ \dots & \dots & \dots & \dots \\ \dots & \dots & \dots & \dots \\ \dots & \dots & \dots & \dots \\ 0 & 0 & \dots & W_n \end{bmatrix}, \mathbf{D} = \begin{bmatrix} \delta_{11} & \delta_{12} & \dots & \delta_{1n} \\ \delta_{21} & \delta_{22} & \dots & \delta_{2n} \\ \dots & \dots & \dots & \dots \\ \dots & \dots & \dots & \dots \\ \dots & \dots & \dots & \dots \\ \delta_{n1} & \delta_{n2} & \dots & \delta_{nn} \end{bmatrix}, \mathbf{A_t} = \begin{bmatrix} a_1(t) \\ a_2(t) \\ \cdot \\ \cdot \\ \cdot \\ a_n(t) \end{bmatrix}, \mathbf{P} = \begin{bmatrix} P_1 \\ P_2 \\ \cdot \\ \cdot \\ \cdot \\ P_n \end{bmatrix}.$$

The vector matrix **H** just stacks up an individual's idiosyncratic preferences H_i. The diagonal matrix **W** has the contribution from each individual i's neighbor to i as described in eqn. (2.31). The matrix **D** is a general representation of the network. Each entry δ_{ij} represents a connection (edge) between agent i and agent j. While a great flexibility of the connection may exist, two assumptions are frequently made. First, either the connection exists or it does not, i.e., there is no *partial* connection, nor is there any difference in the degree of connection. Therefore,

$$\delta_{ij} = \begin{cases} 1, & \textit{if agent } i \textit{ is connected to agent } j, \\ 0, & \textit{otherwise.} \end{cases} \tag{2.34}$$

Second, we assume that the connection (edge), if it exists, is *bi-directional*. Hence,

$$\delta_{ij} = \delta_{ji}. \tag{2.35}$$

In other words, the matrix **D** is *symmetric*. Notice that, according to eqn. (2.31), the matrix **W** is immediately determined by the matrix **D**.

To illustrate, the matrix representations of a ring, star, and fully-connected network with $n = 4$, denoted by $\mathbf{D}_{ring}$, $\mathbf{D}_{star}$, and $\mathbf{D}_{full}$, respectively, are given below from left to right:

$$\mathbf{D}_{ring} = \begin{bmatrix} 0 & 1 & 0 & 1 \\ 1 & 0 & 1 & 0 \\ 0 & 1 & 0 & 1 \\ 1 & 0 & 1 & 0 \end{bmatrix}, \mathbf{D}_{star} = \begin{bmatrix} 0 & 1 & 1 & 1 \\ 1 & 0 & 0 & 0 \\ 1 & 0 & 0 & 0 \\ 1 & 0 & 0 & 0 \end{bmatrix}, \mathbf{D}_{full} = \begin{bmatrix} 0 & 1 & 1 & 1 \\ 1 & 0 & 1 & 1 \\ 1 & 1 & 0 & 1 \\ 1 & 1 & 1 & 0 \end{bmatrix}.$$

With these connections, the corresponding influence matrix representations $\mathbf{W}_1$, $\mathbf{W}_2$, and $\mathbf{W}_3$ are:

$$\mathbf{W}_{ring} = \begin{bmatrix} \frac{W}{2} & 0 & 0 & 0 \\ 0 & \frac{W}{2} & 0 & 0 \\ 0 & 0 & \frac{W}{2} & 0 \\ 0 & 0 & 0 & \frac{W}{2} \end{bmatrix}, \mathbf{W}_{star} = \begin{bmatrix} \frac{W}{3} & 0 & 0 & 0 \\ 0 & W & 0 & 0 \\ 0 & 0 & W & 0 \\ 0 & 0 & 0 & W \end{bmatrix}, \mathbf{W}_{full} = \begin{bmatrix} \frac{W}{3} & 0 & 0 & 0 \\ 0 & \frac{W}{3} & 0 & 0 \\ 0 & 0 & \frac{W}{3} & 0 \\ 0 & 0 & 0 & \frac{W}{3} \end{bmatrix}.$$

The vector $\mathbf{A}_t$ stacks up the binary decisions made by all agents in period t, with each entry described by eqn. (2.29). To trace the dynamics of binary choices, binary decisions are now indexed by t. Initially, $a_i(0) = 0$ for all $i = 1, \ldots, N$. The function g then drives the dynamics of $\mathbf{A}(t)$ based on each agent's optimal decision (2.32). The vector $\mathbf{P}$ is the price charged to each individual. Notice that we have indexed this price by individual i, which indicates the possibility that the price does not have to be homogeneous among agents, considering that we want to study the effect of price discrimination under this framework.

This matrix representation leaves us with great flexibility to deal with various network topologies. As has been demonstrated in Chen, Sun, and Wang (2006), it helps us to deal with the scale-free network. In addition, by varying $\mathbf{D}$, there can be other advantages. Firstly, $\mathbf{D}$ can be *asymmetric*. This is a desirable variation since social influence in general is not symmetric. Secondly, it does not have to be a binary matrix. In fact, δ_{ij} can be any continuous variable between 0 and 1. This can help us to capture different degrees of connection. In sum, a continuous and asymmetric $\mathbf{D}$ provides us with an opportunity to study more complex network topologies that are beyond simple geometric representations.

The Scale-Free Network

The *scale-free network* was first proposed by Barabasi and Albert (1999). The scale free model is based on two mechanisms: (1) networks grow incrementally with the addition of new vertices; and (2) new vertices are attached *preferentially* to vertices that are already well connected.

Let us assume that initially the network is composed of m_0 vertices, and that each is connected to m other vertices ($m < m_0$). Then, at each point in time, a number of new vertices, m_T are added to the network, each of which is again connected to m vertices of the net by the *preferential linking*. It is implemented as follows. At time T, each of the new m_T vertices is randomly connected to a node $i \in \mathbf{V}_T$ according to the following distribution

$$\pi_i = \frac{k_i}{\sum_{j \in \mathbf{V}_T} k_j}, i \in \mathbf{V}_T, \tag{2.36}$$

where $\mathbf{V}_T = \{1, 2, \ldots, \sum_{t=0}^{T-1} m_t\}$. That is the probability of becoming attached to a node of degree k is proportional to k, $\pi(k)$, and nodes with high degrees attracts new connections with a high probability. To avoid redundancy, the random attachment with (2.36) is done by sampling *without* replacement.

In addition to the scale-free network, Chen, Sun, and Wang (2006) also considered something in between the scale-free network and the random network, i.e., a mixture of the two. They allowed each of the incoming m_t agents (vertices) to have a probability, denoted by $1-q$, of being connected to the incumbent agents simply *randomly*, and hence a probability of q being connected to them with preferential linking. By controlling q, one can have a mixture network which is very close to the random network (e.g., $q \approx 0$), and another mixture network which is very close to the scale-free network (e.g., $q \approx 1S$).

It is, therefore, feasible to examine how the emergent properties may change along this spectrum. For simplicity, we shall call this the q-network. Clearly, the q-network is a random network if $q = 0$, and is a scale-free network if $q = 1$.

To study the impact of the network topology on market demand, Chen, Sun, and Wang (2006) define a *penetration rate* as the percentage of buyers in the market, i.e.,

$$\bar{a} \equiv \frac{\#\{i : a_i = 1\}}{N}. \tag{2.37}$$

Since each consumer buys at most one unit, $0 \leq \bar{a} \leq 1$. To make the roles of the price and the network topology explicit, they write $\bar{a}$ as $\bar{a}(P, g)$, where g is the respective network topology,

$$g \equiv \{(i,j) : \delta_{ij} = 1\}.$$

A network topology g is said to *uniformly dominate* other network topologies and is denoted by g^*, if

$$\bar{a}(P, g^*) \geq \bar{a}(P, g), \forall\ P,\ \forall\ g. \tag{2.38}$$

Since the uniformly-dominating network topology may not exist, an alternative measure is to define a *maximum price*, P^*, as follows:

$$P^* \equiv \max_P \{P : \bar{a}(P) = 1\}. \tag{2.39}$$

Again, to acknowledge the influence of the network topology, P^* is also written as $P^*(g)$.

Instead of P^*, we may consider a weighted average market penetration rate ($\bar{a}(P)$) with respect to a distribution of P, and compare the weighted averages among different network topologies:

$$\mu_P = \int_{\underline{P}}^{\overline{P}} \bar{a}(P) f_P(P) dP, \tag{2.40}$$

where $\underline{P}$ and $\overline{P}$ define an effective range of P, such that

$$\begin{cases} \bar{a}(P, g_\emptyset) = 0, & \textit{if } P \geq \overline{P}, \\ \bar{a}(P, g_\emptyset) = 1, & \textit{if } P \leq \underline{P}, \\ 0 < \bar{a}(P, g_\emptyset) < 1, \textit{otherwise}, \end{cases} \tag{2.41}$$

where $g_\emptyset$ denotes the isolated network ($g = \emptyset$). f_P is a density function of P, and when it is uniform over $[\underline{P}, \overline{P}]$, μ_P in a sense can be regarded as a *social welfare measure* if the marginal cost is zero. In this specific context, $\mu_P(g)$ can also inform us how the network topology impacts social welfare.

With the basic framework and analytical concepts, Chen, Sun, and Wang (2006) simulated the agent-based model of network consumption using random networks, regular networks (ring networks), scale-free networks, and mixture networks. To isolate the working of the network topology, they chose the isolated network as the benchmark. In addition, to maintain a degree of heterogeneity among agents, H_i were *uniformly* sampled from the interval $[1, 2]$. With this design of **H**, the corresponding demand curve $\bar{a}(p)$ was, therefore, restricted to the same interval, and was discretized with an increment of 0.02.

Consumer's Surplus

Figure 2.11 is the demand curve $\bar{a}(P)$ under different network topologies. The resultant demand curve is based upon 100 independent runs each with a new initialization of the matrix **H**. So, each single point along $\bar{a}(P)$ is an average of 100 observations. Based on these estimated $\bar{a}(P)$, one can derive the consumer's surplus. Not surprisingly, the world network (fully connected network) and the isolated network provide the two extremes of the consumer's surplus: a maximum of 0.876 is given by the fully connected world network, whereas a minimum of 0.509 is given by the isolated network. The consumer's surplus of the scale-free network and the regular network lie in between, and their differences are not that significant (0.776 vs. 0.766).

Figure 2.11 basically confirms that the consumer's surplus is a positive function of *network density*. Chen, Sun, and Wang (2006) further explored the effect of other network characterizations on the consumer's surplus, specifically the cluster coefficient and average shortest length of a network. To do so, they considered a class of the mixture

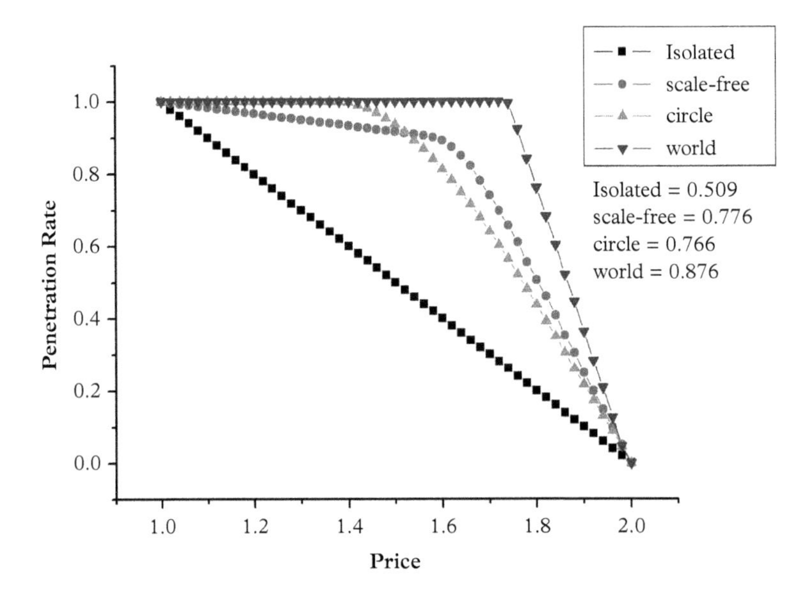

Figure 2.11 *Demand Curves in Various Networks*

Table 2.1 *Consumer's Surplus, Cluster Coefficient, and Average Shortest Length*

$$CS = \alpha_0 + \alpha_1 C + \alpha_2 L$$

Regressors	coefficient	p-value
Constant	0.78019 (α_0)	0.0000
Cluster Coefficients (C)	0.52018 (α_1)	0.0041
Average Shortest Length (L)	–0.0092 (α_2)	0.1112

The R^2 of the above simple linear regression is 0.466 (R^2-adjusted=0.446), and the mean square error is 0.006.

networks by controlling q from 0, 0.1,..., to 0.9, and 1.[16] Five independent runs are conducted for each q-network, and this gives us in total 55 networks.[17] The resulting consumer's surplus of these 55 networks is then regressed against two independent variables, namely, the cluster coefficient (C) and the average shortest length (L). The regression results are shown in Table 2.1. It can be found from Table 2.1, that these two variables C and L can jointly explain almost 50% of the variation in the consumer's surplus. In addition, both regression coefficients have signs consistent with our intuition: the consumer's surplus is positively related to the cluster coefficient, whereas it is adversely affected by the average shortest length.

Avalanche Effect

Another interesting type of behavior of the demand curve is its *jump* or *discontinuity*, known as an *avalanche*, which can be formulated as follows. The demand curve is said to demonstrate an avalanche at price P_v if

$$d_{P_a} = \bar{a}(P_a - \epsilon) - \bar{a}(P_v) > \theta_v. \tag{2.42}$$

The avalanche effect can then be measured by the probability of having an avalanche:

$$Prob_v = Prob(d_{P_v} > \theta_v). \tag{2.43}$$

However, as mentioned in Chen, Sun, and Wang (2006), the avalanche effect may have nothing to do with the network topologies, and can purely come from the homogeneous group of agents. To avoid "*spurious*" avalanches and to disentangle the effect of homogeneity from the effect of the network topology, a great degree of heterogeneity of H_i ($H_i \sim U[1,2]$) is imposed to examine the chance of observing avalanches with respect to different topologies.

[16] Each q-mixture network has a probability of q being a scale-free network and $1 - q$ being a random network.

[17] Only the **D** matrix is regenerated for each independent run. The **H** matrix remains unchanged for all of the 55 matrices.

Table 2.2 *Avalanche Effect*

Topology/p_a	1.98	1.94	1.84	1.74	1.54
Random	0.01	0.04	0.05	0.06	0
q (q = 0.5)	0.07	0.05	0.11	0.03	0
Scale-free	0.06	0.06	0.04	0.09	0
World	0	0.2	0.35	0	0
Regular	0	0	0	0	0

The avalanche effect is defined in eqn. 2.43. θ_a is set to 10%, i.e., 10% of the market capacity. $\epsilon = 0.02$.

A measure of the avalanche effect is defined in eqn. (2.43), which depends on three parameters, namely, a threshold (θ_v),the perturbation size (ϵ), and the evaluation point (P_v). Since it would be meaningless to consider small jumps when talking about "avalanches," Chen, Sun, and Wang (2006), therefore, set θ_v to 0.1, i.e., 10% of the market capacity, and ϵ to 0.02. In other words, if by discounting two cents only, one can suddenly increase sales by 10% of the market capacity, then an avalanche is detected. Finally, since H_i, uniformly distributed over the range [1,2], they choose five evaluation points of P_v from 1.98 to 1.54 (see Table 2.2). This roughly corresponds to the upper half of the distribution of H_i, which should be the ideal place to monitor the avalanche if there is one.

As to the network topology, except for the isolated network, all other partially-connected or fully-connected network topologies are tried in their experiment. 100 runs are conducted for each network topology. The results are shown in Table 2.2. From Table 2.2, they found that except for the regular network, avalanches unanimously exist in all four other types of network, while their structures are different in terms of the distribution of the tipping points (P_v) and the tipping frequencies ($Prob_v$). For example, the world network has more concentrated tipping points ($P_v = 1.94, 1.84$) and a high tipping frequency at these tipping points ($Prob_v = 0.2, 0.35$), whereas the other three network topologies have evenly distributed tipping points, although with much lower tipping frequencies.

Hysteresis Effect

The next issue is a more subtle one in that the demand curve may not be unique and is *state-dependent*. That is, the demand given the price P can depend on what happens before. Has the price before been *higher* or *lower*? This phenomenon, known as the *hysteresis effect*, arises because the demand curve, or equivalently, the penetration rate, derived by decreasing the price is different from the one derived by increasing the price. Formally, hysteresis happens at price P when

$$\bar{a}_u(P) > \bar{a}_d(P), \tag{2.44}$$

where $\bar{a}_u$ and $\bar{a}_d$ refer to the penetration rates derived by moving downstream and upstream respectively. The hysteresis effect of a network topology can then be measured by

$$R \equiv \int_{\underline{P}}^{\overline{P}} (\bar{a}_u(P) - \bar{a}_d(P)) f_P(P) dP. \tag{2.45}$$

$R(g)$ denotes the hysteresis effect of the network topology g.

To examine the *hysteresis effect* as defined by eqn. (2.45), what Chen, Sun, and Wang (2006) did was to first derive the demand curve by running the price downstream, and then by running the price upstream. In this experiment, they considered all network topologies mentioned above. For the q-networks, they only considered the cases $q = 0, 0.5$, and 1. One hundred runs are conducted for each network topology. The result shown in Table 2.3 is, therefore, the average of these 100 runs. Table 2.3, columns two and three, show the consumer's surplus associated with the $\bar{a}_d(P)$ and $\bar{a}_u(P)$. The fourth column of Table 2.3 shows the difference between the two surpluses, i.e., the measure of the hysteresis effect R. From this column, we can see that both the isolated network and the fully-connected network have very little hysteresis effect. As expected, it is identically 0 for the isolated network, and is only 0.013 for the fully-connected network. However, all partially connected networks show some degree of hysteresis. Among them, the scale-free, random, and q ($q = 0.5$) networks are close, whereas the regular network has a strong hysteresis effect.

Network Size

None of the questions discussed so far may be independent of the market size N (the size of the network). It is then crucial to know the limiting behavior as well. For example, would

$$\lim_{N \to \infty} Prob_{v,N} = 0? \tag{2.46}$$

Table 2.3 *Hysteresis Effect*

Topology	Downstream ($r_d(p)$)	Upstream ($r_u(p)$)	R
Isolated	0.510	0.510	0
World	0.876	0.889	0.013
Scale-free	0.776	0.851	0.075
Random	0.738	0.816	0.079
q ($q = 0.5$)	0.758	0.847	0.089
Regular	0.766	0.943	0.177

and

$$\lim_{N\to\infty} R_N = 0? \tag{2.47}$$

If eqns. (2.46) and (2.47) are valid, then in a sense the network topology will not matter when the market becomes thick. In this spirit, we can even ask whether the demand curves associated with different network topologies will be *asymptotically equivalent*.

It is interesting to know whether the property of hysteresis and avalanches obtained above is sensitive to the *size* of the network. In particular, when the network's size becomes very large (ideally infinite), whether these two properties can still be sustained. Chen, Sun, and Wang (2006), therefore, simulated networks with sizes of 1000, 3000, and 5000. The results are shown in Figures 2.12 and 2.13. Shown on the left part of these two figures are the demand curve $\bar{a}(P)$ associated with an isolated network and a scale-free network. By looking at these two graphs visually, we can see that the avalanche

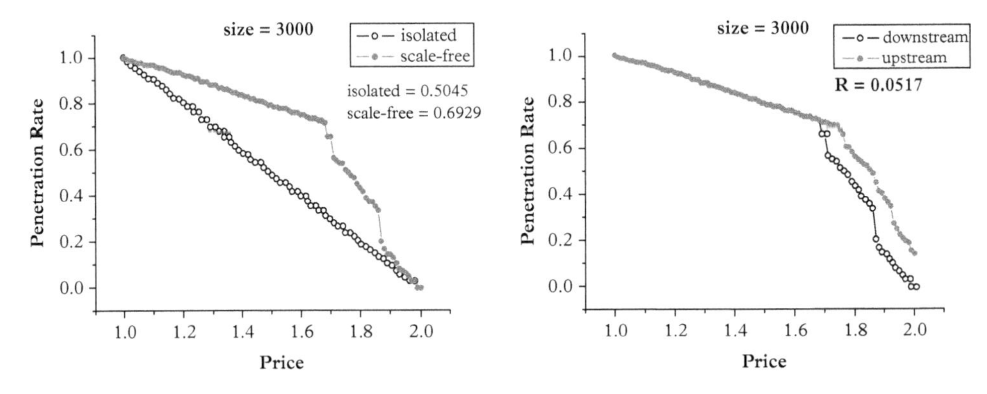

Figure 2.12 *Demand curve with different network topologies (Size=3000)*

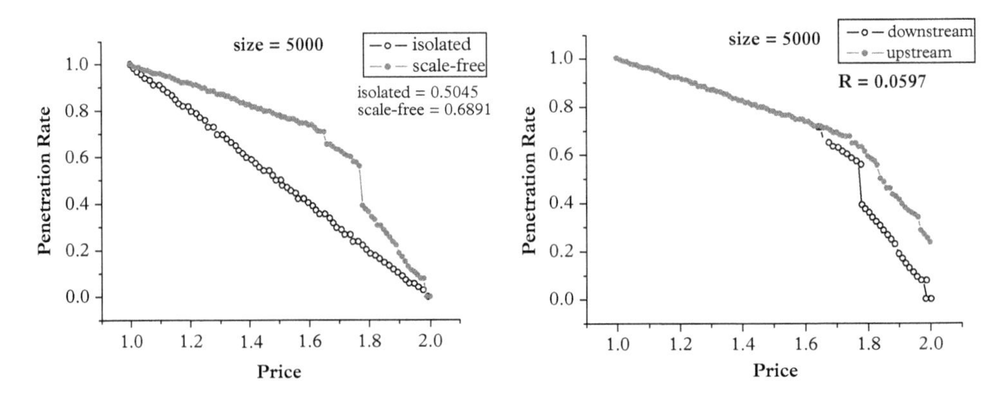

Figure 2.13 *Demand curve with different network topologies (Size=5000)*

effect, characterized by the noticeable jumps in the demand curve, does not disappear as the size gets larger. Furthermore, the right part of the two figures shows the demand curve derived by running downstream and upstream. The bifurcation is clearly there, with hysteresis measures of 0.0517 (N = 3000) and 0.0597 (N = 5000). Compared to the R of the scale-free network from Table 2.3, these two Rs become smaller, but are still quite distinct from the fully-connected network. Therefore, the asymptotic equivalence of network topologies does not hold here as far as these two properties are concerned. Generally speaking, the finding that network topology matters does not depend on network size.

2.3.4 Macroeconomic Stability

While the rapid growth in the literature on social networks indicates the relevance of the network topologies to economic performance (Granovetter, 2005), there is no formal thorough examination of the effect of these network topologies on macroeconomic stability. This challenge is easy to realize, as pointed out in Fogli and Veldkamp (2012), "[w]hile researchers have mapped social networks in schools or online communities..., mapping the exact social network structure for an entire economy is not feasible. (Fogli and Veldkamp (2012), p. 1)" To the best of our knowledge, Chen, Chang, and Wen (2014) is the first work in this direction. Given the daunting nature of this research, they have adopted a strategy that can allow us to at least have a foot in the door before we can move into the hall. In this section, we will introduce their work as a demonstration of the network effect on macroeconomic stability. A few related studies will be given in Section 2.4.

The strategy is to choose the macroeconomic model that is most accessible to the inclusion of different network topologies of economic units. Having said that, we are aware that most agent-based macroeconomic models would have such network topologies, explicitly or implicitly, as part of their models, but not all of them make it easier for the expansions to facilitate a large-scale simulation with a large variety of network topologies. It turns out that the agent-based DSGE (Dynamic Stochastic General Equilibrium) model, as initiated by De Grauwe (2010, 2011) becomes a straightforward choice at this initial stage.

In the standard New Keynesian DSGE model, the representative agent always has rational expectations. De Grauwe relaxed this stringent assumption and started the agent-based version of the DSGE model by replacing the homogeneous rational expectations of output with the heterogeneous boundedly rational counterparts. De Grauwe (2010, 2011) started at the *mesoscopic* level. It distinguishes agents by *type*, the optimist type and the pessimist type, and hence the interaction, learning, and adaptation of agents are operated only at the level of agent types rather than going down to the level of individual agents. Since individuals are not directly involved, social networks, i.e., the connections between these individuals, certainly have little role to play in this model. Chen, Chang, and Wen (2014) on the other hand, start at the microscopic level (individual level), and the interaction, learning, adaptation, and decision-making are all individually based. It is a manifestation of the network-based discrete choice model.

Social networks in these models are obviously indispensable, since they are the driving force behind the subsequent interactions of agents.

Chen, Chang, and Wen (2014) used the well-known Ising model, invented by the physicist Ernst Ising in his PhD thesis in 1924, for interacting agents with regard to their mimetic behavior. This Ising model is operated with different embedded network topologies. They simulated the macroeconomy by using the agent-based DSGE model augmented with the Ising-style market sentiment, which is embedded with different network topologies. They used the agent-based (network-based) DSGE model to generate the time series of two major macroeconomic variables, the output gap, and inflation, and then to study the effects of the network topology on their stability (volatility). They carried out a large-scale simulation based on a large sample of networks with diversified characteristics. An econometric analysis is applied to examine the effect of each network characteristic.

Chen, Chang, and Wen (2014) formally formulated the research question as to the contribution of network topology to economic stability in terms of five major characteristics; specifically, they asked what the relationship is between the economic stability and degree, clustering coefficient, average path length, betweenness centrality, and closeness centrality. A formal econometric examination of the effect of network characteristics on economic stability will involve a review of the possible functional form between the stability variables and the characteristics included. Their baseline model is a linear function of the five network characteristics. They first examined what are the fundamental results that they can have from this basic setup, and then go further to examine whether these fundamental results are sensitive to some possible augmentations. The two augmentations that they considered are the possible influence of the more complex function (*non-linearity*) and the inclusion of other important characteristics (*omitted variables*).

The idea of the Monte Carlo simulation is to have a more extensive sampling so that a thorough examination of the effect of various network characteristics is possible. To achieve this goal, they employed three network generating mechanisms, namely, small-world networks, scale-free networks, and random networks, to generate a sample of 500 networks. Since each network can be characterized by a few parameters, the idea of the random generating mechanism is simply to randomly select a set of parameter values from the parameter space and then generate the network based on the chosen parameter values. In Table 2.4, the parameter space for each type of network is specified. Among the three network families, 437 are sampled from the small-world networks, 36 from the random networks, and 27 from the scale-free networks.

Through this network random generation mechanism, they were able to generate networks with an average degree from 2 to 80, an average cluster coefficient from 0 to 0.8240, an average path length from 1.1919 to 87.0463, a maximum betweenness centrality from 13 to 4681, and a maximum closeness centrality from 0.003 to 1. These ranges, together with other statistics, are shown in Table 2.5. This large range and variation, therefore, facilitates their subsequent regression analysis. A choice of λ that is not large, and also not too small, was made throughout their entire simulation.

Table 2.4 *Parameter Setting for Network Random Generation*

Parameter	Parameter Space	
General		
N	100	number of agents
λ	0.5	intensity of choice
T	300	number of simulation periods
Small-World Networks (437)		
p	[0,1]	rewiring rate
κ	[2,45]	number of neighbors on the left and on the right
Random Networks (36)		
$Prob(\{b_{i,j} = 1\})$	{ 0.5 }	probability of connecting any random pair
Scale-Free Networks (27)		
N_0	[5,90]	initial size (scale-free)
p_0	0.5	initial connecting probability
θ	[0.2, 100]	scaling factor

Inside the parentheses are the numbers generated from the respective networks. There are 437 samples from the family of small-world networks, 36 from the family of random networks, and 27 from the family of scale-free networks.

Table 2.5 *Characteristics of the Randomly Generated Networks: Basic Statistics*

Network Characteristics	Min	Max	Mean	Variance
AD	2	80	39.6134	523.0591
ACC	0	0.8240	0.4288	0.0485
APL	1.1919	87.0463	3.4512	100.9276
MBC	13.0940	4681.677	277.9621	835680
MCC	0.0030	1	0.0090	0.0020

The above table shows the basic statistics of the five network characteristics over the 500 networks, which are randomly generated using Table 2.4. "AD", "ACC", "APL", "MBC", and "MCC" are the abbreviations for the average degree, average clustering coefficient, average path length, maximum betweenness centrality, and maximum closeness centrality, respectively.

To examine the effect of network characteristics on economic stability, they began with eqn. (2.48) and eqn. (2.49), and estimated the coefficients of the two equations *individually*, say, by ordinary least squares (OLS).

$$\begin{aligned} Var(\text{output gap}) = {} & \beta_{y,0} + \beta_{y,1} \times AD + \beta_{y,2} \times ACC + \beta_{y,3} \times APL \\ & + \beta_{y,4} \times MBC + \beta_{y,5} \times MCC + \xi_y \end{aligned} \tag{2.48}$$

$$\begin{aligned} Var(\text{inflation}) = {} & \beta_{\pi,0} + \beta_{\pi,1} \times AD + \beta_{\pi,2} \times ACC + \beta_{\pi,3} \times APL \\ & + \beta_{\pi,4} \times MBC + \beta_{\pi,5} \times MCC + \xi_\pi \end{aligned} \tag{2.49}$$

They estimated these equations as a set of equations, using the familiar *seemingly unrelated regression estimation* (SURE). SURE can be useful when the error terms ξ_y and ξ_π are correlated. Let us rewrite the set of equations (2.48) and (2.49) into a single equation as in (2.50). Eqn. (2.50) is written in a compact form. For all notations used in this compact form, one can find their correspondence in Table 2.6.

$$\Gamma = \boldsymbol{\beta}_0 + \boldsymbol{\beta}\Psi + \Xi \tag{2.50}$$

where

$$\Gamma = \begin{pmatrix} V_y \\ V_\pi \end{pmatrix}, \boldsymbol{\beta}_0 = \begin{pmatrix} \beta_{y,0} \\ \beta_{\pi,0} \end{pmatrix}, \boldsymbol{\beta} = \begin{pmatrix} \boldsymbol{\beta}_y & 0 \\ 0 & \boldsymbol{\beta}_\pi \end{pmatrix}, \Psi = \begin{pmatrix} \mathbf{X} \\ \mathbf{X} \end{pmatrix}, \Xi = \begin{pmatrix} \xi_y \\ \xi_\pi \end{pmatrix},$$

and

$$\mathbf{X}' = \left(X_1\ X_2\ X_3\ X_4\ X_5\right),$$

$$\boldsymbol{\beta}_y = \left(\beta_{y,1}\ \beta_{y,2}\ \beta_{y,3}\ \beta_{y,4}\ \beta_{y,5}\right),$$

$$\boldsymbol{\beta}_\pi = \left(\beta_{\pi,1}\ \beta_{\pi,2}\ \beta_{\pi,3}\ \beta_{\pi,4}\ \beta_{\pi,5}\right).$$

Table 2.6 *Multivariate Regression Analysis: Explained and Explanatory Variables*

Notation	Variables
Explained Variables	
V_y	Output Gap Volatility
V_π	Inflation Volatility
Explanatory Variables	
X_1	Average Degree (AD)
X_2	Average Cluster Coefficient (ACC)
X_3	Average Path Length (APL)
X_4	Maximum Betweenness Centrality (MBC)
X_5	Maximum Closeness Centrality (MCC)

Table 2.7 *The SURE Results*

	Output gap($\times 10^{-6}$)			Inflation($\times 10^{-6}$)		
Intercept	420163	(2507.24)	*	468401	(15015.9)	*
AD	−6.7	(−0.5)		12	(4.75)	*
ACC	824	(0.5447)		−1370	(−5.40)	*
APL	9.713	(0.2297)		213	(141.43)	*
MBC	2.382	(27.59)	*	0.9817	(61.09)	*
MCC	−5160	(−3.05)	*	−9820	(−31.16)	*
R^2		0.67066			0.98559	
Adj R^2		0.66732			0.98544	

The table above shows the SURE regression results. The second and the fifth columns are the estimated coefficients in the output gap equation and the inflation equation. The third and the sixth columns are the t values of the respective coefficients. The fourth and the seventh columns denote the statistical significance at the 5% level with the symbol "*".

Table 2.7 gives the SURE regression results. From this table, we can see that not all network characteristics can have an effect on economic stability. The only characteristic that has a consistent effect on both GDP fluctuations (V_y) and inflation fluctuations (V_π) is *centrality*. Interestingly enough, the two centrality measures have an opposite effect. The betweenness centrality plays a destabilizing role, whereas the closeness centrality plays a stabilizing role. On the other hand, all five characteristics can significantly contribute to the inflation stability or instability. The ones playing the stabilizing role are the cluster coefficient and the maximum closeness centrality, whereas the ones playing the destabilizing role are the average degree, average path length, and maximum betweenness centrality.

Non-Linearity

The effects of network characteristics on economic stability can be much more complex than eqn. (2.50) can represent. Chen, Chang, and Wen (2014) considered a general representation form of the universe of nonlinear models, namely the *polynomial approximation* (the *Taylor approximation*), specifically the linear models with Taylor expansion up to the second and the third order, as shown in eqns. (2.51), (2.52), and (2.53).

$$\Gamma = \boldsymbol{\beta}_0^2 + \boldsymbol{\beta}^2\Psi + \boldsymbol{\gamma}^2\Psi^2 + \Xi_2 \tag{2.51}$$

$$\Gamma = \boldsymbol{\beta}_0^3 + \boldsymbol{\beta}^3\Psi + \boldsymbol{\gamma}^3\Psi^3 + \Xi_3 \tag{2.52}$$

$$\Gamma = \boldsymbol{\beta}_0^{(q)} + \boldsymbol{\beta}^{(q)}\Psi + \boldsymbol{\gamma}^{(q)}\Psi^{(q)} + \Xi_{(q)}, \quad q = 1, 2, 3, 4, 5, \tag{2.53}$$

where

$$\Psi^2 = \begin{pmatrix} \mathbf{X}^2 \\ \mathbf{X}^2 \end{pmatrix}, \Psi^3 = \begin{pmatrix} \mathbf{X}^3 \\ \mathbf{X}^3 \end{pmatrix}, \Psi^{(q)} = \begin{pmatrix} \mathbf{X}^{(q)} \\ \mathbf{X}^{(q)} \end{pmatrix} \quad (q = 1, 2, 3, 4, 5),$$

and

$$\begin{aligned}
\mathbf{X}^{2'} &= \left(X_1^2\ X_2^2\ X_3^2\ X_4^2\ X_5^2\right), \\
\mathbf{X}^{3'} &= \left(X_1^3\ X_2^3\ X_3^3\ X_4^3\ X_5^3\right), \\
\mathbf{X}^{(1)'} &= \left(X_1X_2\ X_1X_3\ X_1X_4\ X_1X_5\right), \\
\mathbf{X}^{(2)'} &= \left(X_2X_1\ X_2X_3\ X_2X_4\ X_2X_5\right), \\
\mathbf{X}^{(3)'} &= \left(X_3X_1\ X_3X_2\ X_3X_4\ X_3X_5\right), \\
\mathbf{X}^{(4)'} &= \left(X_4X_1\ X_4X_2\ X_4X_3\ X_4X_5\right), \\
\mathbf{X}^{(5)'} &= \left(X_5X_1\ X_5X_2\ X_5X_3\ X_5X_4\right).
\end{aligned}$$

These equations are the augmentations of the original linear model (2.50) with the quadratic form (2.51), the cubic form (2.52), or the cross-product forms (2.53). Their SURE of these seven different polynomial models suggest that nonlinear effects do in fact exist.

They found that the nonlinear effects of the network characteristics on the GDP stability are mainly manifested through the two centrality measures, a result that is very similar to the one they had in the baseline model. The two centrality measures are not only significant in the quadratic form and the cubic form, but are also significant in many combined terms. In fact, the combined terms that are significant all have centrality as a part of them. Therefore, for the GDP stability, centrality seems to be the most important characteristic; it impacts the GDP stability not only linearly, but also nonlinearly.[18] The prominent role of centrality in economic stability can also be found in the inflation equation. The two centrality measures are again significant in either the quadratic or cubic forms, and most combined terms having centrality as a part are found to be significant. Nonetheless, the effect of centrality on inflation stability is less certain since the sign of its linear term flips from one equation to the other.

With the presence of these nonlinear effects, they then further examined the robustness of the results obtained from eqn. (2.50) (see Table 2.7). The SURE augmented with the quadratic, cubic, and the cross-product terms are shown in their Table 11 and Table 12. By comparing these tables with their baseline results (Table 2.7), we can see

[18] The nonlinear effect of network characteristics is also found in other studies of economic networks; for example, in a different setting, Gai and Kapadia (2010) found that the network connectivity (degree) has a nonlinear effect on the probability of contagion in the interbank network.

Table 2.8 *The SURE Augmented with Higher Order Terms*

Network Characteristics	Ψ	Ψ^2	Ψ^3	$\Psi^{(1)}$	$\Psi^{(2)}$	$\Psi^{(3)}$	$\Psi^{(4)}$	$\Psi^{(5)}$
				output gap				
AD	–	0	0	1	0	1	1	0
ACC	–	1	1	1	1	0	1	1
APL	+	0	1	1	1	1	1	0
MBC	+	1	1	1	1	1	1	1
MCC	–	0	0	1	1	0	0	0
				inflation				
AD	–	0	0	0	0	1	0	0
ACC	–	1	1	0	1	1	0	0
APL	+	1	1	1	1	1	1	1
MBC	+	0	0	1	1	1	1	0
MCC	–	1	1	0	1	0	1	1

The second column shows the sign of the coefficient of each network characteristic under the baseline model (2.50). The following columns show the robustness of these signs when the liner model is augmented with the quadratic form (2.51), the cubic form (2.52), and the five different cross-product forms (2.53). The "zero" cell indicates that the sign has flipped, whereas the "one" cell indicate that the sign remains unchanged. Details of the SURE results of the nonlinear augmentations can be found in Chen, Chang, and Wen (2014), Tables 11 and 12.

that the signs of some coefficients flip, from either positive to negative or negative to positive. To have a better picture, a summary of these flips is given in Table 2.8. There we denote the term by "0" if the sign of the respective coefficient changes from the baseline model to the augmented model; otherwise, we denote it by "1" if there is no such flip.

For the GDP stability, since only the two centrality measures are significant in the linear model (Table 2.7), their robustness check is, therefore, limited to these two measures. Of the two centrality measures, it is interesting to notice that, even with the augmentation in seven different directions, the effect of the betweenness centrality remains statistically positive, whereas the closeness centrality seems rather sensitive to the augmentations and it flips five times in the seven augmented models. In addition, for the inflation stability, originally in eqn. (2.50), all five characteristics are significant; hence we run through all of these five to see their robustness. Unlike the GDP equation, both centrality measures are somewhat sensitive to the augmentations and flip between positive and negative. The only characteristic that shows the robustness is the average path length. It consistently demonstrates the adverse effect of the average path length on the inflation stability.

Degree Distribution

While the five characteristics examined by Chen, Chang, and Wen (2014) in this study very often appear in the network literature, there is little doubt that they are not exhaustive. Some other characteristics may exist and may play a role. Therefore, they tried to include one more characteristic, i.e., the *shape of the degree distribution*, which may provide us with additional information that is not revealed via the average degree and the other four characteristics. They proposed a measure that has economic meaning, i.e., a measure in line with income distribution or wealth distribution. They assumed that the degree distribution, to some extent, can represent income distribution through the social capital connection. Hence, a measure developed in this way may indirectly enable us to inquire into the relationship between income distribution and economic stability.

If so, then a measure that is frequently used and can be easily calculated is the ratio of the wealthy people's income to the poor people's income, as shown in eqn. (2.54).

$$DIP = \frac{k_{75}}{k_{25}}, \tag{2.54}$$

where *DIP* stands for the degree distribution in terms of percentiles and k_{75} and k_{25} refer to the 75th and the 25th percentiles of the distribution in question. One can also go further to have the measure of the inequality in a more extreme position as in eqn. (2.55).

$$DIE = \frac{k_{max}}{k_{min}}, \tag{2.55}$$

where *DIE* stands for the degree distribution in terms of the extremes and k_{max} and k_{min} refer to the maximum and minimum values of the degree distribution, respectively.

With these two additional characteristics, they rerun SURE (2.50) and the results are shown in Table 2.9. By comparing Table 2.9 with Table 2.7, we can find that the results originally found in Table 2.7 remain unchanged. Those significant variables remain significant with the signs unchanged. This shows that their earlier results are very robust. It seems that the degree of inequality does in fact enhance the instability of the economy, as the coefficients of *DIP* and *DIE* are all positive; however, they are significant only in the GDP equation, and not the inflation equation.

Summing-Up

In sum, Chen, Chang, and Wen (2014) constructed an agent-based New Keynesian DSGE model with different social network structures to investigate the effects of the networks on macroeconomic fluctuations. This simple but fundamental setting allows us to have several findings that are worth summarizing at the outset. First, they found that the network characteristics can have some effects on the economic stability, and different economic variables may be sensitive to different characteristics. Second, they also found, however, that few characteristics are robust to different settings, in particular with the nonlinear augmentations. Third, putting the two together, they did not find any single characteristic that is universally important to both GDP and inflation. While the two centrality measures consistently show their prominence in both the GDP and inflation equations, the signs are not robust across different equations. Fourth, the

Table 2.9 *The SURE Augmented with the Shape of the Distribution*

	Output Gap(x10^{-6})			Inflation(x10^{-6})		
Intercept	417525	(435.19)	∗	46267	(2606.96)	∗
AD	-8.53	(-0.64)		12	(4.72)	∗
ACC	9650	(0.71)		-1360	(-5.38)	∗
APL	9.748	(1.21)		213	(141.30)	∗
MBC	2.384	(27.46)	∗	0.9741	(60.24)	∗
MCC	-5180	(-3.07)	∗	-9820	(-31.11)	∗
DIP	2707	(2.73)	∗	117	(0.63)	
DIE	516	(2.25)	∗	17	(1.29)	
R^2		0.67588			0.98565	
Adj R^2		0.67126			0.98545	

The above SURE is the SURE (2.50) augmented with two additional explanatory variables related to the shape of the distribution, *DIP* and *DIE*. Other variables are defined in Table 2.6. Columns 2 and 5 are the estimated values, and Columns 3 and 6 are the associated t values. The coefficients that are significant at the 5% level are marked by ∗ in Columns 4 and 7, respectively.

effect of the network topology on economic stability is not limited to the five basic characteristics. In addition to them, the shape of the degree distribution is also found to be important. Fifth, more subtly, the effects of the network characteristics can exist in a nonlinear fashion, with quadratic, cubic, or combined effects. Without monotonicity, our understanding and forecasting of the network effects certainly becomes more challenging. Sixth, despite these perplexities, two characteristics quite clearly stand out throughout the analysis, i.e., the (maximum) between centrality on the stability of GDP and the average path length on the stability of inflation. It remains to be seen what underlies these causal mechanisms. This actually calls for a theory of economic stability in terms of network topologies, which deserves an independent study.

2.4 Additional Notes

The beginning lesson of this chapter is that agents' decisions are affected by their neighbors (the neighborhood-based decision rules); obviously, if so, the chain reaction also implies that their decisions are affected by their neighbors' neighbors, and so on. More generally, this neighborhood-based decision rules can be understood as if each agents are, to some extent, mimicking others' behavioral rules and, consequently, may have similar behavioral outcomes. This prompts us to think of the correlation between neighboring agents and when these (positive) correlation disappear. This pursuit actually leads us to *the three degrees of influence*, i.e., the correlation can extend to *three degrees of separation*, to the friends of one's friends' friends (Pinheiro et al., 2014).

As we have seen from Section 2.2 to Section 2.3, the literature development of the networks in complex systems actually underwent two stages, from lattices (cellular automata) to networks. However, solid theory on symbolic dynamics, automata, and complex systems are mainly built upon on the former, but not the latter. For example, the similar theoretical results on cellular automata are generally not available in social networks. Our conjecture is that the chance of obtaining Class-III and IV dynamics are higher when network topologies of lattices are replaced with general networks; hence the chance of get access to computational irreducible systems is higher, even after the application of the coarse-grain method (Israeli and Goldenfeld, 2004, 2006). However, this should be well expected since the social system is generally thought harder than the physical and biological systems (Beckage et al., 2013).

It is, however, a little disappointing to see that the social network literature largely ignores the significance of cellular automata, specifically, the latter as a general theoretical underpinning of the former. For those readers who are interesting in knowing more about cellular automata, McIntosh (2009) gives a rather comprehensive historical review of the origin and the development of the idea of cellular automata and the automata theory. While cellular automata have been studied as a part of the abstract theory of computation since the time that John von Neumann became interested in the possibility of constructing self-reproducing automatic factories, the idea of automata can be even been traced back to the celebrated McCulloch–Pitts model of neural nets (McCulloch and Pitts, 1943), which associated with Alan Turing's work has some influences on the von Neumann's self-reproducing automata.

> McCulloch and Pitts described structures which are built up from very simple elements, so that all you have to define axiomatically are the elements, and then their combination can be extremely complex. Turing started by axiomatically describing what the whole automation is supposed to be, without telling what its elements are, just by describing how it's supposed to function. (Von Neumann (1966), p. 43)

Von Neumann worked out a scheme for an automaton, in terms of a cellular space occupying a two dimensional grid, in which each cell would be found in one of 29 states. Later on, Edgar Codd (1923–2003) worked out a variant that required only eight states per cell, still using the original five cell neighborhood (Codd, 1968). In his review, McIntosh also gives a distinction between automata and cellular automata; the former is characterized by an input–output system, whereas the latter is characterized by a self-organizing or "self-containing" system. The review also extensively covers the influence of many surrounding disciplines, such as semi-group theory, symbolic dynamics, and regular expression. More materials on cellular automata can be found in Voorhees (1996) and Schiff (2008). A more advanced and mathematical account of cellular automata can be found in the Ilachinski (2001).

For the study of the spatial game theory in the framework of automata theory, Nowak (2006), Chapter 9, is a nice companion to Section 2.2.2. The chapter gives a good overview of all the complexity one could ever wish for, such as the evolutionary kaleidoscopes, dynamic fractals, and spatial chaos.

In Section 2.3.1, we assume that social networks can affect the extent of bilateral trade. Empirical studies show that there is an effect of the social network or social distance (cultural distance) on bilateral trade (White, 2007; Tadesse and White, 2010). Bilateral trade can be negatively affected by cultural distance; however, immigrants from countries with different cultures can, to some extent, countervail this negative impact. Leaving the social network or social distance aside and just focusing on spatial distance, there is a large literature on the effect of geographical distance on bilateral trade. These studies show that, despite the advancement in modern telecommunication technology and "the world becomes flat," geographical distance constantly exerts a negative effect on bilateral trade (Leamer, 2007; Disdier and Head, 2008).

Network externalities (Section 2.3.3) can also result in an adverse market selection. A number of historical and theoretical studies published in the 1980s suggested that if network effects and economies of scale are significant, with information goods offered as an example, then free markets could lead to inferior technologies becoming entrenched (Liebowitz and Margolis, 2002). David (1985) claimed that the Qwerty typewriter layout is an example of lock-in to technological inferiority. He asserted that the Dvorak keyboard layout, developed after the Qwerty layout was well established, is technically superior but historical circumstances and market processes have prevented its adoption.

The idea pursued and demonstrated in Section 2.3.3 has attracted quite a few recent studies. Rand and Rust (2011) used the scale-free network to simulate the Bass diffusion curve and compared the simulated diffusion curve with Bass' original results (Bass, 1969). He found that the agent-based simulated diffusion can approximate well the Bass diffusion by appropriately setting a sufficiently large network density.

Delre, Jager, and Janssen (2007) studied the diffusion process using the small-world network topology. The small-world network is parameterized by the rewiring rate, which ranging from zero to one gives us the regular network, the random network, and everything in between. The small-world network has been used frequently in epidemics, and one may assume that what we learn from epidemics may be applicable to the diffusion of new products or knowledge. However, as Delre, Jager, and Janssen (2007) have demonstrated in their study, this presumed equivalence does not hold in a more general setting. They were able to show that diffusion speed is faster when the rewiring rate is neither too low nor too high, but can be slow when the underlying network is regular or random. The result that diffusion rate can be slow when the underlying network is random is different from the usual epidemic dynamics. This is understandable because the behavioral rules that dictate the adoption of a new product or an idea are different from the biological rules that dictate the infection of a disease. In this regard, diffusions of innovation is a process much richer than the spread of diseases.

The hubs or influential agents, as manifested in the scale-free networks, have received some close examinations over the last few years, including the behavior of these hubs, their roles in adoption and diffusion, and the identification of them. It is known that different behavioral assumptions of hubs may actually lead to different results on their role in diffusion (Watts and Dodds, 2007; Goldenberg et al., 2009). For example, Watts and Dodds (2007) based the adoption threshold on the proportion of adoptions in a neighborhood necessary to trigger adoption, whereas Goldenberg et al. (2009) uses a fixed number as a threshold. The latter interestingly argued that "[i]f one person is connected to 10 people and another person is connected to 200, the adoption likelihood is more similar if each of them has 3 friends who adopted than when the first has 3 and the second has 60. (Goldenberg et al. (2009), p. 4)." They, therefore, came up with different conclusion on the role of hubs, tending to support the influentials hypothesis.

While when defining hubs or influential agents normally we mean the agents with a larger number of connections, also known as the *degree centrality*, in social network analysis there exist different measures of centrality and hence different identifications of the influential agents. Kiss and Bichler (2008) used the call data from a telecom company to compare different centrality measures for the diffusion of marketing messages. They found a significant lift when using central customers in message diffusion, but also found differences in the various centrality measures depending on the underlying network topology and diffusion process. The simple (out-)degree centrality performed well in all treatments.

A more practical application of the model introduced in Section 2.3.3 is Katz and Heere (2013). The significance of the scale-free network topology has also been applied to understand the formation of brand communities or tailgating groups. Katz and Heere (2013) conducted an ethnographic study among four tailgating groups of a new college football team during its inaugural season, and identified the leader(s)–followers structure of the groups. The purpose of their study was to gain insight into how individual consumers interacted with each other and how these early interactions contributed to the development of a brand community. These tailing groups were initiated by one major leader with many connected followers, and these connected followers may not have direct interactions with each other for some while, or may form a second group on their own. While the tailgating groups share some features of the scale-free networks, the network formation algorithm is not strictly done through the familiar preferential attachment. If the hubs or leaders can be identified, then direct marketing through them may be more useful than the conventional indirect marketing. Of course, the remaining challenging issue is the way to "discover" or to "learn" who are the influential nodes.

As we mention at the beginning of Section 2.3.4, study on the network topologies and macroeconomic stability is still rarely seen. However, there are some recent attempts made to bring together network topologies, diffusion of knowledge, and economic growth. From a knowledge or technology adoption and diffusion viewpoint, Fogli and Veldkamp (2012) argued that network topologies can matter for economic growth since different network topologies have different implications for the spread of new ideas and innovation and they, in turn, can contribute to economic growth to a different extent. In particular, they asserted that a highly clustered social network and hence a possibly highly fragmental social network may inhibit the spread of ideas required for technology advancement and economic growth, whereas a social network with weak ties may help promote economic growth because it can facilitate the use and the generation of new knowledge.

What makes Fogli and Veldkamp (2012) particularly interesting in their analytic framework is built upon the integration of the epidemic networks and knowledge diffusion networks. Of course, all diffusion processes may share some common channels; however, the network for the diffusion of germs may crucially depend on some kind of geographical proximity, through air, water, or body contact, whereas diffusion of ideas may be wireless. They actually discussed some technical issues involved in the mixture of the two.

In a similar vein, Lindner and Strulik (2014) also related network topologies to knowledge-oriented economic growth. However, unlike Fogli and Veldkamp (2012), they did not deal with a network of individuals and compare how fast knowledge can spread within a country; instead, they chose to work on a network of countries and studied how knowledge can spread from country to country. Their purpose was to endow the neoclassical growth theory with a network underpinning so that divergence and convergence of growth rates among different countries since industrial revolution could be accounted for.

Both Fogli and Veldkamp (2012) and Lindner and Strulik (2014) used small-world networks as the illustrative examples in their models, and mainly focused on the path lengths and cluster coefficients as the main determinant of the diffusion of knowledge. Generally, in additional to path lengths and cluster coefficients, other characterizations of social networks such as network density, betweenness centrality, and associativity may also affect economic growth if they can be shown to have an effect on knowledge spillover.

One point that may be missing in this network-based economic growth theory is how ideas initially originated. Regardless of the individual country model (Fogli and Veldkamp, 2012) or country-wise model (Lindner and Strulik, 2014), some leading countries or leading individuals are assumed to be there at the outset. It ignores the knowledge production process per se; in particular, how one kind of knowledge is used to produce another kind of knowledge, and how network topologies may play a role in facilitating the production. In fact, literature has already shown that knowledge production in terms of the modular structure also require a network (Benkler, 2006). Hence, the social network not only matters for the diffusion and distribution of knowledge but also for the production of knowledge. To make network-based economic growth theory more comprehensive, the latter part should also be taken into account.

3

Collective Dynamics of Adaptive Agents

Our society consists of individuals and the social systems in which they interact. Individuals' behaviors, which depend on the behaviors of others, create the social system of which the individuals are parts. Therefore, social systems could be characterized using feedback mechanisms or *micro–macro loops* between individuals and the whole system.

In this chapter, we study the *collective behavior* of adaptive agents, in which each agent adapts to other agents. When agents adapt to each other, their collective behavior is extremely difficult to predict. This leads to questions about how social systems work and how they can be made to work better. To answer these questions, we may need to identify microscopic (agent) behaviors that produce the macroscopic behaviors of interest.

We are especially interested in answering the following questions: How do agents, who interact and adapt to each other, produce the aggregate outcomes of interest? How do we identify the microscopic agent behaviors that contribute to consistency in behaviors of interest at the macroscopic level?

3.1 Collectives

Most individual activities are substantially free and although people sometimes care about aggregate consequences, personal decisions and behaviors are typically motivated by self-interest. Therefore, when examining collective behavior, we shall draw heavily on the study of individual behaviors.

Billions of people make billions of decisions every day about many things. It often appears that the aggregation of these independent individual decisions leads to a desired outcome. That is, the individuals involved may have a very limited view of the world events but their independent decisions are, to a large degree, coordinated and may produce a desirable outcome at the aggregate level. It is amazing that many economic and social activities generally work well in this way and in the absence of management. The economist Adam Smith (1776) referred to this phenomenon as an *"unseen hand of God"* that brought about implicit coordination among self-interested individuals. This unseen hand is observed behind many market activities and acts as a basic mechanism for allocating limited resources to people who need them.

Agent-Based Modeling and Network Dynamics. First Edition. Akira Namatame and Shu-Heng Chen.

People also constantly interact with each other for different purposes; these interactions show some coherence while generating structures and regularities at the aggregate level. However, there are many systems for which it is difficult to understand both how they work and how to improve them. For instance, many economic and social systems produce inefficient outcomes at the aggregate level without individuals who comprise the system knowing about or being aware of the outcome. Instead, people tend to think that an outcome corresponds to the *intentions* of the individuals who compromise the system. Therefore, when an undesirable outcome occurs, people mainly suspect that it is caused by some components or specific individuals.

Many organisms form aggregations that have strong effects on individual behaviors. Familiar examples include schools of fish and flocks of birds. Auyang (1998) defines such aggregations as *collectives*. Interactions among individuals making up a collective are strong, that is, internal cohesion is strong while external interactions are weak. Collectives also have their own characteristics and processes that can be understood independently of the individuals that comprise the collective.

Another defining characteristic of collectives can be observed in the field of ecology. For example, a school of fish is a collective that emerges from the relatively simple traits of individuals; that is, these traits give rise to individual behaviors that form the collective. In ecological systems, collectives may also exist for longer or shorter times than do the individuals making up the collective. In this way, collectives can be treated as an additional level of organization existing between the individual and the population (Grimm and Railsback, 2005). Individuals belonging to a collective may behave very differently from isolated individuals, therefore, it may be necessary to model different traits for individuals not in a collective. However, the behavior of a collective generally emerges from traits of individuals.

It is generally believed that individual ants do not know exactly how to build the colony in which they live. Each ant has certain things that it does, in coordinated association with other ants, but no single ant designs the whole colony. For example, no individual ant knows whether there are too few or too many ants out exploring for food. Why the collective ant colony works as effectively as it does remains a mystery, but a clue to understanding such collective behavior is hidden in the interactions among ants (Bonabeau et al., 1999).

Representing a collective explicitly does not mean that individuals are ignored. Instead, collectives can be viewed according to: (1) the manner in which individual behaviors affect the collective and (2) how the state of the collective affects individual behaviors. For instance, individuals make decisions about when to disperse, thus affecting the formation and persistence of the collective, but such individual decisions are also based in part on the current state of the collective. Therefore, the collective can only be understood by modeling behaviors of both individuals and the aggregate, as well as the links between them. Repetitive interactions among agents are also a feature of collectives. Therefore, exploration of collective dynamics requires the power of computers to overcome limitations of pure mathematical modeling and analysis.

We use the term *collective system* when it is impossible to reduce the overall behavior of the system to a set of properties that characterize individual agents. An important feature of collective systems is that interaction between agents produces emergent properties at the aggregate level that are simply not present when the components are considered individually. Another important feature of collective systems is their sensitivity to even small perturbations.

At its most basic, a collective system consists of agents and the relationships between them. Thus, a collective system, which consists of many interacting agents, can be described on two different levels: (1) the microscopic level, where the decisions of the individual agents occur; and (2) the macroscopic level, where the collective behavior can be observed. Agents may execute various behaviors that are appropriate for the situations in which they find themselves.

There is no presumption that the self-interested seeking behavior of agents should lead to collectively satisfactory results. For example, each person's enjoyment of driving a car is inversely related to others' enjoyment if too many people drive cars and then become stuck in congested traffic. Such situations constitute a kind of social congestion with the associated problem that there is no way of predicting the behavior of others.

An *externality* occurs when an agent's decision affects the decisions of other agents. Agents' decisions are *contingent* in the sense that they depend on how other agents make choices. In this mode of contingent individual behavior, the resulting collective behavior is often volatile, resulting in far from desirable outcomes. That is, the adaptability of individuals is not related to how desirable is a social situation they collectively create. For instance, in the case of there being too much traffic, individuals are clearly part of the collective problem. In another example, if an individual raises their voice to be heard above a noisy crowd, they add to the noise level, making it difficult for others to be heard (Schelling, 1978).

It might be argued that understanding how individuals make decisions is sufficient for understanding most parts of aggregate behaviors. However, although individual behaviors are important to understand, the resulting knowledge does not explain how a collection of agents arrives at a specific aggregate behavior of interest. For example, when the decision of an agent depends on the decisions of others, such a situations does not usually permit simple summation or extrapolation to the aggregate. To make this connection, it is usually necessary to examine interactions among agents.

Many social and economic systems exhibit *emergent properties*, which are properties of the system that individual components do not have. Emergent properties result from the behavior of individuals as well as the interactions between individuals. Therefore, in the traffic example, the traffic jam is a counterintuitive phenomenon whose mechanisms can only be understood within the framework of the collective system of interacting agents. To view systems in this way, as collectives of interacting agents, requires adopting a new scientific approach that shifts rationale from *reductionism to connectionism*. In the reductionist view, every phenomenon we observe can be reduced to a collection of components, the movement of which is governed by the deterministic laws of nature. In this context, there seems to be no place for novelty or emergence.

The basic view of reductionism in social sciences can be found in *rational choice models*. The rational choice theory posits that an agent behaves to optimize personal preferences; this approach produces relevant and plausible predictions about many aggregate societal phenomena. However, individual motives and their strength are difficult to determine from collective phenomena. Referring again to the traffic example, it is difficult to capture the properties of a traffic jam at the aggregate level without describing the behaviors of individual drivers. Each driver is different from the others and the characteristics of their driving behavior become the rules in the model. When we run such a model it reproduces the traffic jam, but it is necessary to observe closely how the individual drivers interact with each other. We can than inject to see how these interactive behaviors among drivers affect the overall visible properties of the traffic jam.

It is important to look closely at agents who adapt to other agents because the behavior of one agent affects the behaviors of the others. How well agents achieve their goals depends on what other agents are doing. What makes this kind of interactive situation interesting and difficult to analyze is that the aggregate outcome is the focus of study, not merely how the agents behave within the constraints of their own environments. Situations where an agent's adaptation depends on other agents' adaptations usually do not lend themselves to analysis by simple induction or extrapolation as to preference. The greatest promise lies in the development of analyses suitable for studying situations in which agents behave in ways contingent on one another. Such interactive situations are central to the analysis of the linkage between adaptation in individual behavior and the formation of macroscopic behavior.

In summary, a collective of adaptive agents is characterized as a system consisting of individual agents who are continually adapting to each other. Further, agents interact with each other in complex ways and linked agent behaviors are likely highly nonlinear. In order to investigate the performance of a collective system of adapting agents, we need to explore new methods that move beyond conventional equilibrium analysis that tends to emphasize rational aspects of individual agents. In the following sections, we focus on an analysis of the collective behavior of adaptive agents.

3.2 Binary Choices with Externalities

If a system consists of many interacting agents, it can be described on two different levels: (1) the microscopic level, at which the decisions of individual agents occur; and (2) the macroscopic level, at which collective behavior can be observed. In order to connect the two, it is usually necessary to examine interactions among agents.

The most promising approach for determining how the heterogeneous microworlds of individuals generate macroscopic orders of interest and unanticipated outcomes lies in analysis of the linkage between microscopic and macroscopic behaviors. However, understanding the role of this link between two levels remains a challenge.

Human behavior is mainly driven by conscious decision-making. Indeed, many economic models describe how agents behave based on their preferences. An important example is *rational choice theory*, which offers a powerful framework for analyzing how agents make decisions in various situations, whether alone or in interaction with others. The theory also considers whether decisions are made under conditions of uncertainty or with complete information. In our view, the reason for the dominance of the rational choice approach is not that we think it realistic. Nevertheless, it often allows reasonable deduction to explain specific agent choices.

Schelling (1978) classified individual behavior according to two modes: *purposive* and *contingent behaviors*. People usually ascribe individual behaviors as if outcomes were oriented toward their own interests. It is generally assumed that people have preferences and pursue individual goals in order to maximize comfort as well as minimize effort and embarrassment. We might characterize these individual behaviors as purposive behaviors. Purposive behaviors are mostly described in the context of the rational choice model.

However, some have criticized the rational choice approach. The difficulty is that it assumes agents who are sufficiently rational. This is problematic because the interests (purpose) of agents are often directly related to those of other agents or are constrained by an environment that consists of other agents who are pursuing their own goals. For example, economists argue that much individual private consumption is dependent upon other peoples' consumption.

An *externality* can occur when each individual choice affects the choices of others. For example, when deciding which movies to watch or which new technologies to adopt, we often have little available information for evaluating the alternatives. In such cases, we rely on recommendations of friends or simply select the movie or technology favored by most people. Further, even with access to plentiful information, we often lack the ability to interpret this information. When agents' choices are contingent (i.e., dependent on how other agents make choices), they can produce unexpected side effects on the social activities of seemingly unrelated agents; this phenomenon is referred to as externality.

For example, suppose agents face the following *binary choice* problem: buy the product or do not buy the product. Each agent has his/her own *idiosyncratic preferences* concerning the product. For individual idiosyncratic preferences, we assume that each agent (isolated from other agents) would objectively buy the product of interest with preference level which also represents the agent's willingness to pay for the product. In addition, when each agent considers the choices of others, they receive *social influence* from other agents. We define the binary decision variable of agent i as x_i, ($x_i = 1$ if the agent decides to buy, $x_i = 0$ if the agent decides not to buy). By embedding individual idiosyncratic preferences h_i (willingness-to-pay) minus q (product price), and the externality, each agent's utility function, $i = 1, 2, \ldots, N$, is defined as:

$$y_i(t) = h_i - q + \frac{\omega}{k_i} \sum_{j \in N_i} a_{ij} x_j(t) \qquad (3.1)$$

where a_{ij} takes the binary value and if agent i is connected to agent j, $a_{ij}(= a_{ji}) = 1$, otherwise $a_{ij}(= a_{ji}) = 0$. In addition, k_i is the average degree of agent i (the number of neighbors or the cardinality of N_i), ω is some positive constant, and N_i represents the set of neighbors of agent i.

In an isolated case, where each agent does not care about the choices of others, if agent i's willingness to pay h_i is lower than the price q, he/she does not buy the product.

However, the case where each agent cares about the choices of other agents is represented by the third term in eqn. (3.1). If agents receive sufficient positive externality from neighbors' choices, some agents may decide to buy even if their willingness to pay is lower than the price of the product. We set the binary decision variable of agent i as:

$$x_i = \begin{cases} 1 & \text{if } h_i - q + \omega \sum_{j \in N_i} a_{ij} x_j / k_i \geq 0 \\ 0 & \text{otherwise} \end{cases}. \tag{3.2}$$

Further, the binary rule of eqn. (3.2) is written as:

$$x_i = \begin{cases} 1 & \text{if } p > (h_i - q)/\omega \equiv \phi \\ 0 & \text{otherwise,} \end{cases} \tag{3.3}$$

where $p = \sum_{j \in N_i} a_{ij} x_j / k_i$ represents the ratio of the agents to whom agent i is connected.

The rational choice rule of each agent depends on p, the fraction of neighbors who decide to buy. If p is larger than the threshold ϕ in eqn. (3.3), the rational choice of agent i is to buy. In this case, the agent's payoff increases with the number of neighbors who also buy, and positive feedback works in the form of the threshold rule in eqn. (3.3).

We illustrate some properties of agent decision-making according to problems of externality. We consider a collection of agents where each must choose between two alternatives and where the choices depend explicitly on the choices of the other agents. In the social context, most individuals often pay attention to each other. Even when they have access to plentiful information, such as when evaluating new technologies or risky financial assets, they often lack the ability to make sense of the information, and thus rely on the advice of trusted friends or colleagues.

Although an agent's choice depends on how other agents make choices, individual agents are assumed to lack the knowledge necessary to correctly anticipate all other agents' choices. Because no agent can directly observe all other agents' choices, they may instead rely on aggregated information. That is, at any given moment, each agent has the opportunity to observe the proportions of agents that have chosen either of the two alternatives. This class of problems is known generically as binary choices with positive externalities. Both the origins of the externalities and the detailed mechanisms involved in binary choice problems can vary across specific problems. However, in many cases, the decision itself can be considered as a function solely of the relative number of other agents who are observed to choose one alternative over the other; the relevant choice function frequently exhibits a *threshold* nature. This class of problems captures situations in which there is an incentive for individuals to make the same choices as their connected agents.

For modeling purposes, we also consider a collection of interacting agents who make decisions according to several predefined terms. Each agent continuously adapts to the choices of other agents. Suppose that all agents initially make the decision $x_i = 0$. Thereafter, one or a very few number of agents shift to decision $x_i = 1$. It follows that the agents will repeatedly apply the choice rule in eqn. (3.3) and switch to $x_i = 1$ if the proportion of connected agents who have already shifted to $x_i = 1$ is larger than the threshold ϕ.

Table 3.1 *Payoffs to each agent associated with binary choices (coordination game) ($0 \leq \phi \leq 1$)*

agent 1 \ agent 2	A	B
A	$1-\phi, 1-\phi$	0, 0
B	0, 0	ϕ, ϕ

Agents may display *inertia* in switching of choices, but once their personal threshold is reached, the choice of even a single neighbor can shift them from one choice to the other. As simplistic as it appears, a binary decision framework is relevant to surprisingly complex outcomes.

Binary choice problems with positive externalities can also be modeled as games in networks, in which each agent plays a 2×2 game (Table (3.1)). Here, we consider a game-theoretic point of view and scenarios where agent behavior is the result of a rational choice between two alternatives. That is, each agent faces a binary choice problem between A and B. For any agent, the payoff for choosing either A or B depends on how many other agents also choose A or B. Selecting A (or B) could mean, for example, adopting new technology (or continuing to use old technology) or voting for some social issue (or voting against). Agents are identically situated in the sense that every agent's outcome depends on the number of agents who choose either A or B.

The payoff for each choice is given in a symmetric matrix shown in Table (3.1). If two agents' choices are the same, they receive payoffs $1-\phi$ and ϕ for choices A and B, respectively. However, if their choices are different, neither will receive a payoff. Two assumptions that simplify this analysis further are as follows: (1) the payoff function is symmetric, that is, it is the same from every agent's point of view; and (2) every agent's payoff function is linear with respect to the number of agents choosing either A or B. That is, an agent's payoff function is directly proportional to the number of agents choosing either A or B. In this case, the payoff for the agent increases with an increase in the number of neighbors who make the same choice as the agent.

We next define the utility of choice. Let U_i, $(i = A, B)$ represent utilities of choices A and B with the difference in payoff for choosing A or B given as:

$$U_A - U_B = p - \phi, \tag{3.4}$$

where p is the proportion of the agents choosing A. That is, an agent's payoff function is directly proportional to the number of agents making the same choice. In this case, the agent choice rule is specified as a binary threshold rule that is decided by the fraction of neighbors making the same choice and which maximizes the current payoff. The threshold-based choice rule of agent i is given as:

$$x_i = \begin{cases} 1 & if \;\; p \geq \phi \;\; (Choose\; A) \\ 0 & if \;\; p < \phi \;\; (Choose\; B) \end{cases}. \tag{3.5}$$

Here, each agent has to make a choice between two alternatives and the rational choice of each agent depends on p, the fraction of neighbors choosing A, and when p is larger than the threshold ϕ, the rational choice of agent i is to choose A, otherwise the agent chooses B.

A cascading sequence of agents switching to A from B may result when almost all agents in the network come to adopt A by switching from B. The conditions for causing a cascade depend on the threshold ϕ and the underlying network topology. The important question, then, is which network structures favor the rapid spread of new ideas, behaviors, or technologies? For the spread of everything from disease to rumors, a single active agent triggers the activation of his/her neighbors. In this case, hub agents who are connected to many agents can allow for dramatic diffusion of triggers. However, not all propagation results from simple activation of agents after exposure to multiple active neighbors. The propagation process can be modeled as a threshold-based diffusion model, as described in eqn. (3.5). Each agent has an individual threshold that determines the proportion of neighbors required to activate a behavior.

There is another type of externality labeled *negative externality*. Referring again to the traffic example, each person's enjoyment of driving a car is inversely related to others' enjoyment because if too many people drive, they will all become stuck in congested traffic. This is the typical social congestion problem wherein there is no way of knowing what others will do. In our daily life, we face many such problems and the question of how to solve them is an important issue.

Congestion problems always arise when people need to utilize limited resources. Sharing or allocating limited resources in an efficient way requires a different type of coordination of individuals. Division of labor (and other dispersion problems) belongs to this complementary class of social interactions. In this class of problems, agents gain payoffs only when they choose distinct actions according to those of the majority. For instance, each agent chooses a resource to utilize, but the utility of his/her choice depends on the number of other agents who try to utilize the same resource. In this case, the capacity of the resource is limited and there is a preference for agents to be maximally dispersed across the set of possible choices.

The most attractive scientific explanations are brief stories that can be easily understood even by people outside the particular field of study, the core of which, however, constitutes a salient and deeper problem. In this sense, the *El Farol bar problem*, introduced by Arthur (1994) has received much attention as a helpful paradigm for discussing many issues including social inefficiency resulting from rational behavior. The story takes place in an Irish bar by the name of "El Farol" in downtown Santa Fe, New Mexico. The bar has live Irish music every Thursday night. Staff (the agents) from an organization called the Santa Fe Institute are interested in going to the bar on Thursday night to enjoy the live music. All of these agents thus have identical preferences. They will enjoy the night at El Farol very much if the bar is not too crowded but each of them will suffer miserably if the bar does become too crowded. In addition, the bar has limited capacity.

Researchers have used a diverse collection of prediction rules (followed by agents) to investigate the number of agents attending the bar over time. Agents make their choices by predicting whether attendance will exceed capacity and then take appropriate action. What makes this problem particularly interesting is that it is impossible for each agent to be perfectly rational, in the sense of correctly predicting the attendance on any given night. This is because if most agents predict that attendance will be low (and therefore decide to attend) then attendance will actually be high, whereas if they predict that attendance will be high (and therefore decide not to attend) then actual attendance will be low. However, one interesting result obtained by Arthur is that, over time, average attendance at the bar oscillates around the capacity.

The characteristic property of this binary choice problem is that the choice rule is based on the payoff value for each choice, the locality, and the proportion of agents making each choice. The payoff for each agent is given as an explicit function of the actions of all agents, and therefore each agent has an incentive to pay attention to the collective decision. However, the binary decision itself can be considered a function solely of the relative number of other agents who are observed to choose one alternative over the other. This class of binary decision problems is referred to as binary decisions with negative externalities.

For current purposes, we consider a population of N agents, each facing a binary choice between A (go to the bar) and B (stay home). For any agent, the payoff for choosing A or B depends on how many other agents also choose A or B. The payoff for each choice is given in a symmetric matrix in Table (3.2). If two agents' choices are different, they will receive payoffs ϕ and $1-\phi$ respectively. However, if their choices are the same, they will receive no payoff.

Here, we let $U_i, i = A, B$ represent the respective utilities of choices A and B where the difference between payoffs to each agent for choosing A or B is given as:

$$U_A - U_B = \phi - p \tag{3.6}$$

and where p is the proportion of agents choosing A. That is, an agent's payoff function is directly proportional to the number of agents making one of the choices. In this case, the payoff to the agent for each of the two choices decreases as an increasing number of neighbors adopt the same choice. An agent choosing A gains (more or better payoff) if

Table 3.2 *Payoffs to each agent associated with binary choices (dispersion game) ($0 \leq \phi \leq 1$)*

agent 1 \ agent 2	A	B
A	0, 0	$\phi, 1-\phi$
B	$1-\phi, \phi$	0, 0

some other agents choosing A will shift and choose B. The threshold-based choice rule of agent i is given as:

$$x_i = \begin{cases} 1 & \textit{if} \;\; p \le \phi \; (A : \textit{go to the bar}) \\ 0 & \textit{if} \;\; p > \phi \; (B : \textit{stay home}) \end{cases} . \tag{3.7}$$

This rational choice rule implies an incentive for individuals to make choices that are the opposite of choices made by their immediate friends or neighbors.

Both detailed mechanisms involved in binary decision problems and the origins of the externalities can vary widely across specific problems. This is the case when an agent gains a payoff if they make a choice that opposes the choice of the majority. We refer to this agent behavior as being based on the *logic of minority*. Alpern and Refiners (2002) introduced a *dispersion problem* in which agents prefer to be more dispersed (i.e., not as close to one another). The dispersion problem is illustrated by the following typical examples:

1. Location problems: Retailers simultaneously choose their positions within a common space to maximize the area for which they are the closest.
2. Habitat selection: Males of a species choose territories where there are no other males. Animals choose feeding patches with low population density with respect to food supply.
3. Congestion problems: Individuals seek facilities or locations of low population density.
4. Network problems: Travelers choose routes with low congestion levels.

These dispersion problems, in which an agent behaves based on the logic of minority, arise in a large number of domains, including load balancing in computer science, niche selection in biology, and division of labor in economics. Social interactions with negative externalities pose many difficulties not found in social interactions with positive externalities. Most especially, situations where self-motivated agents behave based on the logic of minority need better coordination to achieve efficient agent dispersion and equal benefits for all.

It is important to explore the mechanism by which interacting agents stuck in an inefficient equilibrium can move toward a better outcome. While agents understand that the outcome is inefficient, each independently acting agent may be powerless to manage the collective activity about what to do and how to decide. Self-enforced solutions may be achieved when agents achieve desirable allocation of limited resources while pursuing their own self-interests and without the explicit agreement of other agents; the study of such solutions is of great practical importance.

Effective solutions to social congestion problems or scarce resource allocation problems may need to invoke the intervention of an authority. The central authority may be able to find a *social optimum* and impose that optimal behavior on all agents. While such an optimal solution may be easy to conceive of, implementation is often difficult to enforce in practical situations. For instance, to alleviate social congestion, the central authority often explicitly charges the costs of congestion to users; this solution should eliminate the socially inefficient congestion of a scarce resource. However, this approach often needs to achieve equilibrium solutions in which all agents are fully informed about the structure of the problem and the behaviors of all other agents. Consequently, the relationship between each agent's microscopic behavior and congestion at the aggregate level may be more easily discerned. However, the reliance on information-intensive equilibrium solutions limits the usefulness of this approach in solving many congestion problems.

A more general approach is to employ adaptive choice models with positive or negative externalities structured according to a common framework. We assume that each agent has the same payoff function $u(x, X)$, where the agent's action x is described by a set that is a subset of all real numbers, and the second argument, X, is the average of the others' actions. In this case, others may be a finite number or a continuum of agents. The second argument generates the payoff externality for each agent. The two types of choice problems, one with positive externality and the other with negative externality, are formulated as follows: In the first type, a higher level of activity X by others increases the marginal payoff for each agent and stimulates the incentive to act like others. In this case, each agent behaves based on *the logic of majority*. In contrast, in the second type of choice problem, a higher level of activity X by others reduces the marginal payoff for the agent. For example, each person who visits a popular restaurant increases the waiting time of subsequent other persons.

3.3 Adaptive Choices with Reinforcement

In the current section, we consider the adaptive choices of agents by incorporating two factors: (1) *individual reinforcement*, in which agents reinforce their preferences based on prior choices; and (2) *social reinforcement*, in which agents reinforce choice probability based on the choices of other agents in similar decision environments. We focus particularly on the effects of individual and social reinforcement on macroscopic behavior.

The working assumption of the rational choice approach is that an individual's behavior is guided by the sole motive of maximizing self-interest based on personal preference. While rational choice theory is able to show with great precision how an individual can act rationally according to preference, one shortcoming is that it says little about where individual preferences come from and how they might change. In rational choice theory, preferences are provided exogenously but there is no explanation of how they are formed.

Agent preference is typically fixed, appears to be influenced exogenously, and remains stable across many time horizons. It is surprising how little is known about the effects of external factors on individual preferences. Although psychologists have developed a robust set of facts concerning human preferences and their empirical stability, preferences are far from identical.

Therefore, one of the most challenging issues is to identify the mechanisms of the formation of heterogeneous preferences among individuals. Another important issue is to study the relationship between adaptation of behavior and evolution of preference. In general, the behavioral adaptation of an agent based on preference is much faster than the adaptation of the preference itself. It is often argued that in order to understand how individuals adapt, it is not sufficient to simply observe their preferences.

While researchers usually study the collective behavior of agents with endowed individual preferences, it is worthwhile to try to explain where and how these preferences occur. One natural approach to addressing this issue is to adopt a learning perspective in studying the development of preferences. Accordingly, an agent may change their preference (and choice) after receiving new information about likely outcomes.

For current purposes, we include a framework of preference formation along with a reinforcement learning approach. *Reinforcement* is a stimulus that follows, and is contingent upon, a behavior and increases the probability of a behavior being repeated. The basic premise of reinforcement learning is that the possibility of a behavior occurring in the present increases with the payoff that resulted from past occurrences of that behavior. Therefore, agents may undertake any number of alternative behaviors but will repeat those that previously led to high payoffs. That is, the propensity to exhibit a behavior increases according to the associated payoff. Therefore, it is assumed that agents tend to adopt behaviors that yielded higher payoffs and to avoid behaviors that yielded lower payoffs.

For modeling, it is necessary to examine the validity of agents with fixed preference and to ask deeper questions about the basis for, and plausibility of changing, the structure of individual preference. Agents are generally heterogeneous with respect to certain attributes, a characteristic referred as *idiosyncratic heterogeneity*. With a combined model of learning and specific network structure, interacting agents are generally driven toward heterogeneous individual preferences, even if agents initially appear identical. This characteristic is referred to as *interactive heterogeneity*. Currently, we are concerned mainly with interactive heterogeneity and with showing how a collective of homogeneous agents having similar preferences evolves into a collection of heterogeneous agents with diverse preferences (or the reverse).

We formulate a collective adaptive choice problem in which each agent makes a binary choice in a sequence. The sequential decision model is important for the analysis of *innovation diffusion* or management trends in uncertain environments, such as where many consumers (agents) lack clear objective evidence concerning the merits of alternatives. In the simplest form of our model, agents are born with their own idiosyncratic preference concerning a binary choice. We assume the rationality of each agent in the sense that his/her choices depend on this endogenous preference.

A *logit model* (based on *stochastic utility theory*) is applied to individual rational choice. In stochastic utility theory, an agent is assumed to behave rationally by selecting the option that returns the highest utility. However, individual utility contains some random factors that are treated as a random variable in stochastic utility theory calculations.

We consider a collective of N agents, each of which faces a binary choice, A or B, in a sequence. Let $U_i, i = A, B$ represent the utilities of choices A and B with the probability of choosing A given as:

$$\mu = \frac{1}{1 + exp(-(U_A - U_B)/\lambda)} \tag{3.8}$$

and where λ is the parameter representing the level of rationality.

Next, we let the agent's utility of the choice of A at the time period t be a_t, whereas the utility of choice of B is $b_t = 1 - a_t$. The difference between them, $2a_t - 1$, is normalized to establish a value between -1 and 1. The probability of choosing A in eqn. (3.8) at time period t then becomes:

$$\mu_t = \frac{1}{1 + exp(-(2a_t - 1)/\lambda)}. \tag{3.9}$$

Agent heterogeneity is represented by the parameter a, which represents individual preference level of alternative A over alternative B. We further assume that a is normally distributed with an average of $a = 0.5$. The probability of choosing A is shown in Figure 3.1. For low values of λ, the probability function becomes a step function, whereas for high values of λ, the binary choice approaches a random choice.

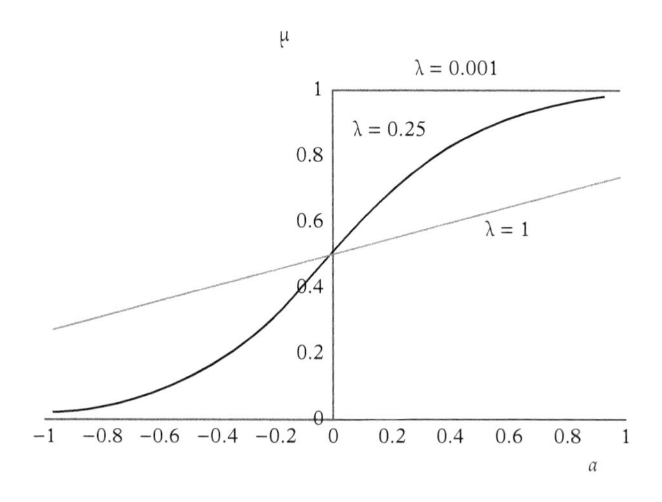

Figure 3.1 *Agent probability of choosing A*

As previously described, reinforcement is a stimulus which follows, and is contingent upon, a behavior and increases the probability of a behavior being repeated. The question then arises as to how agents' preferences are reinforced. The payoff represents the fitness of an agent's choice—that is, to what extent the payoff is at least as high as any alternative. More specifically, the following viewpoints are introduced:

1. the individual favors a repeated choice.
2. repetition leads to increased payoff.

We then formulate a reinforcement learning model by considering the above two factors. An agent increases (or decreases) the utility of his/her choice in eqn. (3.9) based on his/her prior choice as follows:

$$a_{t+1} = \begin{cases} a_t + \delta & if \quad an\ agent\ chooses\ A \\ a_t - \delta & if \quad an\ agent\ chooses\ B \end{cases}. \tag{3.10}$$

The direction of reinforcement is reflected in an agent's choice probability.

There are many social situations in which the behaviors of different agents reinforce each other. Such mutual reinforcement is defined as *social reinforcement*. While there are many different types of reinforcement, for human beings one of the most common is the naturally occurring social reinforcement that we encounter all around us every day. Such reinforcing stimuli from outside sources includes acceptance, praise, smiles, acclaim, and attention from other people.

Researchers have found that social reinforcement can play a vital role in a variety of areas, including health. The influence of people in our social networks can influence the type of health choices and decisions that we make. Indeed, social reinforcement for improving health habits may be more important than how the information is provided. For example, when a person is in the process of quitting smoking, a local community network of friends and neighbors may be more important than a remote celebrity spokesman who encourages smoking cessation. Further, in a famous study example of social reinforcement, Bandura et al. (1963) conducted research with school-age children who typically spent little time studying. When these children were given praise and attention (social reinforcement) for their study efforts, the researchers found that children studied up to twice as much than when they received no social reinforcement.

The concept of *social influence* is also important for the analysis of decision-making in uncertain environments in which agents lack clear objective evidence about the merits of two alternatives. Psychologists contend that when faced with choices, people often use a reasonable heuristic method, such as following a social trend. For example, consider a situation where there are two conflicting social opinions and at different time points an individual has to decide which opinion to support. That is, each agent faces a binary decision in a sequence and makes a choice that reflects both a preference for one of two choices as well as the choices of other agents. These agent choices reflect both personal evaluations of alternatives and social influence from other agents. This formulation resembles that addressed in most contagion models. The important question is how follow-on agents decide between the two opinions.

Bendor et al. (2009) constructed a binary choice model of agents that employs a heuristic method reflecting both individual preference and social influence. Their model considers the following two factors: (1) agents have prior preference for one of the two alternatives; and (2) agents rely heavily on the prior choices of other agents in similar decision-making environments. By combining these two factors, an agent's decision-making is modeled as a stochastic process.

We extend Bendor et al.'s model by incorporating individual preference reinforcement. There are two reinforcement factors underlying individual choices between A and B: (1) individual preference reinforcement; and (2) social reinforcement. Regarding individual preference, we assume that an agent isolated from social influence would objectively choose option A with some probability μ, as given in eqn. (3.9). Thus μ reflects individual relative preference of A over B for each agent. We assume heterogeneity among agents and that choices A and B take some random value [0, 1] in general. If an agent's preference for A is strong enough, then μ is close to 1, and if the two alternatives are nearly interchangeable then μ is close to 0.5.

An agent's probability of choosing A in period $t+1$ is given as the weighted average of individual preference and social reinforcement:

$$p[\textit{agent in period } t+1 \textit{ chooses } A] = p_{t+1} = (1-\alpha)\mu_{t+1} + \alpha S_t, \qquad (3.11)$$

where α $(0 \leq \alpha \leq 1)$ is the weight between the two factors.

Agents increase (or decrease) their utilities in eqn. (3.9) based on the prior choice as follows: $a_{t+1} = a_t \pm \delta$ if an agent chooses A (or B). The probability of an agent choosing A is:

$$\mu_{t+1} = \frac{1}{1 + exp(-(2a_{t+1} - 1)/\lambda)}. \qquad (3.12)$$

Regarding the second factor, social reinforcement, in eqn. (3.11), we assume that the impact of social reinforcement linearly increases in proportion with the number of adopter agents. We let A_t denote the number of agents who have chosen A by period t and B_t denote the number who have chosen B. The social reinforcement factor that works on an agent making their choice at time $t+1$ is given by the ratio $S_t = A_t/(A_t + B_t)$.

We investigate properties of the collective choice when there is a large number of agents ($N = 1000$). To begin with, half of the agents have a stronger preference for A ($\mu > 0.5$), and the other half have a stronger preference for B ($\mu < 0.5$). We examine the two extreme cases with $\alpha = 0$ and $\alpha = 1$, and the mixed cases where $0 < \alpha < 1$:

Case 1. $\alpha = 0$ (Individual Reinforcement Only)

In this case, a collection of heterogeneous agents with diverse preferences eventually evolves into two extreme groups (Figure 3.2): (1) half of the agents have perfect

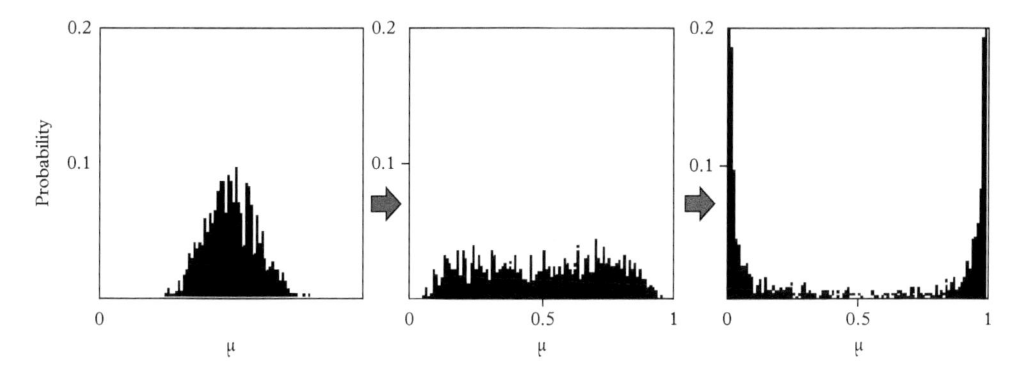

Figure 3.2 *Probability distribution of choosing A* (μ) ($\alpha = 0, \lambda = 0.25$)

preference for A ($a = 1$); and (2) half of the agents have perfect preference for B ($a = -1$).

The El Farol bar problem introduced in Section 3.2 has over time become the prototypical model of a system in which agents, competing for scarce resources, inductively adapt their behavior to the aggregate behavior they jointly create. The bar's capacity can be viewed as a resource subject to congestion, making the El Farol bar problem a stylized version of the problem of efficient allocation of limited economic resources. Real-world examples of this kind of problem include traffic congestion (as previously described) and the congestion of computer networks. In such situations, people hope that resources can be fully utilized (without leaving much idle capacity) while avoiding overuse, which leads to congestion. When it is either undesirable or not feasible to solve such problems by means of central intervention, then they must be solved in a bottom-up manner.

Most research addresses the El Farol problem from the perspective of learning, hence, answers depend on how agents learn. There are two proposed kinds of learning mechanism known as *best-response learning* (Arthur, 1994) and *reinforcement learning* (Bell and Sethares, 1999; Whitehead, 2008), and these show fairly typical results for the El Farol problem. The best-response learning model tends to show fluctuation around the threshold, but the steady state (where aggregate attendance is always equal to maximum capacity) is difficult to reach. The reinforcement learning model, however, shows that almost perfect coordination is possible and that it is, indeed, the long-run behavior to which the system asymptotically converges. However, it is an equilibrium characterized by complete segregation. That is, the agent population splits into two groups: one that frequently or always goes to the bar and one that seldom or never goes. This result has led to a new problem, namely, the inequity issue. The group that does not go to the bar (discouraged by previous unsuccessful attempts) obviously shared very few or no public resources; this may contribute to their being a disadvantaged class in the society.

The El Farol bar problem is a highly abstract model suitable for addressing the fundamental issue of use and distribution of public resources. Early studies of this problem centered mainly on the efficiency aspect of this issue while ignoring the equity problem. Chen (2013) proposed that going to the bar is a social engagement opportunity (to gain more information and social connections) and that not going to the bar can imply social exclusion or isolation. Therefore, application of the El Farol bar problem is not restricted to investigations of the economic distribution problem. It may also be applied in investigations of the social segregation problem characterized by a group of agents who fully occupy an available public resource and a group of agents who are discouraged from participating and are ignored.

Chen (2013) further integrated both efficiency and equity in his social reinforcement model. He showed that the bottom-up mechanism crucially depends on two essential ingredients: social preferences and social networks. He demonstrated that social network topologies matter for the emergence of a good society, that is, a state analogous to the situation where the bar attendance always equals capacity and all agents go to the bar with the same frequency. The introduction of social preference can facilitate the emergence of a good society when groups are sensitive to inequity or have inequity-averse preferences. Thus, emergence of a good society is increasingly likely with an increase in the degree of inequity aversion.

Case 2. $\alpha = 1$ (Social Reinforcement Only)

In this case, the collective choice is unpredictable. Every feasible outcome in terms of penetration is equally likely at every time; that is, the collective choice process is completely blind and completely unpredictable. When the individual choice is purely a matter of social trend, every feasible outcome is equally likely; intuitively, this is a very random process.

The sequential choice based only on social reinforcement resembles the well-known *Polya's urn process*. That is, the expected proportion exactly equals the current proportion when $\alpha = 1$ in eqn. (3.11). It is easy to show that this *martingale property* generally holds; that is, the collective choice process (where each agent makes choices based simply on social trends) is strongly *path dependent* and is expected to stay wherever it is.

Case 3. $0 < \alpha < 1$ (Both Individual Preference Reinforcement and Social Reinforcement)

Here, the sequential collective choice process becomes a mix between the two extreme cases of $\alpha = 0$ and $\alpha = 1$. As shown in Figure 3.3(a), collective adaptive dynamics lead to outcomes that appear to be deterministic and most agents have strong preferences of choosing A ($\mu \doteq 1$) in spite of being governed by a stochastic decision model at the individual level. However, in another trial as shown in Figure 3.3(b), most agents have strong preferences of choosing B ($\mu \doteq 0$). Among these dynamics, when agents make choices under weak social reinforcement, the proportion of agents making a specific choice is determined by the distribution of the agents' preferences; that is, pure *herding* does not occur. When agents make choices under conditions of strong social reinforcement (e.g., according to social trends), the collective adaptation process is not predetermined by the agents' preferences. Even when individual preference for a specific choice is strong, no

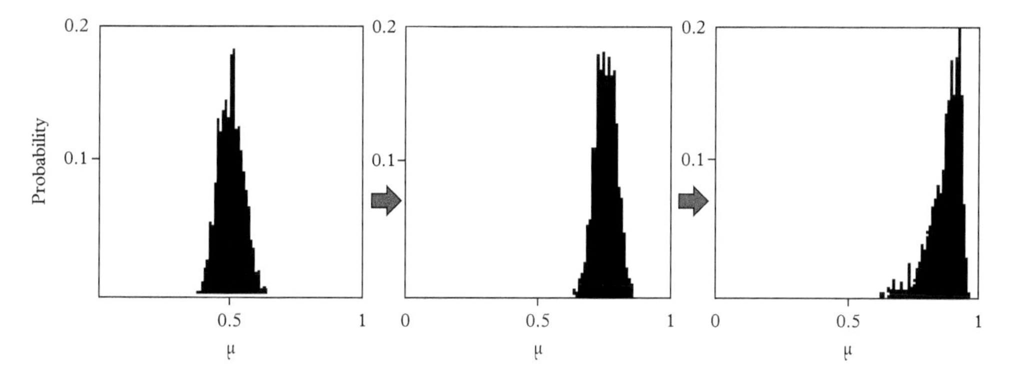

Figure 3.3(a) *Probability distribution of choosing A (μ) (α = 0.5, λ = 0.5)*

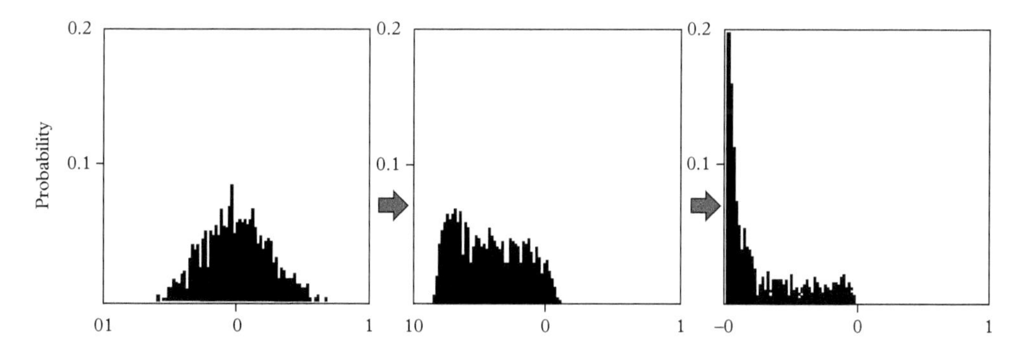

Figure 3.3(b) *Probability distribution of choosing A (μ) (α = 0.5, λ = 0.5)*

sample path of the collective adaptive process exhibits martingale properties. Therefore, the collective outcome is not predetermined by initial circumstances and every possible outcome is equally likely.

To further explain, when individual reinforcement is weak, the choices of most agents converge on a particular alternative and a large-scale cascade occurs. When individual reinforcement is strong, collective outcome is determined by the initial preference distribution. In particular, if agents' preferences nearly split into two groups (one initially favoring *A* over *B* and one initially favoring *B* over *A*), there will be a tendency for two distinct groups of agents to form along these lines.

We are interested in the size of the cascade observed in the collective dynamics and in clarifying the effect of the relationship between individual and social reinforcement on cascade size. When social reinforcement is strong, the evolution of the collective choice is based on a fraction of agents and is not predetermined by objective evidence about the merits of two alternatives—therefore, a large-scale cascade occurs. When individual reinforcement is strong, the adaptive choices are determined by the agents' preferences.

For instance, in the context of fashion, viewing the spread mechanism as a continuing process of collective choice among competing alternatives yields a markedly different picture from that provided by conventional models. These models call attention to the fact that key players (such as innovators and early adopters) are major parts of a collective process that responds to changes in collective taste. The transformation of taste or collective preference results without question from the diversity of experience that occurs in social interaction. However, the idea of individual preference reinforcement may be an important contributor to the analysis of mechanisms underlying collective behaviors such as fashion spread.

3.4 Network Effects on Adaptive Choices

In this section, we consider how collective choice taking place within agent networks is influenced by the topological properties of the networks. Importantly, the choices of individuals can have a clear impact on network topology, such as when conflicting opinions lead to the breakup of social interaction. From an applied point of view, it is desirable to compose an inventory of the types of microscopic dynamics in agent networks and their impact on collective choice at the macroscopic level.

New computerized social networks have changed the ways in which people interact and communicate. Hundreds of millions of internet users now participate in such networks, and are able to establish many social links to new friends. These new networks incorporate new social infrastructures for information sharing, storage, and other internet social applications.

Innovations and spreading of opinions are important processes that take place within social networks and, as such, are influenced by the topological properties of those networks. Individuals' behaviors or preferences spread to their friends, colleagues, and others through social networks as both information and the propagation of messages. Most human interactions today take place with the mediation of information and communications technologies provided by social media. In the past, the set of references, ideas, and behaviors to which people were exposed was restricted according to geographical and cultural boundaries. However, today, many different types of social media are providing more data and extending the boundaries of interdependence.

Currently, we can address a number of phenomena that cannot be modeled well at the whole population level. For example, many individual interactions happen locally rather than at the global level and, consequently, we often form opinions that are aligned with those of our friends and colleagues rather than the decisions of the whole population. When agents perform this type of local interaction, it may be useful to focus on the structure of the network (using graphs) to examine how agents are influenced by their particular network neighbors.

In this way, consideration of individual choices can be merged with explicit network structures. We will next connect these two approaches by exploring some of the decision-making principles that can be used to model individual choices in a social network, thus leading people to align their behaviors with those of their network neighbors. For instance, the process of sequential choices in social networks has become a popular method for measuring online preference aggregation. Consider a social network situation where agents have the option to vote either yes (A) or no (B) on certain specific content posted in an online network. When an agent enters a network to view the content, they are able to view all previous votes and then make their own choice. The social network then provides a tool for aggregating these different voters (agents).

A network $G = \{V, E\}$ can be defined by a set of agents $V = \{1, \ldots, N\}$ and a set of links E. In the present case, the agents represent agents, while a link between two agents $i, j \in V$ represents direct interaction or communication such that $\{i, j\} \in E$. We denote links between agents by the binary variable a_{ij} such that $a_{ij} = 1$ if $\{i, j\} \in E$, and $a_{ij} = 0$ otherwise. The network can be summarized by an *adjacency matrix* A, which is an $N \times N$ matrix. If we restrict the analysis to undirected (or bidirectional) networks, then $a_{ij} = a_{ji}$ and the adjacency matrix is symmetric, that is, $A = A^T$. The neighborhood of agent i is the set $N_i = \{j \in V : a_{ij} = 1\}$. Each agent in the network has a degree of connection that corresponds to the number of links that are connected to the agent. If $p(k)$ denotes the fraction of agents that have degree $k \in K = \{0, \ldots, N-1\}$, corresponding to the probability that an agent chosen uniformly at random has degree k, then the set $P = \{p(k), k \in K\}$ denotes the *degree distribution* of the network.

Next, we introduce several network structures that will be utilized in the social network analysis.

Complete network The basic case would be a *complete network* in which each agent connects to all other agents. A *clique* is a subset of the agents such that every two agents in the subset are connected. Therefore, a clique is a complete subgraph. The term clique comes from social network research modeling cliques of people, that is, groups of people all of whom know each other. Cliques are one of the basic concepts of graph theory and are used in many other mathematical problems and constructions on graphs.

Regular network In a regular network structure, all agents are connected to a constant number, k, of neighbors. The simplest case would be a circle that connects agents to their two nearest neighbors. A random regular network (*RR*) is the special class of regular networks in which each agent is connected to k neighbors, randomly selected from the whole network.

Random network: RND We consider a graph $G(N, p)$ that consists of N agents joined by links with some probability p. Specifically, each possible edge between two given agents occurs with a probability p. The average number of links (also called the average degree) of a given agent will be $z = Np$, and it can be easily shown that the probability $p(k)$ that a vertex with a degree k follows a Poisson distribution.

We can construct a random network by first setting the number of agents equal to N and forming an adjacency matrix A with all entries being zero. We then connected each of these among themselves with constant *linking probability* $p \in (0, 1)$; if i and j are linked, $a_{ij} = 1$ and $a_{ji} = 1$. The resulting degree distribution is binomial with parameters N and p, and in the limits of a large N, it becomes a Poisson distribution where the *average degree* is pN.

Scale-free network: SF(BA) A scale-free network has a power-law degree distribution. A popular mechanism for the generation of scale-free networks is the *preferential attachment* of Barabasi and Albert 1999. This process begins with m_0 initial agents and then continuously adds new agents with degree $m \leq m_0$; each of the m links is in turn connected to the existing agents with a probability that is proportional to the degree of the already existing agents. For this reason, the mechanism is sometimes referred to as "the rich getting richer". When we let $p(k)$ denote the degree distribution, the resulting formula is $p(k) = ck^{-3}$ where c is some constant, and this is called the Barabasi-Albert scale-free network, $SF(BA)$.

Core-periphery network: CRA In realistic networks, core-periphery structures are often observed. Such networks can be constructed by combining a clique network with random attachment and such a network is referred as a *CRA* network.

Figure 3.4 depicts the topologies of these networks. In general, threshold models of networks are described within a specified framework. In considering models for the spread of an idea or innovation through a social network G (represented by a graph), agents are considered to be either active or inactive. Researchers usually focus on motivation, in which each agent's tendency to become active increases monotonically as more of their connected agents become active.

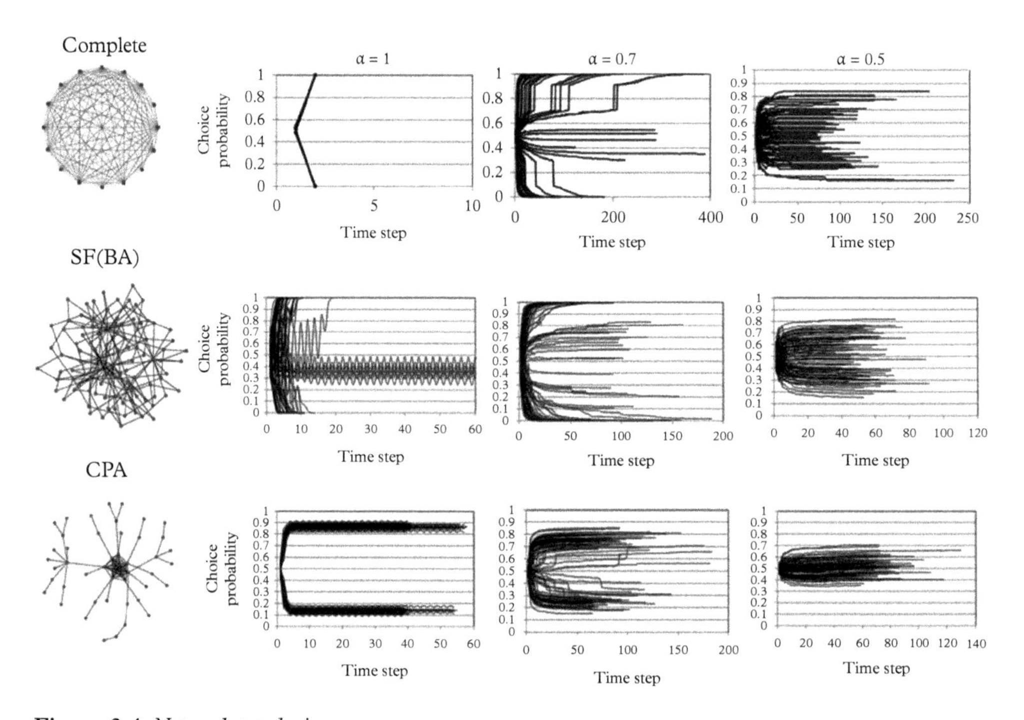

Figure 3.4 *Network topologies*

As discussed in Section 3.3, each agent makes a choice based on both personal preference and social reinforcement. However, by incorporating the agent network, an agent's probability of choosing A in period $t + 1$ is given by the weighted average of individual preference reinforcement and social reinforcement from connected agents:

$$p[\textit{agent chooses } A] = p_{t+1} = (1-\alpha)\mu_{t+1} + \alpha \sum_{j \in N_i} a_{ij} x_j / k_i, \tag{3.13}$$

where the second term $\sum_{j \in N_i} a_{ij} x_j / k_i$ represents the ratio of connected agents who choose A.

In order to implement the stochastic adaptive choice model of eqn. (3.13) on various network structures, the immediate question becomes how to formulate microscopic transition probabilities for individual agents in the network. At the individual level, an agent can either stay in their current state (B) or switch to the alternative state (A). An obvious way of implementing individual transition probabilities would be to posit a transition probability p_i for switching the state of agent i at the individual level, where $\sum_{j \in N_i} a_{ij} x_j$ counts the number of agent i's neighbors that are in the same state. Since an agent is connected to other agents, we would simply implement the transition probability for switching states at the individual levels in the form of eqn. (3.13).

We simulate the collective adaptive choices where a large number of agents ($N = 500$) are placed on the complete network, the scale-free network ($SF(BA)$), and the core-periphery network (CPA). We examine collective choice over time by considering the three extreme cases with $\alpha = 1$, $\alpha = 0.7$, and $\alpha = 0.5$.

Figure 3.4 shows the choice probability of A of each agent and shows the distribution of the choice probability of A on these three networks. With strong social reinforcement ($\alpha = 1$), most agents reinforce their preferences for A or B (strong conformity) when agents are located on the complete network or the scale-free network ($SF(BA)$). However, some diversity is observed when agents are located on the core-periphery network. With mild social reinforcement ($\alpha = 0.5$), most agents are split into two groups, and most diversities are observed when agents are located on the scale-free network ($SF(BA)$) and the core-periphery network.

Changes in opinions or behavior can easily create incentives that shift behavior across the network in ways that were initially unintended. Moreover, such effects are at work whether we are able to see the network or not. When a large group of people is tightly interconnected, members of that group will often respond in complex ways that are only apparent at the agent population level, even though these effects may come from implicit networks that we do not directly observe.

If individuals have strong incentives to achieve good outcomes, then they will appreciate that these outcomes depend on how others behave and will take this into account in planning their own actions. As a result, models of networked behavior must take strategic behavior and reasoning into account. A fundamental point here is that in a network setting, agents should evaluate their actions not in isolation, but with the expectation that the world will react to their actions. This means that cause–effect relationships can become quite subtle.

Many of our interactions with the rest of the world happen at a local rather than a global level. Further, people often do not care as much about the whole population decisions as they do about local decisions made by friends and colleagues. For example, in a work setting, we may choose a technology that is compatible with choice made by people we collaborate directly with, rather than choosing a more universally popular technology. Similarly, we may adopt political views that are aligned with those of our friends, even if their views are in the minority at the whole level. When we perform this type of analysis, the underlying social network can be considered at two conceptually very different levels of resolution: (1) the network may be viewed as a relatively amorphous population of individuals and examined for aggregate effects; and (2) the network may be viewed as a graph to examine how individuals are influenced by their particular network neighbors.

In this way, considering individual choices with explicit network structure merges the models, especially when we examine how people link to others who are like them and in turn can become more similar to their neighbors over time. In the framework of the agent network, we can explore the agent decision-making and network topology that leads agents to become similar to their neighbors: instead, a tendency toward favoring similarity was invoked there as a basic assumption, rather than derived from more fundamental principles.

In contrast to previous research, this section has developed principles that show how, at an aggregate population level, agents tend to make decisions similar to those of their neighbors when they are seeking to maximize the utility of behaviors in particular situations. We saw, in fact, that there are two distinct reasons why imitating the behavior of others can be beneficial: (1) there are informational effects, in that the choices made by others can provide indirect information about what these others know; and (2) there are direct benefit effects, in that there are direct payoffs from copying the decisions of others, such as payoffs that arise from using compatible technologies instead of incompatible ones. We now connect these two approaches by exploring some of the decision-making principles that can be used to model agent decision-making in a social network. These principles may lead people to align their decision-making with that of their networked neighbors.

The well-known saying, "birds of a feather flock together," is a useful analogy for thinking about networks of people. That is, just as birds that look the same fly together, people get together when they have things in common (e.g., similar traits). According to this perspective, the saying is as much about people as it is birds.

4

Agent-Based Models of Social Networks

4.1 Introduction

In this chapter, we shall give a review on how agent-based modeling is applied as a generation process (formation process) of social networks. However, the approach to study the social network formation is not limited to agent-based modeling. Game theory provides another coherent framework to study the formation of social networks. This framework has motivated the use of the human-subject experimental approach to test the game-theoretic prediction of network topologies. The involvement of human subjects in the network games provides inspiration for the use of artificial agents in network formations. While, as we shall see, the network topologies studied by game theory is substantially different from the network topologies studied using agent-based modeling, the observations of the networking behavior of human subjects do provide an impetus for more human-like agent-based models of social networks.

The question that we would like to address in this chapter is how the agent-based models of social networks can be related to the standard model of network formation games, either in the cooperative version or non-cooperative version. Our first observation is that the game-theoretic models of network formation have "trivialized" the more realistic network formation process by assuming that agents know the cost and the payoff of each connection.[1] Networking in this literature does not have a flavor of "adventure" or "chance discovery." The only problem remaining in these studies is whether the "ideal" network configuration (the Nash network, the efficient and stable network) exists and can be formed. The second issue, the formation issue to some extent, has become an "empirical" issue addressed by experimental economists (to be reviewed in Section 4.3), but this literature is not followed by the agent-based research community.[2]

[1] This issue has also been noticed by game theorists. For example, Song and van der Schaar (2013) proposed a network formation game which does not rest upon complete information. They show that the network topologies emerging from their models are in stark contrast to the complete information model.

[2] The few exceptions known to us are Hummon (2000) and Hayashida, Nishizaki, and Kambara (2014). Hummon (2000) is in effect the first one using agent-based simulation to show that some equilibria as

Agent-Based Modeling and Network Dynamics. First Edition. Akira Namatame and Shu-Heng Chen.

Our second observation is exactly related to these experimental studies. While the theoretical work has no restriction on the number of agents, unfortunately all experiments only deal with small number of agents (see Table 4.1). This setup may help us find a convenient way to examine the theoretical result, such as the emergence of the wheel network, the start network, or the fully connected network, but it is not clear as to what extent the result obtained using such a small group of agents has caught the idea of a networking society, and it is particularly not clear whether the result can be scaled up or is robust in a rather large society. Due to this skepticism, it is not even clear whether this "empirical" study is empirically relevant. As Bravo, Squazzoni, and Boero (2012) have remarked, this literature did not look at the importance of the complexity and large scale of social structures.

4.2 Game Theory of Network Formation

4.2.1 Cooperative Game Theory of Network Formation

Before giving a quick review on the network formation game let us introduce a number of notations, some of which have already been seen in previous chapters.

Let **N** be a set of N agents,

$$\mathbf{N} = \{1, 2, \ldots, N\}. \tag{4.1}$$

Let (i, j) denotes the (undirected) link between i and j, and the set of all possible links be denoted by g^N,

$$g^N = \{(i, j) : \forall i, j \in \mathbf{N}\}. \tag{4.2}$$

Let G be all the set of possible networks comprised of the set of agents,

$$\mathbf{G} = \{g : g \subseteq g^N\}. \tag{4.3}$$

Notice that g^N is the fully connected network. Now, we come to the major concept of the theory of network formation games, namely, the value function, v. A value function is a mapping from the set of networks to the set of real values.

$$v : \mathbf{G} \rightarrow R, \quad v(\emptyset) = 0. \tag{4.4}$$

predicted by Jackson and Wolinsky (1996) may not be achievable. They found this much earlier, before the network formation experiment (Section 4.3). The result or the puzzle obtained from Hummon (2000) has evoked some following-up discussions in the literature (Doreian, 2006).

Basically, it informs us of the value of forming a network.[3] Let $\mathbf{V}$ be the set of all possible *value functions*.

$$\mathbf{V} = \{v\}. \tag{4.5}$$

An *allocation rule* Y is a mapping,

$$Y : \mathbf{G} \times \mathbf{V} \to R^N \ni \sum_i Y_i(g, v) = v(g).$$

Given the value function and the allocation rule, it is possible to discuss two fundamental economic properties of a network, namely, its efficiency and stability. A network g is *Pareto efficient* relative to v and Y if there does not exist any g' such that $Y_i(g', v) \geq Y_i(g, v)$ for all i with strict inequality for some i. Notice that Pareto efficiency is conditional on a given allocation rule (income distribution rule). It is possible that one can have a more productive network in terms of its value, but under the given allocation rule not all people will be better off and some may be even worse off. Hence, it is not Pareto efficient. However, through "social reform," one may have a different allocation rule, say, Y', such that the "inequality" issue under Y can be avoided. If such a reform (income redistribution) is always possible, we can have a stronger notion of efficiency. A network g is *efficient* relative to v if $v(g) \geq v(g')$ for all $g' \in \mathbf{G}$.[4] Hence, an efficient network, g^*, is the network with maximum value:

$$g^* = \arg\max_{g \in \mathbf{G}} v(g). \tag{4.6}$$

In game theory, it is often know that an efficient outcome is not necessarily stable, and a stable outcome is not necessarily efficient. Hence, the second important property of network formation is its stability. Like efficiency, two notions of stability have been employed, namely, *pairwise stability* and *strong stability*. Jackson and Wolinsky (1996) define a network to be *pairwise stable* if no pair of agents wish to form a link and no agent individually wishes to sever a link. Formally, g is pairwisely stable if and only if the following two conditions hold.

1. $\forall ij \in g$, $Y_i(g, v) \geq Y_i(g - ij, v)$ and $Y_j(g, v) \geq Y_j(g - ij, v)$, and
2. $\forall ij \notin g$, $Y_i(g + ij, v) > Y_i(g, v)$ then $Y_j(g + ij, v) < Y_j(g, v)$.

[3] This is probably the most problematic part of the entire theory of network formation games. For example, as we shall review later, the value that can be generated from a network does not just depend on who is in the network, but more importantly, also on their coordination, such as the coordination in the *stag hunt game*. In other words, the value function should be generalized to the *value correspondence*. This generalization, however, has not been addressed in the theory of network formation games.

[4] Nonetheless, it is often the case that how the team production is carried out crucially depends on how the gains are distributed among participants. Hence, in this sense, the efficient network only gives us a potential, but may be less practical than the Pareto efficient network.

This notion of stability is based upon a minimal perturbation, i.e., one link at a time, say ij. Hence, what has been tested is restricted only to the pair of the agents involved in the specific link, namely, agents i and j. However, without the assurance of smoothness or monotonicity, it is in general quite likely that agent i can simultaneously take into account many links, say two links, to test the stability of the incumbent network. Hence, for example, let us consider the following two links ij ($\in g$) and ik ($\notin g$), and g is pairwisely stable (both conditions (1) and (2) are satisfied). However, if we replace ij by ik, both agents i and k will feel better off, even though j may be worse off. Clearly, in this slightly broader perturbation with two links, g is no longer stable, and it will be threatened by the altered network, $g+ik-ij$. A notion of strong stability to address this issue is proposed by Jackson and van den Nouweland (2005). To introduce this stronger version of stability, one needs to first consider a perturbation to all possible degrees.

A network $g' \in \mathbf{G}$ is obtainable from $g \in \mathbf{G}$ via perturbation by a *coalition* $\mathbf{S}$ ($\mathbf{S} \subseteq \mathbf{N}$) if:

1. $ij \in g'$ and $ij \notin g$ implies $\{i,j\} \subset \mathbf{S}$, and
2. $ij \in g$ and $ij \notin g'$ implies $\{i,j\} \cap \mathbf{S} \neq \emptyset$.

A network g is *strongly stable* with respect to allocation rule Y and value function v if for any $\mathbf{S} \subseteq \mathbf{N}$, g' that is obtainable from g via perturbation by $\mathbf{S}$, and $i \in \mathbf{S}$ such that $Y_i(g', v) > Y_i(g, v)$, there exists $j \in \mathbf{S}$ that $Y_j(g', v) < Y_j(g, v)$.

Given the notions of stability and efficiency, the next primary issue to be addressed in the theory of network formation games is under what circumstance can networks be both stable and efficient. This issue is very extensive. To approach this issue, Jackson (2005) uses the *symmetric connection model* as an illuminative starting point. The connection model has a simple structure on the benefit and the cost of a link. For agent i, the benefit that he will receive from a link ij is π_{ij}, and the cost associated with the link ij is c_{ij}. The net benefit of a link ij is $\pi_{ij} - c_{ij}$. In addition to the direct benefit and cost, the connection model also allows for the positive externality of connections. Hence, let $d(i,j)$ be the minimal length of the path connecting i and j. If i and j are directly connected, then $d(i,j) = 1$; if i and j are not directly connected, but there is a k such that $ik \in g$ and $kj \in g$, then $d(i,j) = 2$, and so on. If there is no path connecting i and j, then $d(i,j) = \infty$. By taking into account all of these indirect benefits from all connections, then the value allocated to agent i under a connection model is

$$Y_i(g) = \sum_{j \neq i} \pi_{ij}^{d(i,j)} - \sum_{j:ij \in g} c_{ij}. \tag{4.7}$$

One can simplify this general model by assuming that costs and benefits are identical for all agents so that we can omit all the subscripts ij from π and c, and the connection network becomes the symmetric connection network. The allocation rule can then be simplified as follows.

$$Y_i(g) = \sum_{j \neq i} \pi^{d(i,j)} - \sum_{j:ij \in g} c. \tag{4.8}$$

The value function and the allocation rule of the symmetric connection model is, therefore, parameterized by two parameters, namely, π and c. By fixing π, one can focus on the cost of connection, c. It can then be shown that if the cost is high enough, then the unique efficient network is the empty network; if the cost is low enough, then the unique efficient network is the fully-connected network. In between, an interesting topology arises, i.e., a *star* network. More concretely, the result is stated as the following (Proposition 3, Jackson (2005)).

Theorem 4.1 (Efficiency of the Symmetric Connection Network) *The unique efficient network structure in the symmetric connections model is*

(i) *the fully connected network g^N if $c < \pi - \pi^2$,*
(ii) *a star encompassing everyone if $\pi - \pi^2 < c < \pi + \frac{(N-2)}{2}\pi^2$, and*
(iii) *no links if $\pi + \frac{(N-2)}{2}\pi^2 < c$.*

The parameter c also determines the pairwise stability of a symmetric connection model (Proposition 4, Jackson (2005)).

Theorem 4.2 (Pairwise Stability of the Symmetric Connection Network)

(i) *A pairwise stable network has at most one (nonempty) component.*
(ii) *For $c < \pi - \pi^2$, the unique pairwise stable network is the complete network, g^N.*
(iii) *For $\pi - \pi^2 < c < \pi$, a star encompassing all players is pairwise stable, but not necessarily the unique pairwise stable network.*
(iv) *For $\pi < c$, any pairwise stable network which is nonempty is such that each player has at least two links and thus is inefficient.*

By Theorems 4.1 and 4.2, one can see that when the cost of a direct link is too high or too low, the efficient network and the pairwise stable network are reconciled. They both are either the empty network or the fully-connected network. It is when cost c is in the intermediate level that the two properties are in tension. Jackson (2005) provides an example to illustrate this tension. Given $n = 4$ and $\pi < c < \pi + \frac{\pi^2}{2}$, the efficient network is a star network, which is not pairwise stable; on the other hand, the pairwise stable network is empty, which is not efficient. Hence, this example shows clearly the fundamental tension between stability and efficiency. Jackson (2005) then continues to address the fundamental tension from the symmetric connection model to general value functions (vs) and allocation rules (Ys).

4.2.2 Non-cooperative Game Theory of Network Formation

In contrast to Jackson (2005), Bala and Goyal (2000) develop a noncooperative model of the Nash network in which each agent bases his decision to form links on benefits and costs given the action of others. The main idea of the network model in Bala and Goyal (2000) is similar to the connections model. Players earn benefits from being connected to other players and bear costs for maintaining direct links. Benefits result from valuable, non-rival information that flows through the network. Bala and Goyal (2000) distinguish between two different scenarios of information flow. In the first scenario (*the 1-way flow model*), information only flows to the player who maintains the link. In the second scenario (*the 2-way flow model*), information flows both ways. Independent of the information flow, Bala and Goyal (2000) assume that players simultaneously decide with whom to form a direct link, a link being costly to the individual who forms it.

Assuming that information flows through the network with no decay, Bala and Goyal (2000) proved that Nash equilibria of the network-formation game are the following. In the 1-way flow model a Nash equilibrium is either the *empty network*, where no player maintains any connection to any other player, or *minimally connected*, that is, it has a unique component that splits if one link is severed. Analogously, a Nash equilibrium in the 2-way flow model is either the empty network or minimally 2-way connected, that is, it has a unique component, no cycle, and no two individuals maintain a link with each other. Intuitively, a network is Nash in both models if (i) either none or all players are connected and (ii) no redundant links are maintained.

The startling findings of Bala and Goyal are that decentralized agents, via a series of simultaneous decision-making rounds, organize and stabilize at strict Nash network configurations. In a remarkable theoretical result, Bala and Goyal (2000) show that, despite the myopic and naïve behavior of agents, the social communication network evolve toward the Nash equilibrium very rapidly. This result is perhaps best interpreted as a benchmark with respect to the evolutionary capabilities of networks: that, with self-interested and boundedly rational agents, convergence to stable networks is possible.

This completes our brief review of the game-theoretic approach to network formation. For an extended overview of the theoretical literature on network formation, the interested reader is referred to Jackson (2008) and Goyal (2012). However, before we leave this section, it is important to mention that the literature on the network formation games is largely developed with the assumption of *complete information*, i.e., agents know $Y_i(g, v)$, for all $g \in \mathbf{G}$. This assumption has its value and is an inevitable research strategy for an initial attempt to understand this otherwise rather complex system. However, we also know that this assumption is very stringent because it is equivalent to assume that before the network g is formed, agent i already knows the value of it or the share he will be allocated. Therefore, it will be interesting to move a step forward to consider a more realistic situation that agent i can know $v(g)$ or his share in $v(g)$ only after the network is formed.

Let us take the *co-author model* as an example (Jackson and Wolinsky, 1996). The author i can hardly decide what to write with author j without any knowledge of his co-author. He must attend some conferences or read some papers published by author j and may even need to have some personal interactions with agent j to have a shared idea. The environment or embeddedness that can facilitate this probing process has been simply assumed away by the canonical model either in the form of Jackson and Wolinsky (1996) or Bala and Goyal (2000).

In addition to the probing-facilitating process, there is no guarantee that the contact between agent i and j will be successful. Failures are probably more often seen than successes. Hence, it seems to be reasonable to assume that network formation is a dynamic path-dependent process. Whether agents i and j will meet obviously depends on the network already formed before their cross and how their relation will extend also depends on how far they can see the possible paths (sequences) of networks emanating from the link between them. The complexity of this process may make the exiting game-theoretic formulation too primitive to be empirically profound.

To place more realistic network formation under consideration, there are three possible directions to move. First, is to relax the assumption of the complete information and develop game-theoretic models with incomplete information. Presently, the only work known to us in this area is by Song and van der Schaar (2013). Compared to the case of complete information, their results derived from the assumption that incomplete information is more pertinent to agent-based modeling. We briefly summarize their results as follows.

Firstly, it is about the variety of the possibly formed network topologies. When information is complete, the networks that form and persist typically have a star or core-periphery form, with high-value agents at the core. By contrast, when information is incomplete, a much larger variety of networks and network topologies can form and persist. Indeed, the set of networks that can form and persist when information is incomplete is a superset (typically a strict superset) of the set of networks that can form and persist when information is complete. Secondly, even when the network topologies that form are the same or similar, the locations of agents within the network can be very different. For instance, when information is incomplete, it is possible for a star network with a low-value agent in the center to form and persist indefinitely; thus, an agent can achieve a central position purely as the result of chance rather than as the result of merit. Thirdly, even more strikingly, when information is incomplete, a connected network can form and persist even if, when information is complete, no links would ever form so that the final form would be a totally disconnected network.

The second direction to move on is motivated by the Class IV cellular automata or computational irreducibility (Section 2.2.2), using the agent-based simulation model to directly deal with the complex network formation process. The third possible development is to directly work on some interesting networks, such as the buyer—seller network, the labor market network, and to build network formation model directly from there. We shall introduce the development along the second direction in Section 4.4 and then along the third direction in Chapter 8.

4.3 Network Game Experiments

Following the boost in network formation theory, the experimental literature on network formation has been steadily growing in recent years. Most of the experimental analysis of networks has focused on *the endogenous formation of networks.* In a typical network formation experiment, players decide how to form "links" with other players in light of the benefits those links confer. We make no attempt to give a literature review of the network formation experiments. However, we will elaborate on the two earliest studies on the network-formation experiments, Vanin (2002) and Deck and Johnson (2004), to gain some ideas on the further development of agent-based modeling of network formation (Section 4.4).

4.3.1 The Connection Model

Vanin (2002) is the first work on network-formation game experiments. Vanin (2002) considers a connection model, as reviewed in Section 4.2.1. The number of agents, N, is 4. As we mentioned in Section 4.2.1, the connection model is completely determined by the cost parameter c and the benefit parameter π. If these two parameters are arranged in a specific range, the efficient network is a star, but the stable network is empty (Theorems 4.1 and 4.2). This range is:

$$\pi < c < \pi + \frac{N-2}{2}\pi^2. \tag{4.9}$$

Vanin (2002) choose π to be 0.8 and c to be 1; hence the above inequality is held: $0.8 < 1 < 0.8 + \frac{4-2}{2}(0.8)^2 = 1.44$. The resultant network topology is shown in Figure 4.1 (lower panel). Let us first look at the result of the connection model without side payment. The network topology under this design shows two possible topologies, neither empty (the pairwise stable network) nor star (the efficient network), but two in the form of a line (groups A and B) and one in a form of a square (group C). A star is difficult to form since $c > \pi$. Whoever serves as the central node will suffer a loss of 0.6 dollars ($= 0.2 \times 3$).

Since side payment is not allowed, further possible negotiations are impossible. However, the network is not empty. It is either a line or a square. This is because Vanin (2002) allows for open group discussion for the formation of the network.[5] Not surprisingly, if this opportunity is allowed, agents are naturally more interested in the egalitarian rules. However, since side payment is not allowed, the four subjects of Groups A and B came up with an idea of ex ante fairness, i.e., the allocation of value is fair ex ante. This is done by a probabilistic allocation rule. Through random draws, subjects 1 and 4 of the group A and subjects 2 and 4 of the group B are determined to be the two ends of the line network (Figure 4.1, the lower panel, groups A and B). Notice that the allocation of value on these two ends are $0.8 + 0.8^2 + 0.8^3 - 1 = 0.952$, which is higher than the value of the two interior nodes, $0.8 + 0.8 + 0.8^2 - 2 = 0.24$. Distribution inequality between 0.952 and 0.24 clearly presents a challenge for the formation of the network, but by the probabilistic allocation rule, each subject can have a probability of 0.5 to get the high value, 0.952, and a probability of 0.5 to get the low value, 0.24. The expected value is 0.596, which is the same for each agent. As a result, inequity aversion can be successfully achieved by the use of the probabilistic allocation rule. While inequity aversion also applies to the group C, this group did not come up with the idea of the probabilistic allocation rule for achieving ex ante fairness; therefore, the solution turns out to be the

[5] The open group discussion, while a quite commonly social behavior, is rarely seen in current experimental economics, which is dominated by individual-based decisions.

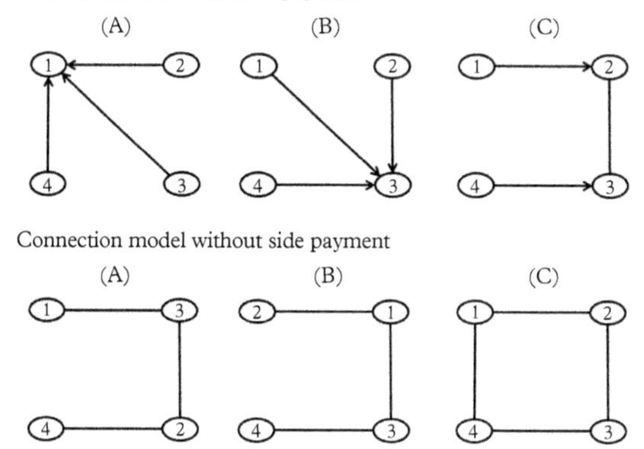

Figure 4.1 *Connection model with and without side payment*

square network (Figure 4.1, the lower panel, group C). In this square network, each subject was allocated a value of $0.8 + 0.8 + 0.8^2 - 2 = 0.24$ only.

In the other series of experiments, side payment is allowed. In this scenario, both groups A and B were able to form the star network. This is so because the central agent, subject 1 in the group A and subject 3 in the group B, can be compensated by the other three peripheral agents. So, originally, the central agent suffered a loss of 0.6, and each peripheral agent can have a gain of $0.8 + 2 \times 0.8^2 - 1 = 1.08$. By an agreement, each of them has to give 0.42 as side payment to the central agent. With this income redistribution, all four subjects ended up with a gain of 0.66. To an extent, the finding in Vanin (2002) is an answer to the question: *will efficient networks be formed if bargaining over the allocation and the network formation are tied together?* The side payment through the group negotiation (bargaining) actually clothes the network game with a flavor of *multi-person ultimatum game* or *multilateral trade*. When the number of subjects (agents) increases, the negotiation cost can become non-trivial, and it cannot be ignored for ascertaining whether a network is efficient.

4.3.2 Spatial Connection Model

An interesting issue arising from the Vanin (2002) is the *trading institution* associated with the allocation rule. In the general framework, reviewed in Section 4.2.1, we do not specify how the allocation rule is determined. Vanin (2002) has shown that collective bargaining can be a way to decide the allocation, which is particularly suitable for achieving the desired goal on both efficiency and equity. Deck and Johnson (2004) in a *spatial connection model* consider a different trading mechanism, namely, auction. As we have noticed, collective decisions can become expensive when the number of agents is large.[6] Hence, an alternative is not to involve collective decision-making, but restore to individual decision-making. Auction is such a form.

In Deck and Johnson (2004), the cost of each possible link, say jk, can be opened for bidding for each agent i, even though $i \neq j, k$. The bid made by agent i for the connection jk is denoted by $bid_{i,jk}$. A link can then be formed if

$$\sum_i bid_{i,jk} \geq c_{jk}. \tag{4.10}$$

If $\sum_i bid_{i,jk} \geq c_{jk}$, agent i only needs to pay the proportion $\frac{bid_{i,jk}}{\sum_i bid_{i,jk}} c_{jk}$. In a sense, this is an alternative form to side payment considered in Vanin (2002). Deck and Johnson (2004) call this auction mechanism the *indirect institution* and propose another two variants and more restrictive versions of auction. One is the *direct institution* and one is *equal spilt institution*. The former does not allow any external agent to bid for a link, i.e.,

$$bid_{i,jk} = 0, \text{ if } i \neq j, k. \tag{4.11}$$

Hence, only agents j and k can bid their connection jk. The latter further requires

$$bid_{j,jk} = bid_{k,jk} = \frac{1}{2} c_{jk}, \tag{4.12}$$

which in effect is an posted-price mechanism.

Deck and Johnson (2004) then considered whether these three different institutions (trading mechanisms) can impact the efficiency of the formed network. The spatial connection model, slightly different from the usual connection model, is characterized by the following three parameter values: $\pi_0 = 1, \pi_1 = 0.9$, and $c = 3$. The benefit of a link is $\pi_0 \pi_1^{d(i,j)-1}$. This connection model is spatial in the sense that the five agents are spatially placed in Figure 4.2; hence, the distance between $d(i, i+1) = 1, i = 1, 2, 3, 4$, $d(i, i+2) = 2, i = 1, 2, 3$, and so on. Since the connection cost is very expensive, the most efficient network is a line, i.e., only geographically consecutive nodes will have a connection between them, as show in Figure 4.2. If this network is formed, the value directly attributed to each subject is as follows.

$$\begin{aligned} v_1(g^*) &= \pi_0 + \pi_0\pi_1 + \pi_0\pi_1^2 + \pi_0\pi_1^3 \\ &= 1 + 0.9 + 0.9^2 + 0.9^3 = 3.439 \\ v_2(g^*) &= 3.71, v_3(g^*) = 3.8, v_4(g^*) = 3.71, v_5(g^*) = 3.439. \end{aligned}$$

[6] While collective decision-making is often seen in real life, agent-based models of collective decision-making is rarely seen. The example of Vanin (2002) is relative simple, because the fair share as a strong focal point can be independent of the number of subjects; hence a consensus on it is easier to achieve by collective decision-making. Nevertheless, when the equity issue is not the only concern or the criterion for a fair share becomes less clear, it can be quite costly to involve a collective action to make a decision.

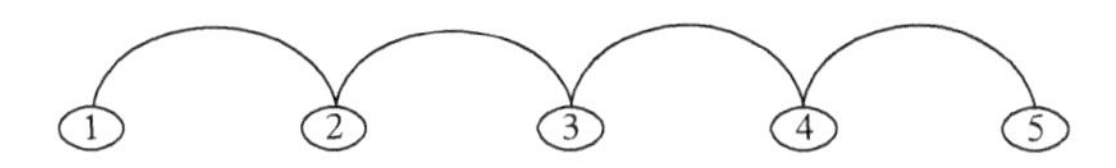

Figure 4.2 *Spatial connection model*

Subtracting the connection costs, which are 12 (= 3 × 4), from the sum of these values ($\sum_i v_i(g^*)$), we can still have a social surplus up to 6.098. However, it is immediately clear that this efficient network cannot possibly be formed under the equal splitting institution. This can easily be seen due to the fact that agent 2 has no incentive to assume a cost of 1.5 to get a reward of only 1 from agent 1 (Figure 4.2). Hence, the link (1, 2) cannot sustain (not pairwise stable), but if link (1,2) is severed, then (2,3) will not remain for the same reason. It turns out that the only pairwise stable outcome is the empty network.

With more flexible auction rules in either the direct or indirect institution, the form of the efficient network is possible, but this efficient network was not observed in a human subject experiment. The realized efficiency is even negative. Since in this case, the benefits of the efficient network are not evenly distributed among the five agents, it is not hard to see that, if we allow them to have multilateral bargaining, a rule that each subject pays the connection fee in proportion to his received benefits should easily be worked out as a focal point.[7] If so, then obviously subject 3 should pay more, followed by subjects 2 and 4, and subjects 1 and 5 pay the least. In this case, it turns out that the auction mechanism is poor at coordinating agents' individual actions.

Through the two earliest network-formation game experiments, we can see that in the formulation of Jackson and Wolinsky (1996), the network-formation game experiments can be read as an N-person ultimatum game. N persons, through some institutional arrangement, can collectively decide whether they would agree with the share of costs for forming a specific network.

4.3.3 Non-cooperative Game Experiments

Other human subject experimental studies on network formation are only briefly reviewed here (also, see Table 4.1). Theoretical studies of star networks have shown that under certain conditions star networks emerge as efficient and stable equilibria. Experimental studies that use the Bala-Goyal model (Bala and Goyal, 2000) as the network formation protocol, however, did not succeed in discovering star networks (Callander and Plott, 2005; Berninghaus et al., 2007; Falk and Kosfeld, 2012). Callander and Plott (2005) considered various conditions that differed in terms of the linking cost, as well as the value of information. They also examined the impact of having network agents

[7] Of course, there is always the possibility that some agents will be greedy and are not satisfied with their "fair" share, as we have often seen in the ultimatum game.

Table 4.1 *Network Experiments*

Experiment	Number of Subjects
Vanin (2002)	4
Deck and Johnson (2004)	5
Callander and Plott (2005)	6
Berninghaus et al. (2006)	6
Berninghaus et al. (2007)	6
Corbae and Duffy (2008)	4
Goeree, Riedl, and Ule (2009)	6
Di Cagno and Sciubba (2010)	6
Falk and Kosfeld (2012)	4
Rong and Houser (2012)	4
Van Leeuwen, Offerman, and Schram (2013)	4,8
Charness, Feri, Melendez-Jimenez, and Sutter (2014)	10
Van Dolder and Buskens (2014)	[9,15]
Elbittar, Harrison, and Munoz (2014)	4

with heterogeneous payoff structure, an issue unaddressed by the Bala-Goyal model. Their main finding was that star networks did not consistently emerge under theoretical conditions, and that even introducing payoff heterogeneity did not lead to systematic formation of star networks. Consequently, they reported that "significant and persistent inefficiency" was a feature of all of their network environments. Berninghaus et al. (2007) focused on the comparison between discrete and continuous time environments. In the discrete environments, their results showed that players have a tendency to reduce network distance over time. However, the overall average frequency of star networks found in their data is only 11.33%.

The experiment of Callander and Plott (2005) was inspired by the network-formation model of Bala and Goyal (2000). However, they only considered the 1-way flow model and, in this model, only those treatments where the *wheel* is the unique efficient strict Nash network. Falk and Kosfeld (2012) presented a general analysis of the Bala-Goyal model, where both 1-way and 2-way flow networks were studied and several treatment conditions were implemented yielding different theoretical predictions. They did observe equilibrium networks in their treatments with "1-way flow", i.e., where information only flows in the direction of the subject who maintains the link. In contrast with "1-way flow", they found that when information flows two ways, the network fails to converge to a star. Instead, Falk and Kosfeld (2012) found equilibrium wheel networks to emerge. They concluded that the need for asymmetric strategies combined with inequality aversion might contribute to the difficulty in realizing star networks.

In addition, Falk and Kosfeld (2012) hardly found any convergence to Nash networks unless they allowed for pre-play communication between participants. This holds when players are homogenous in their value to others. With heterogeneity in values, star networks form more often (Goeree, Riedl, and Ule, 2009). The non-occurrence of Nash equilibria has mainly been attributed to inequity aversion (Goeree, Riedl, and Ule, 2009; Falk and Kosfeld, 2012). In environments where the equilibrium payoff differences between core and periphery players are large, Nash networks are typically not observed. The role of inequity aversion is highlighted in Goeree, Riedl, and Ule (2009), who estimated the parameters from the Fehr-Schmidt model (Fehr and Schmidt, 1999) and found that subjects experience envy, but no guilt. Berninghaus et al. (2006) also highlight the role of inequity aversion in network formation. In their network formation experiment in continuous time, participants often rotated being the core of the star in order to equalize payoffs. Falk and Kosfeld (2012) also observed rotating, but only if they allowed participants to communicate prior to the experiment. Schram, Van Leeuwen, and Offerman (2013) did not observe any behavior consistent with rotating the core position in their data.

Schram, Van Leeuwen, and Offerman (2013) contributed to this literature in a number of interesting ways. First of all, they highlighted the role of social benefits and payoff asymmetries in network formation and showed that the introduction of social benefits increases the rate of convergence and the overall efficiency. Second, they showed that superstars may form in the (finitely) repeated game equilibrium, but that superstars with a fixed core player can only be part of a repeated game equilibrium with social benefits. Third, they found that superstars are formed with social benefits when group size increases and players jockey to be in the core. Fourth, they show that the core-periphery networks predicted by the Galeotti-Goyal model (Galeotti and Goyal, 2010) are observed in the presence of social benefits. Finally, they showed that stable star networks may form when the value of a player is endogenous. This extended the work by Goeree, Riedl, and Ule (2009) who showed that star networks form under exogenous value-heterogeneity but not with homogeneous values.

Corbae and Duffy (2008) analyzed network effects on equilibrium selection in 2×2 coordination games. While they also considered endogenous networks, their main aim was not to study the emergence of particular network structures but the impact of different interaction networks on subjects' play in the coordination game. The authors found that coordination on an inefficient, risk-dominant Nash equilibrium is more likely to occur if players are embedded in a network of local rather than global interaction.

From the observations of human subject experiments, one can see a variety of human actions involved in network formations. It can be a completely independent action requiring no efforts from the others; it can be a bilateral bargaining; it can involve group discussion and decision. The motives behind the decision and action can be purely self-interested, free-riding, but may be also other-regarding, altruistic, or inequity aversion. The game in the repeated form has been done in many cases so that presumably learning of human subjects can happen. Human subjects are heterogeneous. They can be myopic or far-sighted. Callander and Plott (2005) provided the first analysis of the possible strategies used by human subjects in their networking decisions. Interesting enough, the best-response decision rule as suggested by Bala and Goyal (2000) was rejected; instead, their data support a simple strategic rule indicating that subjects are forming links in counterclockwise fashion.

4.4 Agent-Based Modeling

4.4.1 Origin

Agent-based modeling of social networks can be regarded as the application of the games, conventionally studied in small groups (two persons, three persons), to a grand large society with various social constructs (social norms, reputations, social punishment, taxes, law), and then to see how these games played in a decentralized fashion can constantly reshape the topologies of social networks, which may further cause the change of agents' strategies and behavior. Hence, while quite distinct from the game-theoretic approach of network formations (as reviewed in Section 4.2), agent-based modeling of social networks does rely heavily on the use of various games, such as prisoner's dilemma game, stag hunt game, ultimatum game (bargaining game), trust game, public good game, or various social dilemma, to provide mechanisms (algorithms) for the evolution (dynamics) of social networks. The primary goal of these studies, in a nutshell, is to have a view of the coevolution of individuals and social networks, and create models that are more true to the life fraught with structural changes.

The idea of playing games in a lattice or in a graph already existed in the 1980s. Axelrod (1984), Albin (1992), and Nowak and May (1992) placed the spatial prisoner's dilemma (PD) game in a checkerboard and studied the PD games from the perspective of cellular automata (Section 2.2.2). These studies became pioneering work in the field known as *spatial prisoner's dilemma game* (Section 2.2.2). The spatial prisoner's dilemma game provides a possible route to sustain cooperation. A survey of various models of spatial games can be found in Nowak and Sigmund (2000).

The spatial game was later generalized into the network game. Hence, the games are played in an exogenously given social network (Abramson and Kuperman, 2001), as a part of the literature reviewed in Chapter 2. When the literature has advanced to this stage, it is natural to expect that the game played in endogenous social networks is the next step to move. Skyrms and Pemantle (2000) is one of the earliest studies toward this direction. At the time when their article was written, "the idea of simultaneous evolution of strategy and social network appears to be almost completely unexplored. (Skyrms and Pemantle (2000), p. 9340)" In the following section, we shall begin with a review of their model as a preparatory step for the later developed models.

4.5 The Skyrms-Pemantle Model

4.5.1 Two-Person Games

The idea of the social network that Skyrms and Pemantle (2000) started to work with is a weighted directed network.

> Most current research on theory of network formation takes the point of view that networks are modeled as graphs or directed graphs, and network dynamics consists of making and breaking of links. ... In taking an interaction structure to be a specification of probabilities of interaction rather than a graphical structure, we take a more general view than most of the literature ... It is possible that learning dynamics may drive these probabilities to zero or one and that a deterministic graphical interaction structure may crystallize out, but this will be treated as a special case. We believe that this probabilistic approach can give a more faithful account of both human and non-human interactions. (Skyrms and Pemantle (2010), p. 278)

As in the notation used in Section 4.2, each agent i, $i \in \mathbf{N}$, has to decide with whom he or she would like to connect. The payoff of each connection is determined by an underlying two-person game. It is assumed that each agent can only play one game at a time, i.e., to have only one connection at a time. This restriction makes the problem become a multi-armed bandit problem, in which agent i has to make a choice out of the $N-1$ possible choices (connections).

The standard approach to model the decision for the multi-armed bandit problem is the stochastic choice model. Let $p_{i,j}(t)$ be the probability that agent i chooses to connect with agent j, and the vector of the choice probabilities is given as

$$p_i(t) = \{p_{i,j}(t)\}_{j=1}^{N}, \tag{4.13}$$

where $p_{i,i}(t) = 0, \forall t$.

The vector $p_i(t)$ is constantly updated depending on the payoffs that agent i interacts with agent j. This decision-making can be, and is frequently modeled as, a stochastic choice model. Let us assume that each agent i has a repertoire of his experience of interacting with agent j, characterized by a weight vector $w_i(t)$

$$w_i(t) = \{w_{i,j}(t)\}_{j=1,j\neq i}^{N}, \tag{4.14}$$

where

$$w_{i,j}(t+1) = \begin{cases} \phi w_{i,j}(t) + \pi_{i,j}(t), \text{ if } (i,j) \text{ at time } t,\ i \neq j, \\ \phi w_{i,j}(t) + \pi_{j,i}(t), \text{ if } (j,i) \text{ at time } t,\ j \neq i, \\ \phi w_{i,j}(t), \text{ otherwise.} \end{cases} \tag{4.15}$$

The payoff $\pi_{i,j}(t)$ depends on the strategy that agents i and j use when they met at time t. For example, in the context of the prisoner's dilemma game, the strategy can be to cooperate or to defect; in the context of the stag (hare) hunt game the strategy can be to hunt stags or to hunt hares. Here, we assume that the payoff matrix is fixed (time-invariant), but the strategy used by the agents can be time-variant, as indicated by the time index t. The parameter ϕ is a decay factor $(0 \leq \phi \leq 1)$. This decay parameter characterizes the influence of agent i's past experience on his choice, also known as the *recency effect*, and $\phi < 1$ indicates that the agent i's memory decays with time.

At each time t, one of the agents, say i, will be randomly chosen, and he will then make a choice of connection by following the logistic distribution,

$$p_{i,j}(t) = \frac{\exp^{\lambda w_{i,j}(t)}}{\sum_{k=1,k\neq i}^{N} \exp^{\lambda w_{i,k}(t)}}. \tag{4.16}$$

One can further add a noise ϵ $(\epsilon > 0)$ to the choice probability so that agent i's decision may deviate from what his experience may suggest,

$$p_{i,j}(t) = \frac{\epsilon}{N-1} + (1-\epsilon)\frac{\exp^{\lambda w_{i,j}(t)}}{\sum_{k=1,k\neq i}^{N} \exp^{\lambda w_{i,k}(t)}}. \tag{4.17}$$

Given eqns. (4.13) to (4.17), the general question addressed by Skyrms and Pemantle (2000) is the dynamics of the social network, characterized by the connection probabilities of all agents, $\mathbf{P(t)}$,

$$\mathbf{P(t)} = (p_1(t), p_2(t), \ldots, p_N(t))'. \tag{4.18}$$

Skyrms and Pemantle (2000) show that their model is rich enough to generate a great variety of structures. Depending on the payoff function, the memory decay rate, and noises, different topologies of social network can emerge. The limit of $\mathbf{P(t)}$ can be symmetric or asymmetric; the limit vector $p_i(t)$ can be a random vector, a uniform vector, or a degenerate vector with only one entry being non-zero (one). There are other interesting topologies such as the pairing component where i only connects with j and vice versa, or the star networks where a center player appears.

4.5.2 Multi-Person Games

Pemantle and Skyrms (2004) extend their model with two-person network games into the one with multi-person games. A three-person *making friends game* and a three-person *stag hunt game* are studied. The stag hung game is a story that became a game. The story is briefly told by Rousseau (1984). This game has been further well illustrated in Skyrms (2004). Consider a two-person stag hunt game. Denote the action "hunt a stag" by "1", and denote the action "hunt a hare" by "0". In its normal form, the stag hunt game has the following payoff matrix:

$$\left| \begin{array}{ccc} & 1 & 0 \\ 1 & \pi_{11} & \pi_{10} \\ 0 & \pi_{01} & \pi_{00} \end{array} \right|, \tag{4.19}$$

where $\pi_{11} > \pi_{10} = \pi_{00} > \pi_{01}$. Assume that hunting a stag requires good team work, i.e., requiring both players to choose the action "1". If they both do so, they will successfully hunt a stag with a payoff π_{11} for each. However, if one player is not faithful to this commitment and changes to hunt a hare, then he will get a hare characterized by a payoff π_{10}. Nonetheless, in this situation, the one who is still hunting a stag will be doomed to fail and receive a payoff π_{01}. If both decide to hunt for a hare, they both get a hare with a payoff π_{00}. Hunting a stag is more valuable ($\pi_{11} > \pi_{10} = \pi_{00}$), but is risky because its success depends on the cooperation of the other player. On the other hand, hunting a hare is riskless because the job can be accomplished alone, but the payoff is just a hare.

Consider a H-person game, where $2 \leq H \leq N$. The decision for i to make in each of his iterations is to form a team by finding other $H-1$ players. Pemantle and Skyrms (2004) assume that the agent i's choice is team-based, instead of individual-based. For i, there are a total of $\binom{N-1}{H-1}$ possible ways to form a team, and the reinforcement is made on these $\binom{N-1}{H-1}$ possible options. Eqn. (4.15) is extended from one link to multiple links as in eqn. (4.20).

$$w_{i,j}(t+1) = \begin{cases} \phi w_{i,j}(t) + v_i(g_i)(t), \text{ if } j \in g_i \text{ at time } t, j \neq j, \\ \phi w_{i,j}(t) + v_i(g_j)(t), \text{ if } i \in g_j \text{ at time } t, j \neq i, \\ \phi w_{i,j}(t), \text{ otherwise.} \end{cases} \tag{4.20}$$

Notice that we have replaced the payoffs obtained from the two-person game $\pi_{i,j}$ by the value function v introduced in Section 4.2. This does not mean that we want to change the game from a non-cooperative game to a cooperative game. The use of the value function v and the distribution rule Y simply enhance our expressive power. This general form allows us to includes the interesting case that the payoffs are collectively generated by a team (team-production) as indicated by the value function $v(g_i)$. It also makes the involved distribution problem explicit as indicated by the appearance of the distribution rule $Y_i(g_i, v)$ $(v_i(g_i))$.[8] In non-cooperative games, one can get $v_i(g_i)$ by reading the payoff matrix directly.

The value function used in eqn. (4.20) is different from the one used in Section 4.2 at two places. First, we have indexed the team g by i, as g_i, to indicate that the "invitation" to join the team g is sent by agent i, i.e., the decision is made by agent i. Alternatively, agent i may be invited by other agents, say, agent j, to his "hosted" team, g_j. This additional subscript simply facilitates the presentation of propensity updating and may be less important.

Second, we have added the time index t to shows that the value function is not fixed. The multiplicity of value is quite normal in non-cooperative games, since each entry of the matrix implies one possible value (production) of the team. The team production then depends on how well the team coordinates its team members. Do they all cooperate, or do they all defect? Do they all hunt stags, or do some hunt hares? However, in

[8] In eqn. (4.20), we denote $Y_i(g_i, v)$ more compactly by $v_i(g_i)$.

cooperative game theory, the multiplicity of value is rarely seen, since the value function is normally deterministic and unique. This fine difference shows that v in a more general situation is unknown; agent i does not know what they may gain by attending a club or how they can contribute to the club. Therefore, they have to learn from their experience as to what $v(g)$ and $Y(g, v)$ are, and then make their networking decision accordingly. That is why reinforcement learning plays an important role in the Skyrms-Pemantle Model.

The choice probabilities for the possible $\binom{N-1}{H-1}$ teams are given in eqn. (4.21).

$$p_{i,g_h}(t) = \frac{\exp^{\lambda \prod_{j \in g_h} w_{i,j}(t)}}{\sum_{\{s|g_s \in G^H\}} \exp^{\lambda \prod_{j \in g_s} w_{i,j}(t)}}, \tag{4.21}$$

where

$$G^H \equiv \{(i, i_j)_{j=1}^{H-1} : i_j (j = 1, \ldots H-1) \text{ are distinct and } i_j \in \mathbf{N}, \forall j\}. \tag{4.22}$$

Eqn. (4.22) indicates the set of all possible $\binom{N-1}{H-1}$ teams that agent i can consider. The basic structure is the same as in eqn. (4.16), except that we take a multi-linear form (the product form) of the propensity connecting each of the agents constituting the team g_h.

Reinforcement learning with a decay factor ϕ can result in the so-called *trapping states*. This has been shown in the theory of urn processes.[9] In the stag hunt network game, starting with the equal number of stag hunters and hare hunters, initially they randomly make attempts to connect to each other in forming a three-person stag hunt game, and, then in the limit, *cliques* of different sizes are formed.[10] Agents (hunters) will then only have a positive probability of hunting with the agents in the same clique. Typically, stag hunters will only hunt with stag hunters, and their choice probability for hunting with hare hunters is zero, even though hare hunters may still choose to connect with stag hunters.

Needless to say, the number $\binom{N-1}{H-1}$ can increase very fast and makes the choice become increasingly overloading. For example, consider the case where $H = 6$. When $N = 10$, there are a total of 252 options, but when $N = 100$ the number of options already expands into 75,287,520, which may be beyond the practical boundary of reinforcement learning. In fact, the number of iterations required to make reinforcement learning sensibly work is also exceedingly large.[11] This choice-overloading problem restricts the Skyrms-Pemantle model to only a small number of N, comparable to the size of most network experiments. For example, most of the simulations presented in their paper, N

[9] For the mathematical background of reinforcement learning with a decay factor, please see the review in Pemantle and Skyrms (2004).

[10] See Pemantle and Skyrms (2004), Theorem 4.1, p.320.

[11] Obviously, given the astronomical numbers of options, human life is so short that only a very limited number of possible networks (teams) may have been visited in his lifetime. Many opportunities or challenges are just left there without even giving them a glimpse.

and H are set to six and three only, respectively. This setting leaves agents with only 20 options to consider simultaneously.

4.5.3 Strategy Revision

The essential purpose of the Skyrms–Pemantle model is to study the evolution of individual behavior and social structure together. Hence, something has to be said on the evolution of strategies. The review and revision of strategies is done through a kind of social learning, popularly known as imitation, or simply, "copy the best". In each point in time, there is a small probability, say, q, that agent i will be granted a chance to review and revise his incumbent strategy. When this time comes, he will change his strategy (action) to whichever strategy (action) that was most successful in the previous round. Hence, the Skyrms-Pemantle model simultaneously applies both individual reinforcement learning and social learning to the adaptive behavior of agents, reinforcement learning for interaction structure and imitation for strategy revision, which is also called *reinforcement-imitation* model.

Skyrms and Pemantle (2010) align the *reinforcement-imitation* model into a *double reinforcement* model; hence, reinforcement learning is applied to both interaction structure and strategy revision.

> In this Double Reinforcement model, each individual *has two weight vectors*, one for interaction propensities and one for propensities to either Hunt Stag or Hunt Hare. Probabilities for whom to visit and what to do are both gotten by normalizing the appropriate weights. Weights are updated by adding the payoff from an interaction to both the weight for the individual involved and to the weight for the action taken. Relative rates of the two learning processes can be manipulated by changing the magnitude of the initial weights. (Skyrms and Pemantle (2010), p. 159; Italics added.)

Let $\mathbf{A}$ be the action (strategy) space with a finite number of actions (strategies); $a_s \in \mathbf{A}$, $(s = 1, 2, \ldots, S)$. Let $w_{i,a_s}(t)$ be the propensity of strategy a_s at time t for agent i. Then the reinforcement made on the strategy a_s can be written as eqn. (4.22).

$$w_{i,a_s}(t+1) = \begin{cases} \phi w_{i,a_s}(t) + \pi_{i,j}(t), \text{ if } a \text{ is applied at } t \text{ by } i, \\ \phi w_{i,a_s}(t), \text{ otherwise,} \end{cases} \tag{4.23}$$

$s = 1, 2, \ldots, S$. The choice probabilities are then given in eqn. (4.23).

$$p_{i,a_s}(t+1) = (1 - 1_q) + 1_q \left(\frac{\epsilon}{S} + (1-\epsilon) \frac{\exp^{\lambda w_{i,a_s}(t)}}{\sum_{l=1}^{S} \exp^{\lambda w_{i,a_l}(t)}} \right), \tag{4.24}$$

where 1_q is an indicator function with a probability q being one and $1 - q$ being zero. Notice that in the reinforcement for strategy revision, we basically apply the same parameters previously used for the interaction structure, but this is of course only suggestive.

Maybe the point is not so much on the learning scheme per se, but on the parameter q in both the reinforcement-imitation model and the double reinforcement model. This parameter q controls the relative speed of the evolution of social structure and the evolution of individual behavior. This feature is the essence of the study of the endogenous formation of social networks. When q is small, the evolution of social structure toward one steady state can proceed with less perturbation, whereas when q is large the evolution of social structure may be constantly disturbed by the change of individual behavior. For example, in their two-person stag hunt game, by simulating the stag hunt game for 1000 time steps, they found that

> When $q = 0.1$, we found that in 22% of the cases all hunters ended up hunting stag, while in 78% of the cases, all hunters hunted rabbit. Thus there was perfect coordination, but usually not to the most efficient equilibrium. On the other hand, when $q = 0.01$, the majority (71%) of the cases ended in the optimal state of all hunting stag, while 29% ended up all hunting rabbit. (Skyrms and Pemantle (2000), p. 9346)

Generally, the essential question here is that, when an agent is not satisfied with his current payoffs, will he change his interaction structure (network) first or change his strategy first. For example, considering an agent playing a network prisoner's dilemma game and when he is not satisfied with his payoffs, will he attribute his frustration to the partners, to the strategy that he applied, or to both? If he has habitual inertia in networking structure, then he may change his strategy first or more frequently, and vice versa. Of course, to address the question above, the models discussed in these sections need to be further modified. It will be more realistic by allowing for agents' heterogeneity in their degree of habitual inertia, either in networking behavior or strategy-playing behavior.

4.6 The Zimmermann-Eguiluz Model

Earlier we mentioned that the game-theoretic approach does not directly deal with the society with a large number of agents. While the theory is general and elegant and does not restrict its application to a large society, when it is tested in the laboratory, only small societies are considered (Table 4.1). A similar problem can also be found in the Skyrms-Pemantle model, in which we have argued that the straightforward application of this model to a large society introduces a choice-overloading problem (Section 4.5.2). In fact, the famous *social brain hypothesis* provides a non-trivial cognitive constraint for the agent when he is placed in a large crowd.[12] Of course, one can argue that this parameter may be captured by the linkage cost; however, in agent-based modeling, an alternative way is to directly place a ceiling on the number of connections that agents can build. In this section, we introduce the second agent-based model of social network, which exactly has

[12] By the social brain hypothesis (Dunbar, 1998; Dunbar and Shultz, 2007), humans' brains are capable of managing a maximum of just 150 friendships. A related argument is given in Chen and Du (2014), who question whether reinforcement learning can be effectively applied to the situation with a large number of options.

this feature and can effectively address network formation in a large society. This model, called the *Zimmermann-Eguiluz model*, has been introduced by Zimmermann, Eguiluz, and San Miguel (2004), Zimmermann and Eguiluz (2005), and Eguiluz et al. (2005).

The Zimmermann-Eguiluz model is the earliest prominent example showing how a social network can be developed through a prisoner's dilemma game. Each agent i ($i \in \mathbf{N}$) is randomly assigned a set of neighbors, say, $\mathbf{N}_i$ ($\mathbf{N}_i \subseteq \mathbf{N}$). The number of the connections for i, i.e., the degree of $\mathbf{N}_i$, is randomly determined by a Poisson distribution. The relationship with each neighbor j stands for a undirected link $(i,j) \in g^N$. Agent i will then play a *two-person game* with each of his neighbors. Let $\pi_{i,j}$ be the payoffs that agent i obtained from the game, and the total payoff of agent i's game playing with all his neighbors is

$$\pi_i(t) = \sum_{j \in \mathbf{N}_i(t)} \pi_{i,j}(t). \tag{4.25}$$

Notice, we have indexed both the payoffs and the set of neighbors by time t because both of these two sets may change over time. For the former, player i and/or j may change their strategies (actions) in the course of the game, whereas for the latter agent i may change some of his neighbors. To make this point explicit, we rewrite $\pi_{i,j}(t)$ as a functional relationship with players' employed strategies,

$$\pi_{i,j}(t) = \pi\,(a_i(t), a_j(t)), \tag{4.26}$$

where $a_i(t) \in \mathbf{A}_\mathrm{i}$ and $a_j(t) \in \mathbf{A}_\mathrm{j}$. $\mathbf{A}_\mathrm{i}$ is the action (strategy) space of agent i. At each time all agents play with all of their neighbors simultaneously. After that, agents learn and adapt. Learning and adaptation includes two parts: first, the strategy, and, second, the neighborhood.

For the strategy update, Eguiluz et al. (2005) take the same approach as the Skyrms-Pemantle model, i.e., to imitate and copy the best. However, different from the Skyrms-Pemantle model, the best in Eguiluz et al. (2005) is defined locally not globally. Upon learning, agent i will review the payoffs earned by his neighbors; he then copies the strategy employed by the neighbor with the highest payoffs, as shown in eqns. (4.27) and (4.28):

$$a_i(t+1) = a_j^{i,*}(t), \tag{4.27}$$

where

$$a_j^{i,*}(t) = \arg\max_{a_j(t)} \{\pi_j(t) : j \in \mathbf{N}_i(t)\}. \tag{4.28}$$

For the neighborhood update, if a certain condition is satisfied, then there is a probability that agent i may sever a link and replace the severed link by randomly choosing a new neighbor. The condition considered by Eguiluz et al. (2005) is

$$a_i(t+1) = a_j^{i,*}(t) = D \text{ (Defection)}. \tag{4.29}$$

This condition (4.29) basically says that if the agent learns from his neighbor j to defect, then he has a probability of replacing neighbor j with a randomly chosen new neighbor k. This network rewiring rule is kind of domain-specific; it certainly can work well in the case of the network PD game. However, it does not have a good behavioral explanation, and hence it is unclear how one can generally apply this rewiring rule to other games.

The intuition to support this rewiring rule is that agent i may have already anticipated that $(a_i(t+1), a_j(t+1)) = (D, D)$ is not an lucrative outcome; he, therefore, takes a preemptive action to sever the link and not to play with j. However, if he can do so, he may also like to sever the links with other neighbors who may also apply the strategy D. The behavioral rule does not explain why the latter cases are not considered. Also, since j may switch to C (cooperate) in the next period, the preemptive action taken by agent i is based on his prediction, for example, the naïve adaptive expectation, If he is able to do that, it is not clear why this naïve adaptive expectation does not apply to others of his neighbors. In sum, this rewiring rule is somewhat ad hoc, even though the results obtained from this setting are rather interesting, to which we now turn.

4.6.1 Emergent Social Complexity

Leaving the technical weakness aside, from the viewpoint of network generation, Eguiluz et al. (2005) did make significant progress in agent-based models of social networks. Earlier, we mentioned the game-theoretic approach to social network formation. However, the network topologies are not exciting in the sense that most works done in reality (the laboratory) are small networks, not really large complex networks. On the other hand, the theoretical underpinnings of the laboratory works actually contribute to this smallness, since they seem to be interested in some specific network topologies, such as star or ring networks, which occupy only a small portion of the empirical complex network literature (see Chapter 8). The small-world network, the scale-free network, and the core-peripheral network that receive great attention in empirical observations are not at the center of the game theory of network formations. As to how these networks can be formed socially, it has get to be discovered.

Using the network PD game with the behavioral rules above, Eguiluz et al. (2005) can generate social networks with a number of properties which seem to be empirically relevant. First, the model is able to generate a social network that can demonstrate an interesting social hierarchy, a leader–follower hierarchy. There are only a few leaders but many followers standing in different levels of the hierarchy. Each agent in this hierarchy is followed by some agents at his immediate lower level unless he is the one at the bottom level. These leaders and followers are not given initially, but emerge from their own adaptation process. Psychologists may argue that leaders have some personal traits that may distinguish them from the rest. However, this agent-based model shows that this property is not a necessary condition. As experimental economics have shown, the personal traits may play a determining role in accounting for the results of PD games, but none of these personal traits have been included in the Zimmermann-Eguiluz model.

A direction for further research is to augment this model with agents' personal traits and individualize agents' behavioral rules.

Even without the inclusion of personal traits, the system has already generated a leader–follower network. This shows that the agent-based model has the potential to replicate the self-organization property of social structure, sometimes called *a spontaneous division of labor.*[13]

Second, it generates a distribution of social capital (number of connections) and economic gains. While the society initially starts with a Poisson distribution of the number of connections, the end distribution can be exponential, depending on the payoff structures and rewiring rules.

4.6.2 Comparison with the Skyrms-Pemantle Model

While the basic idea that a group of agents play games in an evolving social network is shared by both the Zimmermann–Eguiluz (ZE) model and the Skyrms–Pemantle (SP) model, these two models differ in several interesting ways, which actually offer two different kinds of agent-based modeling of social networks.

First, the game played in the network is different. For the ZE model, it is the prisoner's dilemma game, whereas for the SP model, it is the stag hunt game. Both games are important for our general understanding of cooperative behavior in the society and, technically, they differ only in the parameters of the payoff matrix; hence this difference may be seemingly less important.

Second, nevertheless, there is a non-trivial difference between the two. The games played in the SP model are network games (multi-person games), such as the three-person stag hunt as demonstrated in Skyrms and Pemantle (2010), but the games played in the ZE model are two-person games. In other words, in the SP model, agents play with their neighbors *altogether in one game*, while in the ZE model agents play with each neighbor individually in multiple games.

Third, in addition to the payoffs, this difference also affects the concept of the social structure. As we mentioned in Section 4.2, in economics what matters is the team production captured by the value function in the cooperative game theory. This idea has been well presented in the SP model, but is absent in the ZE model. This contrast prompts us to think more deeply about the purpose of networking for individuals. Clearly, simultaneously having many bilateral relations has a value, such as having a coauthor in each of the working papers, but it is not at the same level as hunting a stag.

[13] This term has been used in literature on the adaptive evolutionary networks (Gross and Blasius, 2008). This term originates from economics. From the viewpoint of productivity, the formation of economic networks is exactly driven by the division of labor. One can consider a *joint production game*, a more general version of the stag hunt game, where each agent not only has to decide what to produce but also with whom to produce (two degrees of freedom) or what to produce given a fixed team (one degree of freedom). We suspect that if we have a network joint-production game, then we may be able to generate social network evolution simultaneously with the division of labor.

4.7 The Zschache Model

4.7.1 Network Public Good Games

When network topology is no longer fixed but can be determined endogenously, what interests us are not just strategies employed by the agents and the ecology of the strategies, but a subject defining what evolutionary game theory is. Networking itself also becomes a part of the strategy which can be modified as time goes on. The resultant coevolutionary system of strategies and network topologies can be highly complex; hence, an easier kickoff is to restrict the attention to *two-person network games* only, as done in the Zimmermann-Eguiluz Model (Section 4.6). Even though the Skyrms-Pemantle model has been extended to the N-person game, only a three-person game has actually been demonstrated (Section 4.5.2). However, after some basic understanding of the coevolution of the network topology and the strategy ecology has been gained in the context of N*-way two-person games*, it is natural to generalize such a framework into the *one-way* N*-person game*, and the Zschache model (Zschache, 2012) is the first model that gives a fully-fledged version of the one-way N-person network game.

> The main differences are that, in an n-way Prisoner's Dilemma, each actor directly interacts with everybody else in 2-person games, and, hence, the payoffs depend on actions that are directed towards a particular player. In contrast, a public good is jointly produced, and nobody can be excluded from its benefits. Thus, the public goods game describes a different, yet often occurring, social situation. (Zschache (2012), p. 539)

To give an introduction to this model, we begin with a fundamental question on the role of the network in the game. As we have generally reviewed earlier in this book, a network can have a number of important functions pertinent to economic behavior. First, it facilitates the information flow. This has been fully demonstrated in the network-based decision models. Agents' behavior and decisions are generally dependent on the information that they received from their neighbors. The development of social economics is based on this pillar. Second, it enlarges the scale of production, such as in the stag hunt game (Section 4.5). However, neither of these two roles are the one considered in the Zschache model. The Zschache model introduces the third role, namely, the *social norm*. The network helps to form the social norm.

Social norm is a set of behavioral or decision standards. The compliant agents will be rewarded, whereas the defiant agents will be punished. It has been considered as a mechanism to promote cooperative, altruistic, or prosocial behavior, and has become part of the evolutionary game theory. However, what has been ignored in the literature is how this social norm is formed in the first place and how it can perform the function of social control. Indeed, it is difficult to address these issues sensibly without taking into account the social network underpinning the social norm. What Zschache did in his model is to use the personal network to define a norm, a local norm.

As before, let $g(t)$ be the social network at time t, $g(t) \subseteq g^N$. Let $\mathbf{N}_i(t)$ be the personal network of agent i at time t, i.e.,

$$\mathbf{N}_i(t) \equiv \{j : (i,j) \in g(t)\}.$$

Let $\mathbf{A}$ be the action space. Without losing generality, we, as before, assume $\mathbf{A}$ to be one-dimensional and binary, i.e.,

$$\mathbf{A} = \{-1, 1\}.$$

Zschache introduces a utility function related to the local norm (the personal network norm), $\pi_i^{\aleph}$, as:

$$\pi_i^{\aleph}(t) = \sum_{j \in \mathbf{N}_i(t)} a_i(t) a_j(t). \tag{4.30}$$

By eqn. (4.30), if at time t, agent i's action is the same as the action taken by the majority of his neighbors, then he will have a positive gain; otherwise, he will suffer a loss, even psychologically. Hence, eqn. (4.30) basically says that the majority of $\mathbf{N}_i(t)$ determines the norm applied to agent i.

Generally speaking, agent i will be happier if his action is closer to the centroid of the actions taken by his neighbors, and unhappier if his action is further away from the centroid. By our conventional understanding of the social norm, one cannot do anything about changing the social norm, no matter how dissatisfied he is with it. Partially this is because the conventional analysis of the social norm does not leave room for social networks, and the default (implicit) setting is the global network (fully-connected network). By that assumption, agent i cannot insulate himself by "escaping to an island" or making himself disappear in the world. However, when a network is formed locally, agent i can have a choice of his own neighbors.

The simple structure characterized by eqn. (4.30) renders the network formation process a strong flavor of Schelling's segregation model (Section 2.2.1). Basically, it is well expected that cliques of agents who take same action will be formed; hence, type-"1" agents will be fully connected and be segregated from type-"-1" agents who themselves are also fully connected as a clique, a result similar to the clique of stag hunters and hare hunters (Section 4.5.2). From the sociological viewpoint, this result can be explained as a consequence of *homophily*, similar to the cultural dissemination model introduced by Axelrod (1997).

The real effect of the social norm is that it causes agents to change their behavior or action. Hence, in this regards, the Zschache model differs from the Schelling model in the sense that agents in the Schelling model cannot change their natural ethnicity (action), and the choice left for them is to migrate (rewire), whereas agents in the Zschache model, in addition to rewiring, can also consider changing their actions so as to comply with the majority. Therefore, even though cliques may form in both models, it is this fine difference that makes this model more similar to the Axelrod model (Axelrod, 1997).

The option to migrate (rewire, severe, and reconnect) or change the action can further depend on other considerations. From the game-theoretic viewpoint (Section 4.2), severing a link or adding a link can both be costly. In addition, based on the notion, severing a link can be a unilateral decision, whereas adding a link may require mutual consensus. If we take into account these technical conditions with other behavioral assumptions, agents may prefer to change the action, instead of re-networking. The initial configuration plays a role here. Consider the case that a type-"–1" agent is surrounded by N_i type-"1" agents. Currently, his utility $\pi_i^{\aleph}$ is simply $-N_i$, and if he can simply switch his action from "–1" to "1", his utility will be N_i, an increment of $2N_i$. Alternatively, he can sever all links and replace each of them with a connection to the same type of agent to gain a utility of N_i. If the cost of changing the action is relatively low, say zero, and the cost of deleting/adding a link is relatively high, say c, then the agent may decide to change the action, instead of reconnecting. On the other hand, if the cost of reconnecting is free but there is a substantial cost of changing the action, then the agent may prefer to get rewired.

Even though the cost is not a concern, adding a new link needs mutual consensus, and the other end of the link may decline the request if such a link does not bring in an improvement in utility. In other words, the condition of pairwise stability also needs to be satisfied during this re-networking process.

Now, we have been back to the fundamental issue raised by Skyrms and Pemantle (2000). Is it more likely for the agent to change the strategy or more likely to change the structure (network)? Different games may imply different costs. For example, in the stag hunt game, the switch from the hare hunter to the stag hunter may involve substantial profession training and hence involve a high cost of switch. Zschache (2012) considers a different game, known as the *public goods game*. The public goods game has been pursued in the literature, specifically, in the experimental economics, for more than three decades.[14] From the theoretic viewpoint, the provision of public goods has long been considered suboptimal given the potential threat of the free-rider problem. Nevertheless, in reality, contributions to public goods in various forms are often seen; for example, knowledge and information sharing through the social media networks is the rule not the exception. Therefore, it becomes crucial to understand the factors affecting the willingness of individuals to contribute to the provision of public goods.

In the linear public goods game[15], agent i shall divide his given endowment z_i into two parts. One part, z_i^{pr} is for his own *private goods* consumption, and the other part, z_i^{pu}, is for the public goods consumption;

$$z_i = z_i^{pr} + z_i^{pu}.$$

[14] We make no attempt to give a survey here. The reader is referred to two meta-analysis of the game: one is on threshold public goods experiments (Croson and Marks, 2000) and one is on liner public goods experiments (Zelmer, 2003).

[15] The threshold public goods game is different from the linear public goods game in the minimum received contribution for the provision of public goods. For the linear public goods game, such a threshold is zero; hence, no matter how small Z^{pu} (eqn. 4.31) is, the provision of public goods is Z^{pu}. For the case of the threshold public goods game, if Z^{pu} does not meet the minimum, the provision of the public goods will be zero.

Since the public goods are not exclusive, all agents can consume all provision of public goods contributed by other agents. Therefore, the total consumption of agent i, the value allocated to agent i, is

$$Y_i = \alpha z_i^{pr} + \sum_{j \in \mathbf{N}} z_j^{pu} = \alpha z_i^{pr} + Z^{pu}, \tag{4.31}$$

where $\frac{1}{\alpha}$ is called the *marginal per capital return*. This parameter, characterizing the value of the public good relative to the forgone private good, is an important determinant of voluntary contributions to public goods. Isaac, Walker, and Thomas (1984) demonstrates the influence of this parameter on the provision of linear public goods. In public goods experiments, α is set to be greater than one (Zelmer, 2003). The linear public goods game considered by Zschache (2012) is a special case of eqn. (4.31) in the sense that z_i^{pu} (z_i^{pr}) is binary, either 0 or 1 (1 or 0). In other words, the action in the binary set **A** is either "to contribute" or "not to contribute", i.e.,

$$z_i^{pu} = \begin{cases} 1, & \text{if } a_i(t) = 1, \\ 0, & \text{if } a_i(t) = -1. \end{cases} \tag{4.32}$$

Hence, eqn. (4.31) can be rewritten as

$$Y_i(t) = \sum_{j \in \mathbf{N}} z_j^{pu}(t) + \alpha(1 - z_i^{pu}(t)) = \sum_{j \in \mathbf{N}} z_j^{pu}(t) + \alpha - \alpha z_i^{pu}(t). \tag{4.33}$$

Since α is a constant, it does not affect the decision to contribute or not; therefore, we can normalize eqn. (4.33) by removing α:

$$Y_i(t) = \sum_{j \in \mathbf{N}} z_j^{pu}(t) - \alpha z_i^{pu}(t). \tag{4.34}$$

Eqn. (4.35) is the payoff function of the standard linear public goods game without the network effect. By taking the network effect into account (eqn. 4.30), the payoff function becomes

$$Y_i(t) = \sum_{j \in \mathbf{N}} z_j^{pu}(t) - \alpha z_i^{pu}(t) + \sum_{j \in \mathbf{N}_i(t)} a_i(t) a_j(t). \tag{4.35}$$

Given eqn. (4.35), how agent i will modify his personal network and his action, i.e., the mapping

$$(\mathbf{N}_i(t), a_i(t)) \rightarrow (\mathbf{N}_i(t+1), a_i(t+1))$$

including how he decides whether to accept a request for a connection from some other agents, is an open-ended issue in agent-based modeling. From simplicity to complexity, from naïvety to deliberation, there is no definitive answer for it. For example, one can work with Simon's satisficing assumption and assume that agent i will basically do nothing if he is satisfied with his recent payoffs (Simon, 1955). On the top of this, one can also apply the *prospect theory* and introduce a *reference point* upon which agent i can decide whether the current situation is satisficing (Kahneman and Tversky, 1979). By the above procedure, if a criterion for making a change is met, then a minimal (incremental) adjustment will be made, which includes switching the action, severing a link with an agent with a different type, and adding an agent with the same type. Then watching for the results of this small change for some periods of time, if the result is not satisficing, a further incremental change will be made.

Zschache (2012) proposes two different behavioral assumptions for agents, one is the myopic optimizer (the forward-looking agent), and one is the adaptive agent (the backward-looking agent). He then studies the provision of public goods under these two behavioral assumptions and other technical parameters of the models, such as α, c, initial degree of segregation between type-1 agents and type-"–1" agents, and the density (degree) of the network. In fact, the endogenously formed network topology is not a focused issue in his study, even though his model can serve that purpose as well.

4.7.2 An Extended Connection Model

The network public goods game considered by Zschache (2012) is very much driven by social preferences or social conformity. The psychological gain from social conformity is almost equally important as the material gains from the consumptions of public goods. There are other versions of network public goods games which, while they are not necessarily agent-based, are more economically (materially) motivated. Cabrales, Calvo-Armengol, and Zenou (2011) is a case in point. To some extent, Cabrales, Calvo-Armengol, and Zenou (2011) can be read as an *endogenous extension* of the Jackson-Wolinsky connection model (Jackson and Wolinsky, 1996). The main difference is that in the original connection model, the gains from the connections are exogenously given, and the only question left to decide is whether to connect. The connection model itself already encapsulates the idea of the public goods since information per se flowing in this network can be regarded as a public good. The contribution made by each individual is to provide the infrastructure (the connection). In Cabrales, Calvo-Armengol, and Zenou (2011), the (information) gain or the public good which each agent (node) can possibly provide is endogenously determined by the effort (investment) that each agent would like to make.

> The decision makers in the model are parents. Each parent is altruistic and cares about the future educational outcome of his child. Each parent exerts two types of costly effort: *productive effort* with the child (i.e. doing homework with the child, doing sport activities together, driving him to different activities, and so on) and *socialization effort* related to education (going to parental evenings, birthday parties, or any activity that involves other parents). (Cabrales, Calvo-Armengol, and Zenou (2011), p. 340; Italics added)

Consider a knowledge (information) network, where agent i can spend his resources in two main stays: either investing in connections (investing in others) or investing in knowledge discovering (investing in himself). The knowledge discovered by agent i becomes a public good for those who are either directly or indirectly connected to agent i. In addition to investing in himself, agent i can also invest in social activities and get connected to many other agents who are able to, directly or indirectly, transmit knowledge. By doing so, agent i also helps distribute knowledge.

The connection cost associated with (i,j) needs to be distributed between agent i and agent j through negotiation and bargaining. In this setting, certain interesting strategic behavioral pattern can be expected. For example, agent i can fully invest in knowledge discovery and contribute nothing to social engagement, and make others who want to connect to him pay the cost and, at the same time, use these "free-riding" connections to get access to the spillover of the knowledge discovered by others. At the other extreme, agent i can spend zero cost on knowledge discovery and invest fully in social engagements, making himself stand in an important position in the connection model to acquire bargaining power for getting more connections.

It is then interesting to see how agents in this "generalized" connection model distribute their limited resources to knowledge discovery and social engagement, and how this network will coevolve with the ecology of investment portfolios.

4.8 The Bravo-Squazzoni-Boero Model

4.8.1 Trust Games

The agent-based network models introduced in the following two sections are all based on trust games. Hence, we begin with a review of *trust games*. The trust game was first introduced by Partha Dasgupta (Dasgupta, 1988) in the context that a customer wants to buy a car from a salesman, but is not guaranteed whether the salesman is honest so that he will get a reliable car, or is dishonest so that he will get a lemon. Dasgupta introduced a game tree which has a structure very similar to the investment game introduced later in experimental economics.[16]

Trust and reciprocity have been studied by experimental economists since 1995, when Berg, Dickhaut, and McCabe (1995) published a paper where they reported the results of an *investment game* that became the prototypical trust game in the subsequent works. In this two-stage game, the two players are endowed with \$10 each. In stage 1

[16] In fact, there are many other games which, while named differently, share similar ingredients of the trust game or the investment game. For example, Rosenthal (1981) introduced a game known as the *centipede game*. The centipede game is a two-person finite-move extensive-form game in which each player alternately gets a turn to either terminate the game with a favorable payoff to himself, or continue the game, resulting in social gains for the pair. Specifically, the game consists of a sequence of a hundred moves with increasing payoffs. Other similar games include the gift-exchange game (Fehr, Kirchsteiger, and Riedl, 1993), the peasant–dictator game (Van Huyck, Battalio, and Walters, 1995), and the donor-recipient game (the help game) (Seinen and Schram, 2006).

the first mover decides how much money to pass to an anonymous second mover. All money passed is tripled. In stage 2 the second mover decides how much to return to the first mover. In this original experiment, out of 32 first movers, 30 sent positive amounts and only 2 sent 0, whereas, out of 28 players who received amounts greater than $1, and 12 returned $0 or $1, 12 returned more than their paired player sent them. So, the results clearly departed from the Nash equilibrium outcome that would be reached by perfectly rational and selfish players. This experiment has been replicated many times since then, showing that these results are, from a qualitative point of view, quite robust. It is now widely accepted that trust and reciprocity are fundamental aspects of human social behavior.

However, most studies on trust games are *one-shot* only; studies on *repeated trust games*, while relatively few, also exist. Repeated trust games enable us to have additional treatments to study the effect of subjects' learning or experiences, their interaction patterns and employed strategies, and the significance of reputation, information provision, and other related institutional designs. Repeated games are further divided into *finite games* and *indefinite games*. They differ in the determination of the duration of the game; for the former, it is fixed at the outset, whereas, for the latter, it is stochastically determined. The repeated game can also be distinguished by *whom* the subject is matched to. In the *fixedly-pairing game*, he is matched to the same opponent through the entire duration of the game, whereas in the *randomly-pairing game*, his opponent is randomly determined in different rounds.

To some extent, the behavior observed in the repeated trust games can be useful for our design of artificial agents. For example, the well-known *reciprocity hypothesis* (Fehr and Gachter, 1998) shall predict that the amount sent by the trustor in the current period is positively related to the proportion returned by the trustee in the previous period, and the proportion returned by the trustee is positively related to the amount sent by the trustor. This hypothesis, which has been examined and well accepted by Cochard, Nguyen Van, and Willinger (2004), will be taken as the basis upon which the behavioral model of artificial agents is built. At this point, we can switch to the Bravo-Squazzoni-Boero (BSB) Model.

4.8.2 Experiment-Like Model

Earlier we mentioned the network formation experiments, which are well motivated by game theory of network formation. These experimental studies have little effect on the agent-based models of network formation. This is because what concerns the agent-based modeling literature is not links and the value of links per se, but the interpersonal relations characterized by a game. However, in the BSB model, human subject experiments and agent-based modeling are well integrated, as shown in Figure 4.3. The construction of the BSB model has two stages. In the first stage, the repeated trust games are carried out to get behavioral observations (Figure 4.3, the left block). These observations are then used to calibrate two parametric decision functions in the trust game, namely, the investment function (4.38) and the return function (4.39). Both decision functions (behavioral rules) are formulated based on the reciprocity hypothesis; in fact,

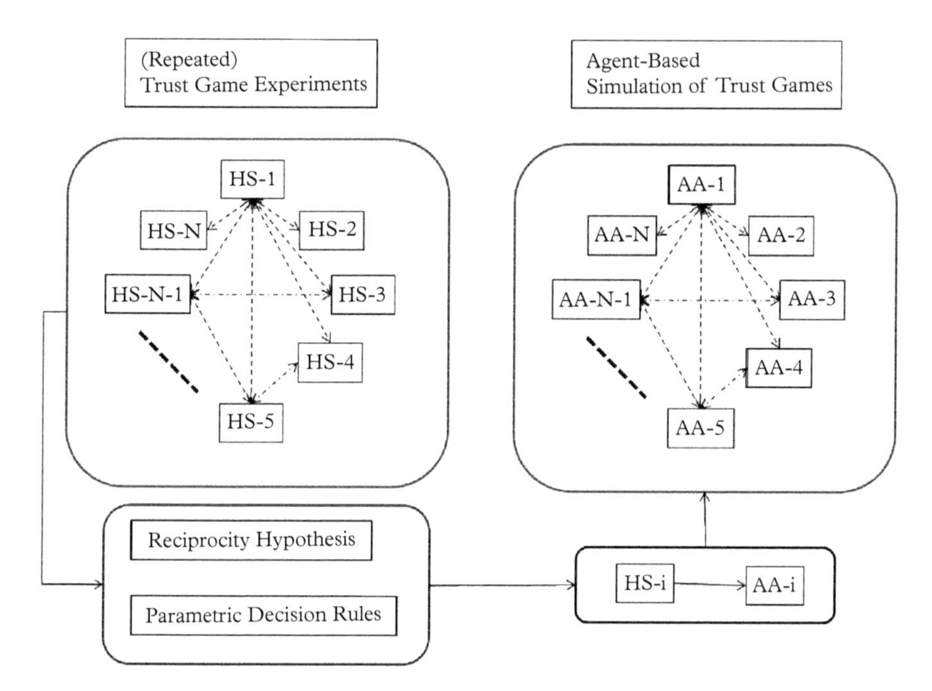

Figure 4.3 *The Experiment-Like Model*

"HS" stands for the human subject, and "AA" stands for the artificial agent.

as we shall see, they are formed as a *linear* reciprocal function. Each individual subject's data are used to calibrate a pair of parametric functions, which in turn are used to build an artificial agent. In this way, we have N avatars of N human subjects. The agent-based model of the trust game is then built upon these N avatars (artificial agents) (Figure 4.3, the right block).

Therefore, the distinguishing feature of the BSB model is the integration of the human subject experiments with agent-based modeling and simulation. The major decision rules of the trust game or investment game are statistically derived from the human subject experiments. Basically, there are two rules involved, namely, the *investment rule* and the *kickback rule* (Figure 4.4). Let $k_{i,j}(t)$ be the investment from agent i (the trustor) to agent j (the trustee), and $y^i_{i,j}$ is the return from agent j to agent i on agent i's investment, or the so-called *kickbacks* from j to i. The notation $y^i_{i,j}$ is motivated as follows. First, by the essence of the trust game (another form of joint production),

$$y_{i,j} = \tau k_{i,j}. \tag{4.36}$$

τ is also called the *investment multiplier*. In the literature on experimental economics, τ has been frequently set as a constant of three. Then the generated returns $y_{i,j}$ needs to be distributed between agents i and j, and this decision is made by the trustee j.

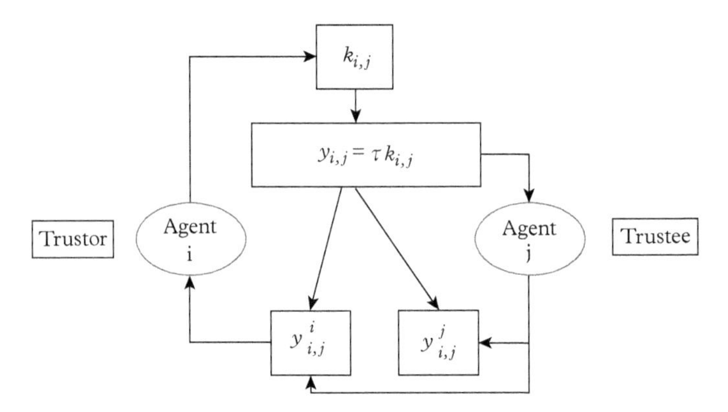

Figure 4.4 *Trust Game: Investment and Returns (Kickbacks)*

Hence, we introduce the notation $y^i_{i,j}$ as the returns allocated to agent i as kickbacks and $y^j_{i,j}$ as the returns allocated to agent j as his "brokerage" fee. In this way,

$$y_{i,j} = y^i_{i,j} + y^j_{i,j}. \tag{4.37}$$

With the notation above, the investment rule and the kickback rule are given as follows:

$$k_{i,j}(t) = \alpha_i + \beta_i[y^i_{i,j}(t-1) - k_{i,j}(t-1)]. \tag{4.38}$$

The coefficient β_i indicates how much trustor i modifies his investment each period as a function of the difference between the amount invested ($k_{i,j}(t-1)$) and the amount received by the trustee j in the previous round. Eqn. (4.38) takes into account the fact that trustor i could have an individual constant propensity to trust represented by the individual intercept α_i, but also the capability of reaction upon past experience represented by the coefficient β_i.

$$y^i_{i,j}(t) = \omega_j y_{i,j}(t) = \omega_j \tau k_{i,j}(t). \tag{4.39}$$

Eqn. (4.39) indicates that trustee j is supposed to react mainly against what he received from truster i. ω_j is the average amount returned by trustee j as proportion of the amount received from truster i in the current round. ω_j represents an estimate of truster j's trustworthiness.

In Bravo, Squazzoni, and Boero (2012), 105 subjects' data from trust game experiments were used to estimate the behavioral parameters α_i, β_i, and ω_i for each individual subject i. Over these set of estimates,

$$\{\alpha_i, \beta_i, \omega_i\}_{i=1}^{105},$$

they further found that there was a significant coefficient between α_i and ω_i, which led them to conclude that "*trusting subjects were also trustworthy: they tried to cooperate and responded by cooperating to others' cooperation.* (Bravo, Squazzoni, and Boero (2012), p. 484; Italics added)".

These 105 artificial agents are then placed into an agent-based model of network-based trust games (Figure 4.3), which is an N-way 2-person trust game. Given a personal network $\mathbf{N}_i$, agent i plays a two-person trust game with each of his neighbors. In the typical trust game experiment, subjects can play the role of either a trustor or a trustee, but not both in the same round. Bravo, Squazzoni, and Boero (2012) introduce a *dual-role design*, i.e., in each round, agents play the role of both a trustor and a trustee with their neighbors. Hence, both behavioral rules, (4.38) and (4.39), are applied simultaneously. Bravo, Squazzoni, and Boero (2012) considered two versions of network-based trust games. In the first version, for each agent i, $\mathbf{N}_i$ is exogenously given. They experimented with different network topologies, including the fully-connected network, the small-world network, and the scale-free network. In the second version, they left $\mathbf{N}_i$ to be determined by each agent. This requires a separate rule for the *networking* or *rewiring* which one cannot have from the usual two-person trust game experiment.

The behavioral rule that they used is similar to the one considered in Zschache (2012). Basically, at the end of time t and the beginning of time $t + 1$, agent i will review the *degree of satisfaction* that he received from each link (i,j), $j \in \mathbf{N}_i(t)$, based on the returns $y^i_{i,j}(t)$. If $y^i_{i,j}(t)$ does not meet a reference point, say $R_{i,j}(t)$, then (i,j) will be severed. In Bravo, Squazzoni, and Boero (2012), $R_{i,j}(t)$ is set to $y^i_{i,j}(t-1)$, if $j \in \mathbf{N}_i(t-1)$; otherwise, $R_{i,j}(t)$ is set to 0. This behavioral rule is widely used in behavioral economics, and can be extended or enriched in many directions. For example, $R_{i,j}(t)$ can be designed by taking into account the historical average of $y^i_{i,j}$, the "peer standard", and even the "social norm" of giving returns. Also, the decision on the cut of the unsatisfied links can be deterministic and stochastic. These varieties of extensions can allow for more tests for the network formation under the trust game.

For the severed link, Bravo, Squazzoni, and Boero (2012) consider two algorithms for its replacement.[17] The first algorithm does not allow the cut link to be rewired unless the agent i will become isolated. In this case, the agent i, who is to be isolated, will be randomly paired to another agent h. The second algorithm allows one of the two agents involved in a broken link to be connected to a new agent.[18]

To see the difference between these two algorithms, notice that both algorithms allow agents to punish "non-cooperative" agents by severing the current link. However, the cut-without-replacement algorithm has made the punishment become *costly* since,

[17] Unlike what we shall see in Section 4.9, Bravo, Squazzoni, and Boero (2012) did not allow agents to *add* more links. Only severed links can be replaced by new links.

[18] Notice that in Bravo, Squazzoni, and Boero (2012), both cutting and adding a link can be done by an agent alone and is not conditioned on the consent of the other agent; of course the other agent, if not satisfied with the link, can cut it later.

without replacing the served links, agent i may have lost some investment opportunities, and money capital is either lost or no longer productive. The cut-with-replacement algorithm can mitigate the effect of severing, and the cost of punishment is alleviated.[19] Bravo, Squazzoni, and Boero (2012) showed that these two different algorithms can lead to very different network topologies and economic outcomes.

The cut-with-replacement algorithm essentially allows agents to select their partners, since they are given a 50% chance to do a replacement. This "norm" actually leads to a network topology that is more similar to the core-peripheral network that we saw in Eguiluz et al. (2005). The degree distribution becomes more skewed, which indicates that leaders emerge, as also demonstrated in Eguiluz et al. (2005). It is further found that those who can be leaders are those who tend to be more trustful and trustworthy (a higher value of α_i and γ_i), which are typical pro-social characteristics. In addition, those agents with these pro-social characteristics also performed better financially than those agents with lower values of the two parameters.

It is interesting to note that the trustfulness and trustworthiness, characterized by the parameter α_i and γ_i, are the calibrated parameters of human subject i, who attended the earlier trust game experiment before the agent-based simulation. Therefore, the repertoire of the 105 behavioral parameters correspond to 105 different personalities in terms of α_i and γ_i. In other words, each artificial agent in the BSB model has their own personality and are heterogeneous. It is then interesting to examine the relationship between the personality and the resultant personal networks as well as the resultant earnings. This is the only model reviewed thus far that examines the role of personality in social network formation.

4.9 Network-based Trust Games

The network games that we have reviewed up to this point, including the prisoner's dilemma game, the stag hunt game, the public good game, and the trust game, share a common feature, i.e., the payoff function of the game is constant, regardless of being the N-way two-person version or one-way N-person version (Figure 4.5). Clearly, when g is exogenously given, there is no point in considering the payoff-dependence on g. However, in this chapter, g is endogenously formed, the dependence of the payoff function on the formed g becomes relevant. As we shall introduce in Section 4.9.1, a more general setting of the network dynamics should not only allow for the *g-dependent payoffs*, but more fundamentally, the *g-dependent games*. We move one step forward toward this more general setting: while the game played by the agents is the same, the payoff function is no longer g-independent.[20]

[19] The cost of punishment has not been taken into account by Bravo, Squazzoni, and Boero (2012); the reference point and the severing decision of agent i are independent of this cost.

[20] For the cooperative game theory, the g-dependent payoff becomes trivial, because it always begins with the value of the entire network g first, $v(g)$, and then proceeds to the possible distribution of network gains, Y_i. Doing that, the g-dependent payoff is automatically incorporated into $v(g)$.

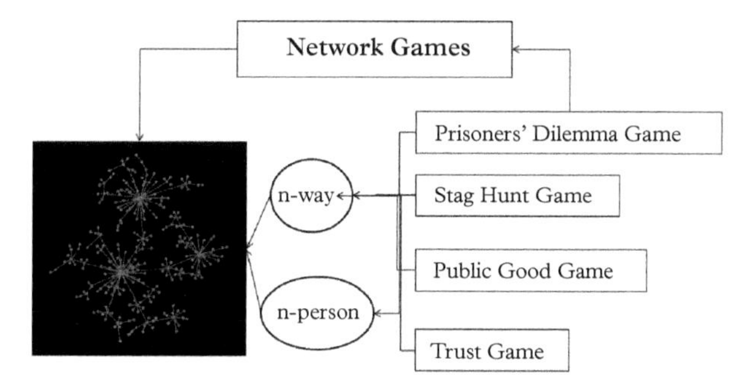

Figure 4.5 *Game-Playing and Network Formation: Who Plays with Whom*

4.9.1 Co-Author Network

To motivate this more general setting, let us consider the *co-author network* introduced in Jackson and Wolinsky (1996). Without losing generality, let us assume that all papers are co-authored by, at most, two authors, i.e., agents i and j. The joint productivity comes from the synergy effect of the *effort* made by agents i and j, which is the *time* that they can devote to the joint paper. Assuming that each agent has one unit of time and assuming that the time is distributed evenly to each paper in progress, the joint production coming from the connection (i, j) has the following form:

$$Y_i(g) = \sum_{j:ij \in g} \left(\frac{1}{n_i} + \frac{1}{n_j} + \frac{1}{n_i n_j} \right) \tag{4.40}$$

where n_i is the number of joint papers that agent i is working on. From eqn. (4.40), we can see that the benefit of joint production for agent i comes from the effort made by agent j, i.e., $\frac{1}{n_j}$ and the joint effect $\frac{1}{n_i n_j}$. Now, consider this as an N-way two-person (co-authoring) game carried out in a network, then the payoff of agent i is no longer independent of the specific formed network, g. For example, if $g = \emptyset$, then the payoff of agent i by cooperating with agent j is 3; if $g = \{(j, h)\}$, the payoff becomes 2; if $g = \{(j, h_1), (j, h_2)\}$, then the payoff is 0.44. However, most network games studied in the agent-based models does not have this feature. Although each payoff function does depend on the network $\mathbf{N}_i$, such as the stag hunt game or the public good game, it does not depend on g in general. Nevertheless, in the co-author network, in addition to $\mathbf{N}_i$, the payoff function also depends on $\mathbf{N}_j, j \in \mathbf{N}_i$; hence, more generally, it can depend on g. The next agent-based model of network formation to be introduced in this chapter is the model proposed by Chen, Chie, and Zhang (2015), abbreviated as the CCZ model, which is also based on the trust game.

4.9.2 Networking and Investment

Like the BSB model (Section 4.8), the CCZ model is a hybridization of both the repeated trust games and the network games. It is outlined as follows. N agents engage in a repeated trust game with T repetitions (T rounds). In each round, each agent has to make a two-stage decision: *networking* and *investment*.

1. (*Partner Selection*) In the first stage, which is called the *network formation stage* or the *partner selection stage*, the subject i ($i = 1, 2, \ldots, N$), acting as a *trustor*, has to decide whom he would like to choose to be his *trustees*, say j ($j \neq i$). Denote this decision by δ_{ij}.

$$\delta_{ij} = \begin{cases} 1, & \textit{if } i \textit{ chooses } j, \\ 0, & \textit{otherwise}. \end{cases} \tag{4.41}$$

   A link between i and j is said to be formed only if either $\delta_{ij} = 1$ or $\delta_{ji} = 1$. That is, all links are *undirected*.

2. Based on the first-stage decisions of all agents, a network topology is determined by a set of links, g,

$$g = \{\bar{ij} : \delta_{ij} = 1 \ \ or \ \ \delta_{ji} = 1, 1 \leq i < j \leq N\}. \tag{4.42}$$

   The *neighbors* of agent i, denoted by $\mathbf{N}_i$, are defined as follows.

$$\mathbf{N}_i = \{j : \delta_{ij} = 1 \ \ or \ \ \delta_{ji} = 1, j = 1, 2, \ldots, N \ \ and \ \ j \neq i\}. \tag{4.43}$$

3. In the second stage, a standard trust game is implemented on each pair connected by a link $\bar{ij}$. This will separate agent i's neighbors into two sets: the trustees of i (to whom agent i will send money, $\delta_{ij} = 1$) and the trustors of i (from whom agent i will receive money, $\delta_{ji} = 1$), denoted as $\mathbf{N}_{i,S}$ and $\mathbf{N}_{i,R}$, respectively. Obviously, $\mathbf{N}_i = \mathbf{N}_{i,S} \cup \mathbf{N}_{i,R}$, but $\mathbf{N}_{i,S} \cap \mathbf{N}_{i,R}$ may be nonempty.

4. (*Investment*) Then, for all those links in which the agent i plays the role of a trustor ($\delta_{ij} = 1$), he/she has to make a decision on the investment in each link, $k_{i,j}$ ($j \in \mathbf{N}_{i,S}$), constrained by his/her endowment or wealth,

$$\sum_{j \in \mathbf{N}_{i,S}} k_{i,j} \leq K_i, \tag{4.44}$$

   and send the money. In the meantime, agent i's trustors js ($j \in \mathbf{N}_{i,R}$) also make decisions on their investment in i and send the money to him. As to the investment decision per se or the determination of $k_{i,j}$, this will be discussed in Section 4.9.3.

5. (*Social Cohesiveness Hypothesis*) All the investment $k_{i,j}$ will then be associated with a multiplier $\tau_{i,j}$, which depends on the network topology. This leads to a major

novelty of the CCZ model. The *multiplier*, intuitively, is related to productivity. The idea is to fully acknowledge the significance of the network size or the scale effect on productivity. In the investment game, all business relations are simply *dyadic*, but the dyadic relation is not isolated. Instead, it is embedded within a large network where many other dyadic relations coexist. Chen, Chie, and Zhang (2015) first assume that the *cohesiveness* of this social embeddedness functions as an *infrastructure* which can be productivity-enhancing, and then use the size of the *clique* containing the specific dyadic relation as a measure of the size. A clique is a completely (fully) connected subnetwork of a given network. Let $g^*_{\bar{ij}}$ be the *clique* (the largest fully connected subnetworks) that $\bar{ij}$ or, equivalently, $\bar{ji}$ belongs to:

$$g^*_{\bar{ij}} = \{(i', j') : \delta_{i'j'} + \delta_{j'i'} \neq 0, \;\; if \;\; \delta_{uv} + \delta_{vu} \neq 0, u \in \{i, j\}, v \in \{i', j'\}\}. \tag{4.45}$$

The *network cohesiveness* is then defined as the degree of $g^*_{\bar{ij}}$ and is denoted by $N_{\bar{ij}}$. By eqn. (4.45), we are searching for the maximally fully connected subnetworks within which the business relationship between i and j is embedded. Intuitively, if the business between i and j is run within a well-connected society instead of a fragmentally isolated small group, then we expect a larger scale effect.

6. (*State-Dependent Multiplier*) Consequently, Chen, Chie, and Zhang (2015) assumed that the multiplier $\tau_{i,j}$ is monotonically increasing in $N_{\bar{ij}}$ (the size of $g^*_{\bar{ij}}$). They set the investment multiplier as a linear function of the cohesiveness of the social embeddedness of the partner relation (i, j), i.e.,

$$\tau_{i,j} = 1 + \alpha \left(\frac{N_{\bar{ij}}}{N} \right), \tag{4.46}$$

where α is a constant. Notice that when the cohesiveness comes to its maximum, i.e., $N_{\bar{ij}} = N$, $\tau_{i,j} = 1 + \alpha$. By setting $\alpha = 2$, we then have the usual setting of having a multiplier of three, frequently used in experimental economics. The production function and the total return received by the trustee is

$$y_{i,j} = \tau_{i,j} k_{i,j}. \tag{4.47}$$

By eqn. (4.46), $\tau_{i,j} = \tau_{j,i}$; hence, $y_{j,i} = \tau_{j,i} k_{j,i}$.

7. (*Kickbacks*) Then, as the usual second stage of the investment game, agent i has to make his decision on the share of the yield $y_{j,i}$ ($j \in \mathbf{N}_{i,R}$) that he would like to return to his trustors j. By following the same notations used in Section 4.8,

$$y^i_{j,i} + y^j_{j,i} = y_{j,i}. \tag{4.48}$$

In the meantime, he also receives money from his own trustees, $y^i_{i,j}$ ($j \in \mathbf{N}_{i,S}$). The details of the decision on kickbacks will be fully developed in Section 4.9.3.

8. This finishes one round of the network-based investment game. An end result is the net income earned by agent i:

$$K_i(t+1) = K_i(t) + \sum_{j \in \mathbf{N}_{i,S}} (y^i_{i,j}(t) - k_{i,j}(t)) + \sum_{j \in \mathbf{N}_{i,R}} y^i_{j,i}(t). \tag{4.49}$$

9. We then go back to step (1). Each subject renews the network formation decisions, and they together form a (possibly) new network topology. The trust game, steps (3) to (8), is then played with this renewed social network with additional links being added or deleted.
10. The cycle from steps (1) to (8), as described in (9), will continue until the maximum number of rounds, T, is achieved.

4.9.3 Heuristics

Section 4.9.2 provides a general description of the network-based trust game model, which is more sophisticated than most network games that we have reviewed so far. In this section, we address the behavioral aspects of the CCZ model. Based on the description in Section 4.9.2, there are three major behavioral aspects that need to be addressed, namely, the decisions on *trustee selection* (Step 1), *investment and portfolios* (Step 4), and *kickbacks* (Step 7).

Trustee Selection

First, we begin with the trustee selection. The starting question is how to characterize an appropriate set of alternatives for agents. We can make no restriction on the set of candidates, i.e., the agent can always consider everyone in the society except himself $\{1, 2, \ldots, N\} \setminus \{i\}$; nonetheless, how many trustees can he choose at each run of the game? One obvious setting is as many as he wants. However, in considering all costs associated with communication, search, computation, or, simply, transaction costs, it seems to be reasonable to assume an incremental process for the upper limit of the number of trustees that an agent can choose. This upper limit is primarily restricted by the cost affordability of the agent. Here, without making these costs explicit, we indirectly assume that the affordability depends on the wealth of the agent, i.e., K_i. Hence, in a technical way, we assume that the additional number of trustees (links) can be available if the growth of the wealth increases up to a certain threshold. For example, an additional link becomes possible if he has positive growth in wealth, and vice versa.

$$l_{max}(t) = \begin{cases} l_{max}(t-1) + 1, \text{if } \dot{K}_i(t) > 0, \text{ unless } l_{max}(t-1) = N-1 \\ l_{max}(t-1) - 1, \text{if } \dot{K}_i(t) \leq 0, \text{ unless } l_{max}(t-1) = l_{min} \end{cases} \tag{4.50}$$

where

$$\dot{K}_i(t) = \ln \frac{K_i(t)}{K_i(t-1)}. \tag{4.51}$$

Note that Equation (4.50) serves only as a beginning for many possible variants, but the idea is essentially the same: each agent starts with a minimum number of links, say $l_{min} = 1$, and gradually increases the number of links associated with his good investment performance, and vice versa. The rule (4.50) leaves two possibilities for the agent to change at each point in time: either adding one link (if he has not come to the maximum) or deleting one link (if he has not come to the minimum). For the former case, he will choose one from those who were not his trustees in the last period, i.e., the set $\mathbf{N} \backslash \mathbf{N}_{i,S}(t-1)$; for the latter case, he will choose one from his last-period trustees, i.e., the set, $\mathbf{N}_{i,S}(t-1)$. Let us assume that, for both cases, his main concern for this one-step change is *performance-based* or *trust-based*. Call this the *trust-based selection mechanism*, which basically says that the agent tends to add the most trustful agent and delete the least trustworthy agent. To do so, Chen, Chie, and Zhang (2015) define the *effective rate* of return on the investment from agent i to j, measured in terms of its kickbacks, as

$$\kappa_{i,j}(t-1) = \begin{cases} \frac{y^i_{i,j}(t-1)}{k_{i,j}(t-1)}, & \text{if } k_{i,j}(t-1) > 0, \\ 0, & \text{if } k_{i,j}(t-1) = 0. \end{cases} \tag{4.52}$$

Then the frequently used *logistic distribution* can be used to substantiate the trust-based selection mechanism as follows.

$$Prob(j \mid j \in (\mathbf{N} \backslash \mathbf{N}_{i,S}(t-1))) = \frac{\exp(\lambda_1 \theta_{j,i}(t-1))}{\sum_{j \in \mathbf{N} \backslash \mathbf{N}_{i,S}(t-1)} \exp(\lambda_1 \theta_{j,i}(t-1))} \tag{4.53}$$

where

$$\theta_{j,i}(t-1) = \frac{k_{j,i}(t-1)}{\sum_{j \in \mathbf{N} \backslash \mathbf{N}_{i,S}(t-1)} k_{j,i}(t-1)},$$

and

$$Prob(j \mid j \in \mathbf{N}_{i,S}(t-1)) = 1 - \frac{\exp(\lambda_1 \kappa_{i,j}(t-1))}{\sum_{j \in \mathbf{N}_{i,S}(t-1)} \exp(\lambda_1 \kappa_{i,j}(t-1))}. \tag{4.54}$$

Eqns. (4.53) and (4.54) are familiarly known as the stochastic choice model, and the parameter λ_1 is known as *the intensity of choice*. Chen, Chie, and Zhang (2015) call it the *degree of relative reciprocity*. Eqn. (4.53) above applies to the situation where agent i can add a link, whereas eqn. (4.54) applies to the situation where agent i needs to delete a link. By eqn. (4.53), agent i tends to favor more those agents who have good trust in him and invest in him generously, i.e., $j \in \mathbf{N}_{i,R}(t-1)$ and $k_{j,i}(t-1)(\theta_{j,i}(t-1)) \gg 0$, than those who did not, i.e., $j \notin \mathbf{N}_{i,R}(t-1)$ $(k_{j,i}(t-1) = 0)$. By eqn. (4.54), agent i will most likely cut off the investment to the agent who offers him the least favorable rate of return, i.e., the lowest κ.

Investment and Portfolios

Once the new set of trustees ($\mathbf{N}_{i,S}(t)$) is formed, the trustor then has to decide the investment portfolio applied to them, i.e., how to distribute the total wealth, $K_i(t)$ over $\mathbf{N}_{i,S}(t) \cup \{i\}$. We assume again that this decision will also be *trust-based*. The idea is that agent i tends to invest a higher proportion of his wealth in those who look more promising or trustworthy, and less to the contrary. Technically, very similar to the decision on the trustee deletion (eqn. (4.54)), let us assume that agent i will base his portfolio decision on the effective rate of return $\kappa_{i,j}(t-1)$. Those who have reciprocated agent i handsomely in the previous period will be assigned a larger fund and vice versa. Then a trust-based portfolio manifested by the logistic distribution is as follows:

$$w_{i,j}(t) = \frac{\exp(\lambda_2 \kappa_{i,j}(t-1))}{\sum_{j \in \mathbf{N}_{i,S}(t) \cup \{i\}} \exp(\lambda_2 \kappa_{i,j}(t-1))}, \quad \forall j, j \in \mathbf{N}_{i,S}(t) \cup \{i\} \tag{4.55}$$

where $w_{i,j}(t)$ is the proportion of the wealth to be invested in agent j; consequently,

$$k_{i,j}(t) = w_{i,j}(t) K_i(t). \tag{4.56}$$

Two remarks need to be made here. First, part of eqn. (4.55) is self-investment, i.e., $w_{i,i}(t)$, and

$$w_{i,i}(t) = \frac{\exp(\lambda_2 \kappa_{i,i}(t-1))}{\sum_{j \in \mathbf{N}_{i,S}(t) \cup \{i\}} \exp(\lambda_2 \kappa_{i,j}(t-1))}. \tag{4.57}$$

Like the typical trust game, agent i can certainly hoard a proportion of the wealth for himself; however, based on the rules of the trust game, this capital will have no productivity and its effective rate of return is always 1, $\kappa_{i,i}(t) = 1, \forall t$. Therefore, by eqn. (4.55), hoarding becomes more favorable when an agent suffers general losses on his investment, namely, $\kappa_{i,j}(t-1) < 1$ for most j. Of course, when that happens, the social trustworthiness observed by agent i is lower and he can then take a more cautionary step in external investment.

Second, for the new trustee ($j \notin \mathbf{N}_{i,S}(t-1)$), $\kappa_{i,j}(t-1)$ is not available. We shall then assume that it is $\kappa_{i,0}$, which can be taken as a parameter of agent i's trust in *new trustees*. The culture or the personality that tends to have little trust for new partners, being afraid that they will take all the money away, has a lower κ_0 and zero in the extreme. The introduction of this parameter then leaves us room to examine how this initial trust (discrimination) may impact the later network formation.

Kickbacks

Finally, they consider the decision related to kickbacks. When investing in others, agent i also plays the role of a trustee and receives money from others $k_{j,i}$ ($j \in \mathbf{N}_{i,R}$). In the end, the total revenues generated by these investments are given in eqn. (4.58).

$$Y_i(t) = \sum_{j \in \mathbf{N}_{i,R}(t)} y_{j,i}(t) = \sum_{j \in \mathbf{N}_{i,R}(t)} \tau_{j,i}(t) k_{j,i}(t). \tag{4.58}$$

Let us assume that the total fund available to be distributed to agent i himself and all of his trustors is simply this sum, $Y_i(t)$. That is, agent i will not make an additional contribution from his *private* wealth to this distribution. Furthermore, they assume that the decision regarding kickbacks is also *trust-based*. They assume that agent i tends to reciprocate more to those who *seem* to have a higher degree of trust in him and less to those who *seem* to have less. This subjective judgment is determined by the received size of investment, $k_{j,i}(t)$. Hence, a straightforward application of the logit model leads to the proportions of kickbacks allocated to each trustor of agent i.

$$\omega_{i,j}(t) = \frac{\exp(\lambda_3 \phi_{j,i}(t))}{\sum_{j \in \mathbf{N}_{i,R}(t) \cup \{i\}} \exp(\lambda_3 \phi_{j,i}(t))}, \quad \forall j, \; j \in \mathbf{N}_{i,R}(t) \cup \{i\}, \tag{4.59}$$

where $\omega_{i,j}(t)$ is the proportion of $Y_i(t)$ that will be returned to agent j as kickbacks, and

$$\phi_{j,i}(t) = \frac{k_{j,i}(t)}{\sum_{j \in \mathbf{N}_{i,R}(t) \cup \{i\}} k_{j,i}(t)}.$$

Hence,

$$y^j_{j,i}(t) = \omega_{i,j}(t) Y_i(t). \tag{4.60}$$

Note that part of eqn. (4.59) is the reserve that agent i keeps for himself. In fact,

$$\omega_{i,i}(t) = \frac{\exp(\lambda_3 \phi_{i,i}(t))}{\sum_{j \in \mathbf{N}_{i,R}(t) \cup \{i\}} \exp(\lambda_3 \phi_{i,j}(t))}. \tag{4.61}$$

By eqns. (4.56) and (4.57), the self-investment is

$$k_{i,i}(t) = w_{i,i}(t) K_i(t), \tag{4.62}$$

and the "retained earnings" are

$$\sum_{j \in \mathbf{N}_{i,R}(t)} y^i_{j,i}(t) = \omega_{i,i}(t) Y_i(t). \tag{4.63}$$

Then the behavioral interpretation of eqn. (4.61) is that an agent who has a large hoarding size tends to be more selfish in the sense that he keeps a large proportion of the fund as "retained earnings," reserved for himself. Such people invest a small share in others, but keep a large share to themselves. These people are, therefore, less social and less cooperative. The parameter which dictates this behavior is κ_0.

4.9.4 Trust, Growth, and Distribution

In this article, they addressed the impact of the technology potential on the network formation. They found that when the technology potential, characterized by the technical parameter, i.e., α (the upper limit of the investment multiplier), can facilitate the networking of the society; hence, a society with a higher technology potential is denser, in terms of network topology, and wealthier, in terms of physical or financial capital. The average degree, average path length, and average clustering coefficient of the society with the high technology potential are consistently larger than the one with low technology potential. Their model, therefore, complements the literature on economic development by not just showing the pivotal role of knowledge in personal productivity but also its enhancement to social cohesiveness and networking production.

They also analyzed the role of the degree of relative reciprocity. Reciprocity seems to be the primary factor in all behavioral models of the trust game experiment, and relative reciprocity, captured by λ, is the extension of this fundamental behavioral parameter into the N-person version of the game. It is found that this parameter can be unfavorable for economic and social growth. A society with a lager degree of reciprocity ends up with a slower development; it is less dense, poorly connected, and less wealthy compared to the one with a negligible degree of relative reciprocity. This result is rather intriguing, but not entirely surprising. The reciprocity behavior in the context of N-person game tends to lead to a more concentrated investment and leaves the gains from networking to a smaller group of agents. The consequence is not just unfavorable for growth, but also unfavorable for income or wealth distribution.

4.10 Toward a General Model of Network Dynamics

In this chapter, we have seen the agent-based network formulation using prisoner's dilemma games, stag hunt games, public goods games, and trust games. From this series of illustrations, one can sense agent-based modeling of social networks in a more general setting. Each game with the respective payoff function can be regarded as a formal representation of the interpersonal relations. These interpersonal relations are potential, and are triggered by the underlying games. While all models consider only one game, a more general model should allow for multiple games.

In real life, different games appear at different times and in different spaces which causes the network dynamics to become even more complex. An agent can come across residential communities, schools, and workplaces. Each space in time may introduce a different game (different interpersonal relations or economic and social dilemma) for the agent; over his lifetime he has to learn what these games are, what the payoff functions are, and what networking strategies to adopt. This grand picture is what we try to paint in this chapter. Obviously this grand picture is far more complex than any single agent-based model can deal with and is far more than what the game theory of network formation can capture.

What is not addressed in this chapter, but should not be ignored, is that the games are not fixed in time–space, but actually evolve with the networks. Therefore, in addition to the coevolution of networks and behavioral rules (strategies) under a given game, there is also a coevolution of networks, strategies, and games. Finally, considering that networks are made possible only if they can be supported by a proper level of technology, the underpinning technology level should also be taken into account. However, technology as a result of R&D also requires an appropriate network to support the "joint production" or the "team production," as has been shown in Section 4.9. Therefore, technology, networks, games, and strategies are actually coevolving. In other words, each model that we have seen so far only gives a partial view of network formation. A full view of network formation needs to place the study of networks in the context of the evolution of technology, interpersonal relations (games), and, finally, strategy behavior. This general understanding of network dynamics in the form of agent-based modeling and simulation has not been pursued in the literature, and is a research subject for the future.

4.11 Additional Notes

Perc et al. (2013) take an approach very similar to the organization of some parts of this book; specifically, they provide a splendid review of the evolution of games over lattice, networks, and the evolutionary networks, while the adopted methodology is statistical mechanism, not the agent-based models. They also focus more on the N-personal game, specifically, the public goods games, rather than the usual pairwise games, and give it a different name, *group interactions*. "Most importantly, group interactions, in general, cannot be reduced to the corresponding sum of pairwise interactions." In this sense, they can be read as a companion to Section 4.7.

Despite the fact that various machine learning, artificial intelligence or statistical learning algorithms have been used in modeling artificial agents' learning and adaptive behavior, imitation as a form a social learning remains to be the most rudimentary and popularly used assumption for learning behavior. In fact, while social learning has a much broader meaning than just imitating or copying others, when applied to agent-based models, it is normally carried out in a way of *merely* copying, as we have seen in Sections 4.5 and 4.6. Some of recent studies seem to support this simple form of social learning (Rendell et al., 2010). Nevertheless, there are also issues addressing the potential limitations and risks of "just copying others" (Boyd, Richerson, and Henrich, 2011; Mills, 2014; Rahwan et al., 2014).

> Nevertheless, social learning has its limitations. For example, extensively copying the behavior of successful or prestigious models is a low-cost way to acquire successful behaviors—but it often comes at the potential cost of not understanding the reasons why these behaviors were successful in the first place. In other words, social learners can be prone to blindly copying the behavior of their models, without acquiring the causal knowledge or the reasoning processes that were responsible for this behavior.... (Ibid, p. 1)

This series of discussions motivates us to reflect upon the form of social learning and imitation, specially, what to imitate, how much to imitate, and why? We believe that this discussion can be useful for the implementation of social learning in agent-based modeling. It can also advance our understanding of its relation to individual learning, as an alternative to modeling agents' adaptive behavior.

5

Agent-Based Diffusion Dynamics

Information, ideas, opinions, and many other things spread in communities and can influence many individuals. Diffusion is a daily social phenomenon of propagation within a society. In this chapter, we will explore the dynamic process of diffusion. We characterize the diffusion scenario, which defines whether promotion or prevention of diffusion will occur.

5.1 Social Diffusion

The study of *social diffusion* examines how epidemic disease, information, opinion, and the latest trends spread within a society. For decades, social scientists, economists, and physicists have been interested in how these things spread through a society (Buchanam, 2007).

The struggle with preventing the spread of epidemic disease has been a major problem throughout human history. When we talk about epidemic disease, we think of contagious diseases spreading from person to person. Epidemics can pass explosively through a population, and a single disease outbreak can have a significant effect on a whole society. Understanding how computer viruses move through a computer network is also an important area in computer science.

Individuals pass information to other individuals. An individual who knows a piece of information will try to share this information with friends or colleagues. The fundamental question is under what conditions the information will diffuse throughout a society. How information will diffuse depends primarily on the probability that it will be passed from one individual to others.

The information diffusion process can be viewed as a virus spreading by a contact process between individuals. However, the biggest difference between biological and information diffusion lies in the process by which one person infects another. In the case of information diffusion, the method of information handling by individuals affects information diffusion, and the diffusion process is sufficiently complex and unobservable at the person-to-person level. On the other hand, in the case of epidemic diffusion, there is a lack of information processing in the transmission of the disease from one person to another, and it is most useful to model the diffusion process at the person-to-person

Agent-Based Modeling and Network Dynamics. First Edition. Akira Namatame and Shu-Heng Chen.

level as a probabilistic diffusion. That is, we will generally assume that when two people are directly linked in the contact network, and one of them has the disease, there is a given probability that he or she will propagate it to the other.

Innovation diffusion is the process by which new products or technological practices are invented and successfully introduced into a society. Numerous studies on the diffusion of innovations have been conducted, and many common features, such as the *S-shaped diffusion curve*, have been empirically verified. One basic question posed by innovation diffusion is why there is often a long lag time between an innovation's first appearance and the time when a substantial number of people have adopted it (Chakravorti, 2003). Why is diffusion sometimes slow? Why is it faster in some regions than others? Why do rates of diffusion differ among different types of innovation? What factors govern the wide variation in diffusion rates?

An extensive amount of theoretical and empirical literature has been devoted to this phenomenon and the mechanisms behind it. Rosenberg (1972) observed two characteristics of the innovation diffusion process: the overall slowness of the process, on one hand, and the wide variation in the rates of acceptance of different inventions on the other. There are many other reasons for the slow pace of technological change. In our world, markets occasionally accept innovations very slowly despite technological advances. New ideas, products, and innovations often take time to diffuse, a fact that is often attributed to the heterogeneity of human populations. Global markets are now larger and more complex than ever before. As such, the amount of information available to consumers has considerably increased; consumers must spend a great deal of time, thought, and consideration before making a decision. Thus, consumers require more time to make personal decisions about matters such as whether to use a certain new product or to adopt the latest technology.

Among research concerning many different types of social diffusion, the diffusion of diseases over social networks has received the most attention (Colizza et al., 2006). There are clear connections between epidemic disease and the diffusion of ideas. Both diseases and information can spread from person to person, across similar types of networks that connect people, and in this respect, they exhibit very similar structural mechanisms. The conditions under which an epidemic or information spreads depends on the diffusion rate and the network structure. Once we estimate these parameters, we can infer the conditions for large-scale diffusion triggered by a few individuals.

Gladwell illustrates certain aspects of social diffusion. In his book (Gladwell, 2002), he synthesizes various theories from the fields of sociology, psychology, epidemiology, and business to show that ideas and messages working their way through the public consciousness behave in much the same way as a contagious disease. In this social diffusion, a specific mechanism works such a *tipping point*, which is the pivotal threshold at which ideas and behaviors spread uncontrollably throughout a society, is reached. The tipping point is the point at which a dynamic diffusion process becomes self-sustaining. A tipping point in adoption is the point at which new ideas or behaviors begin to spread rapidly and experience a sudden increase in popularity. For social diffusion, the tipping point is the level of adoption by the population at which diffusion becomes *self-sustaining*, such that each additional adoptive agent leads to one or more additional adoptive agents, until the diffusion saturates a society.

Networks such as the Internet, the World Wide Web, power grids, and transportation systems are part of our everyday lives. These networks are constructed as interaction networks. For example, social networks represent interactions between friends. In social networks, individuals can be represented as agents, and relations or information flows between individuals can be represented as links. These links can represent different types of relations between individuals including exchange of information, transfer of knowledge, collaboration, and mutual trust.

It becomes important to investigate the influence mechanism that emerges in social networks. Today many people use social media to interact with each other, and information often diffuses virally in social networks. Social networks play a fundamental role in the diffusion of information, opinions, idea, recommendations, and new products. The flow of information through a large social network can be modeled as the spread of an epidemic. Individuals become aware of information, ideas, or new products and pass them on to their friends, colleagues, and other related people.

The phenomenon of social diffusion whereby information, opinions, and even behaviors can spread through networks of people the way that infectious diseases do is intuitively appealing. When individuals have many colleagues, this represents many potential sources for spreading information. The relationships between individuals form the basis of social networks. One of the cardinal rules of human behavior is "*birds of a feather flock together*," which means that people of similar interest tend to become friends and this property fosters the development of dense clusters of connections throughout social networks.

Many other stylized facts on the role of social network structure on information transmission are well understood in the literature on social diffusion. For instance, the dynamics of information transmission is also a *threshold phenomenon*, where, below a certain value, the information will spread only within a limited group. However, above the threshold it will spread throughout the entire society.

5.2 Diffusion Models

In this section, we introduce some simple probabilistic diffusion models in networks. In the spread of an epidemic, some agents initially become infected through exogenous sources, and consequently some of these agents' neighbors are infected through their contact network.

Two major diffusion models, the Susceptible-Infected-Recovered (*SIR*) and the Susceptible-Infected-Susceptible (*SIS*) models, are well investigated for the study of epidemic diffusion. In these models, it is assumed that each agent can interact with others uniformly, which is called *homogeneous mixing*. In the *SIR* model, as shown in Figure 5.1(a), the population of agents is classified into three categories: susceptible (*S*), infected (*I*), and removed (*R*). Each agent is initially susceptible to epidemic diffusion and can become infected by infected neighbors. Once infected, an agent continues to

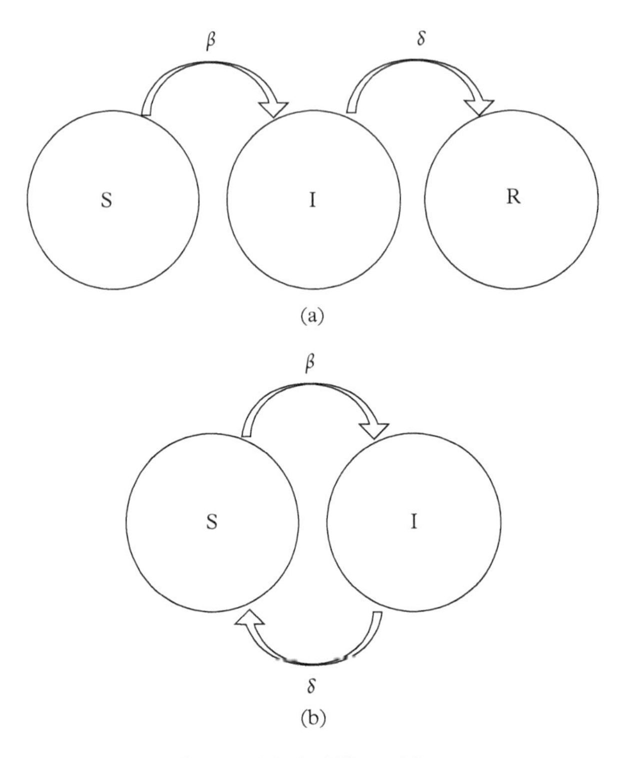

Figure 5.1 *(a) SIR model. (b) SIS model*

infect its neighbors until it is randomly removed from the population. The *SIR* model presumes that, once infected, an agent will eventually infect all of its susceptible neighbor agents. Infected agents can either recover and stop transmitting the pathogen or die and completely disappear from the population. The infected agent is removed from the population with some probability, and removed agents do not contact other agents after removal. Therefore, the number of agents who will be infected decreases with time, until there are eventually no infected agents in the population. Another widely used model is the susceptible-infected-susceptible (*SIS*) model, in which infected agents can randomly recover, but, after recovery, they are once again susceptible to infection, as shown in Figure 5.1(b). The *SIS* model corresponds well with many real-world viral infections that cause agents to transition back and forth between health and illness.

The diffusion of innovations can be modeled after the spread of infectious diseases (Meyers et al., 2005). Epidemiological models describe infection in terms of transmissibility and susceptibility. However, such models cannot adequately describe diffusion processes that usually involve individual information processing. People are usually affected by *word-of-mouth*, which is also referred to as *social influence*. What influences people to participate in the diffusion process? How do consumers choose which items to buy? These questions are constant in the minds of many scientists in different fields.

Various models of human behavior span the extremes from simple and easily influenced behavior to interdependent and complex behavior, where nothing is easily controlled or predicted (Garcia and Jager, 2011).

For the diffusion of a new product formulated using the *SIR* model, at an initial time period, most consumers have neither awareness nor preference for a newly launched product (susceptible). Being exposed (infected) to information about the product through word-of-mouth from peers, each consumer becomes aware of the product with some probability. On the other hand, the consumer with awareness may forget or lose awareness (removed) with some probability. Consumers with awareness transmit information to their peers with a certain probability. For simplicity, this probability is assumed to be constant.

There are basically two types of diffusion processes. Suppose each agent has one of two internal states, A or B. A diffusion process is *progressive* if, once an agent switches from state A to state B, the agent remains in state B in all subsequent time steps. The other type of diffusion is a *non-progressive* process in which, as time progresses, agents can switch from state A to state B or from state B to state A, depending on the states of their neighbors. The *SIR* model describes progressive diffusion, and the *SIS* model describes non-progressive diffusion.

Many diffusion phenomena can be modeled as contact processes on networks, including the spread of infectious diseases, computer virus epidemics on the Internet, word-of-mouth recommendations, viral marketing campaigns, adoption of new products, and behaviors. A contact process on networks is simply a diffusion of activation on a graph, where each infected agent can infect his/her neighbor agents with some probability. The effects of network topology on diffusion dynamics have been widely studied, and one of the more important results is the existence of an *epidemic threshold* (tipping point), which is determined from the contact network topology and transmission probability between two agents.

Infectious diseases spread through populations via networks formed by physical contact among agents. The patterns of these contacts tend to be highly heterogeneous. However, the traditional *SIR* or *SIS* model assumes that population groups are fully mixed and every agent has an equal chance of spreading the disease to every other. In these models, the estimation of the fundamental quantity called the *basic reproductive number*, which is the number of new disease cases resulting from a single initial case, is important. If this number is above one, most outbreaks will spark a large-scale epidemic diffusion.

The modeling of an epidemic diffusion process was pioneered by Daniel Bernoulli in the eighteenth century. Mathematical modeling of epidemics has evolved and a refined methodology has been applied in other scientific fields. The epidemic diffusion mechanism determines how the infection is transmitted based on the epidemic statuses of individuals and the transition probability between two individuals. Although epidemic diffusion is a stochastic process, deterministic models are frequently used. This model approximates the mean of the random process by assuming a fully mixing pattern.

Kermack and McKendrick established a deterministic epidemic model (*KM* model) with a fixed population of N agents (Kermack and McKendrick, 1927). Each agent has three states: susceptible (S), infected (I), and removed (R). Denoting the fraction of susceptible agents by $x(t)$, the fraction of infected agents by $y(t)$, and the fraction of removed agents by $z(t)$ at time t, we have:

$$\begin{aligned}\frac{dx(t)}{dt} &= -\beta x(t)y(t)\\ \frac{dy(t)}{dt} &= \beta x(t)y(t) - \delta y(t)\\ \frac{dz(t)}{dt} &= \delta y(t)\end{aligned} \tag{5.1}$$

where β denotes the pairwise rate of infection, and δ is the curing or removal rate.

A major outbreak occurs if $\frac{dy(t)}{dt} > 0$, which is equivalent to:

$$x(0) > \delta/\beta. \tag{5.2}$$

Therefore, whether a major outbreak will occur depends on the initial condition $x(0)$, the fraction of susceptible agents at the start of the epidemic. The dependency of the spread on the initial condition is a specific feature of the *SIR* model, and this constitutes a benchmark for a range of epidemic models.

An important parameter for an epidemic is the basic reproductive number R_0, which shows the number of secondary infected agents generated by one primary infected agent. If $R_0 > 1$, the number of infected agents is larger than that of recovered agents, so the virus will survive and spread to the population.

Contact network models attempt to characterize every interpersonal contact that can potentially lead to epidemic transmission in a community. These contacts may take place within households, schools, workplaces, etc. In general, the community consists of several classes classified by ages, gender, schools, and each class has different diffusion rates. The *KM* model in eqn. (5.1) can extend to the stratified population, which involves interaction between classes.

For the *SIR* model in a stratified population with m classes, where x_i denotes the fraction of susceptible agents, y_i denotes the fraction of infection, and z_i denotes the fraction of removals in class i, we have:

$$\begin{aligned}\frac{dx_i(t)}{dt} &= -x_i(t)(b_{1i}y_1(t) + \cdots + b_{1m}y_m(t))\\ \frac{dy_i(t)}{dt} &= x_i(t)(b_{1i}y_1(t) + \cdots + b_{1m}y_m(t)) - \delta_i y_i(t)\\ \frac{dz_i(t)}{dt} &= \delta_i y_i(t)\end{aligned} \tag{5.3}$$

where b_{ji} is the pairwise rate of infection between class j and i, and δ_i is the removal rate of class i.

The matrix $B = (b_{ij})$ is defined as the *interaction matrix* of the stratified population. Although estimating the interaction matrix is very difficult, once it is possible to explore influence among each class, the epidemic diffusion in very large-scale communities can be analyzed with high accuracy.

Although the modeling of diseases is an old discipline, the epidemic theory was first applied to computer malware spread by Kephart and White (1991), which is known as the *KW* model. The *KW* model is a homogeneous *SIS* model, which incorporates the concept of a network-based approach to epidemic diffusion. The malware is spread at the same speed in every part of the network and agents are cured from infection with the same frequency.

In the *KW* model, the number of infected agents in the network at time t is denoted by $I(t)$. If the total number of agents of the network N is sufficiently large, we can convert $I(t)$ to $y(t) \equiv I(t)/N$, the fraction of infected agents. The rate at which the fraction of infected agents changes is determined by two processes: (i) the curing of infected agents and (ii) the infection of susceptible agents. The ratio $1-y(t)$ denotes the probability that agent i is susceptible, and $ky(t)$ denotes expected value of infected neighboring agents when agent i has k neighboring agents. For process (i), the cure rate of a fraction of infected agents $y(t)$ is denoted by $\delta y(t)$. The fraction of infected agents will change if (i) susceptible agents become infected and (ii) infected agents become susceptible. The rate at which the fraction $y(t)$ grows in process (ii) is proportional to the fraction of susceptible agents, $1 - y(t)$.

Combining all contributions yields the time evolution of $y(t)$, the change of infected agents can be formulated as:

$$\frac{dy(t)}{dt} = \beta k y(t)(1 - y(t)) - \delta y(t). \tag{5.4}$$

This differential equation is defined as the *KW* model, which has the following solution:

$$y(t) = \frac{y(0)y(\infty)}{y(0) - (y(\infty) - y(0))e - (\beta k - \delta)t} \tag{5.5}$$

where $y(0)$ is the initial fraction of infected agents, and $y(\infty)$ is the fraction of infected agents at the steady-state.

From eqn. (5.4), if $\beta/\delta < 1/k$, $y(t)$ converges to zero, and otherwise converges to $y(\infty)$.

In epidemic modeling, the *infection strength* or *effective infection* parameter is important, and indicates the ratio of the infection rate to the cure rate. If the infection strength is higher than a certain threshold, which we define as the *diffusion threshold*, then the epidemic spreads through the population and persists in the long run.

The diffusion threshold of the *KW* model is given as:

$$\tau_c = \frac{1}{k} \tag{5.6}$$

and the infection strength $\tau = \beta/\delta$ is larger than the inverse of $1/k$, i.e., the number of neighboring agents, then a large-scale diffusion is triggered.

Such epidemic diffusion is known as a *threshold phenomenon*, where there is some parameter such as infection strength which, when below a certain threshold results in the epidemic spreading to only a limited number of agents, but above which results in the epidemic spreading throughout the network.

The *KW* model assumes a homogeneous network in which each agent has an average of k neighboring agents. This assumption was later shown to be inadequate for the spread of malware. Pastor-Satorras and Vespignani (2001) discuss the discrepancy between the data for the spread of a virus on the Internet and the theoretical results for the *KW* model and introduced a model that underlines the influence of the degree of heterogeneity in the network. The number of links emanating from an agent, i.e., the number of contacts of each agent, is called the *degree* of an agent. The distribution of the numbers of contacts, i.e., the *degree distribution*, becomes a fundamental quantity.

In order to take into account the heterogeneous contact process in a social context, Watts and Strogatz (1998) consider modeling epidemics in more general networks. A contact network is the substrate in which the spread of an epidemic occurs and is composed of nodes representing agents and links representing the contact between any pair of agents. The simplest representation of a contact network is a binary network in which the contact level takes two values, either "1" when a pair of agents is in contact, or "0" otherwise. The binary network, or graph, can be represented by an *adjacency matrix*, $A = (a_{ij})$, $1 \leq i, j \leq N$, which is a symmetric matrix, where $a_{ij} = a_{ji} = 1$ if two agents i and j are connected, or $a_{ij} = a_{ji} = 0$ if they are not connected.

The diffusion threshold of a scale-free network is obtained as follows (Pastor-Satorras and Vespignani, 2001):

$$\tau_c = \frac{< k >}{< k^2 >} \tag{5.7}$$

where $< k >$ is the average degree, and $< k^2 >$ is the average of the square of the degree. When the size of the network is sufficiently large, $< k^2 >$ becomes very large, and the critical infection rate τ_c approaches 0 for scale-free networks. This result may come as a surprise, because it implies that any epidemic, even for an infection strength of approximately zero, $\tau = \beta/\delta \simeq 0$, can spread on scale-free networks. This is evidence that the network topology affects diffusion dynamics.

Wang and Chakrabarti (2003) proposed a spectral analytical approach for studying epidemic diffusion in networks. Consider a network of N agents, where the contact is determined by the *adjacency matrix* $A = (a_{ij})$. Agent j is a neighbor of agent i, which is denoted by $j \in N$. If agent j is a neighbor of agent i then $a_{ij} = 1$, otherwise $a_{ij} = 0$. In the *SIS* model, the state $x_i(t)$ of agent i at time t is a random variable, where $x_i(t) = 0$ if agent i is susceptible, and $x_i(t) = 1$ if agent i is infected. The *curing process* for infected agent i is a Poisson process with cure rate δ. The *infection process* for susceptible agent i in contact with infected agent j is a Poisson process with infection rate β. The competing infection processes are independent. Therefore, a susceptible agent effectively becomes

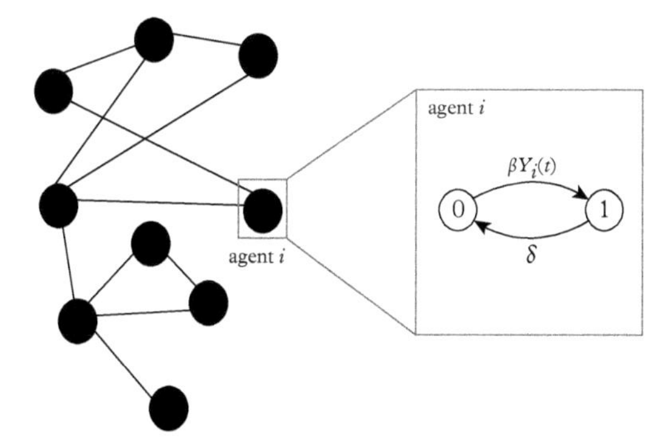

Figure 5.2 *Schematic diagram of a contact network along with the agent level stochastic transition diagram for agent i according to the SIS epidemic diffusion model. The parameters β and δ denote the infection rate and the curing rate, respectively, and $Y_i(t)$ is the number of neighbors of agent i that are infected at time t*

infected at rate $\beta Y_i(t)$, where $Y_i(t) = \sum_{j=1}^{N} a_{ij}x_j(t)$ is the number of infected neighbors of agent i at time t. A schematic diagram of the *SIS* epidemic diffusion model over a network is shown in Figure 5.2.

The advantage of the spectral approach is the incorporation of an arbitrary contact network, which is characterized by the adjacency matrix. This generalizes the *KW* model, which assumes a homogeneous *mixing pattern*, and the only network characteristic considered is the average degree. Through the agent-based model, we can study the spread of epidemics on any type of network by deeper analysis of epidemic diffusion mechanisms.

We can also view information propagation using the *SIS* model. An informed agent propagates the information to another agent in a single step with probability β, while at the same time an informed agent may forget or lose interest with probability δ. The arrival of information along each link and the forgetting process of an agent are assumed to be independent Poisson processes with rate β and rate δ, respectively. The effective infection ratio β/δ and the network topology determine whether information diffusion occurs.

Mieghem et al. (2009) develop a more precise agent-based *SIS* model as an *N-intertwined model*. Each agent is in either a susceptible (*S*) or an infected (*I*) state. If an agent is infected, the agent can recover. However, after recovery, the agent once again returns to a susceptible state and may be reinfected. With an N-intertwined model, we can separately study the state of each individual during the infection process, revealing the role of an individual's centrality properties in spreading the infection across the network.

At time t, agent i is in either a susceptible state $x_i(t) = 0$ or an infected state $x_i(t) = 1$. Each agent changes its state depending on the states of his/her neighbors, to whom the agent is connected. We denote the state vector of all agents at time t by the column vector $X(t)$. The state vector $X(t + \Delta t)$ at the next time $t + \Delta t$ is given by the stochastic matrix P and $X(t)$ in the next equation:

$$X(t + \Delta t) = PX(t)$$
$$P = \begin{pmatrix} 1 - \beta\Delta t & \beta\Delta t \\ \delta\Delta t & 1 - \delta\Delta t \end{pmatrix}. \tag{5.8}$$

The transition probability from the susceptible state to the infected state of each agent during time Δt is assumed to be given by the linear function $\beta\Delta t$ when the time Δt is sufficiently small. The continuous time N-intertwined *SIS* model in eqn. (5.8) describes the global change in the state probabilities of the networked agents. This model, however, is limited to relatively small networks due to the exponential divergence in the number of possible network states with the growth of the network size.

Mieghem et al. analyze the average state dynamics of eqn. (5.8). The connection network of N agents are represented by an undirected network, which is characterized by the adjacency matrix $A = (a_{ij})$, which is symmetric, $A = A^T$. Denote the infection probability of the i-th agent by $v_i(t) = Pr[X_i(t) = 1]$, and the infection probabilities of N agents by the column vector by $V(t) = [v_1(t), \ldots, v_N(t)]$. Actually, the exact Markov set of differential equations has 2^N states. They used a first-order mean-field type approximation to develop the model as a set of ordinary differential equations:

$$\frac{dv_i(t)}{dt} = (1 - v_i(t))\beta \sum_{j=1}^{N} a_{ij}v_j(t) - v_i(t)\delta \tag{5.9}$$

which represents the time evolution of the infection probability for each agent. Summarizing eqn. (5.9), we have the vector form:

$$\frac{dV(t)}{dt} = \beta AV(t) - diag(v_i(t))(\beta AV(t) + \delta u) \tag{5.10}$$

where u is the column vector with all elements 1 and $diag(v_i(t))$ is the diagonal matrix with each element $v_i(t), 1 \leq i \leq N$.

When the diffusion rate of each agent $v_i(t)$ is small, the nonlinear term $\beta A diag(v_i(t))V(t)$ in eqn. (5.10) is negligible, then the changes of the infection probabilities are approximated as follows:

$$\frac{dV(t)}{dt} = (\beta A - \delta)V(t). \tag{5.11}$$

This differential equation can be solved as follows:

$$V(t) = exp((\beta A - \delta I)t)V(0). \tag{5.12}$$

Using the eigenvalue decomposition $A = U\Lambda U^T$, where the $\Lambda = (\lambda_i)$ denotes the diagonal matrix in which the diagonal elements are eigenvalues of the adjacency matrix A, and U denotes an orthonormal matrix in which the i-th column corresponds to the eigenvector of the i-th eigenvalue. The term $exp((\beta A - \delta)t)$ in eqn. (5.12) can be written as:

$$V(t) = \sum_i exp((\beta\lambda_i(A) - \delta)t)x_i y_i^T V(0) \tag{5.13}$$

where $\lambda_i(A)$ is the i-th eigenvalue of the adjacency matrix A, x_i is the corresponding right eigenvector (row vector), and y_i is the corresponding left eigenvector (column vector).

In epidemic modeling, the epidemic threshold is important. If the relative diffusion probability is $\tau = \beta/\delta$, i.e., the ratio of infection probability and the curing probability is higher than a certain value τ_c, which is defined as the epidemic threshold, then the epidemic spreads through the network and persists in the long run. For a single graph representing the contact network of the population under consideration, the epidemic threshold τ_c turns out to be dependent on the network topology.

From eqn. (5.13), the influence of the maximum eigenvalue is dominant. If the infection strength $\tau = \beta/\delta$ is less than the threshold value τ_c, the inverse of the maximum eigenvalue of the adjacency matrix A:

$$\tau_c = \frac{1}{\lambda_1(A)} \tag{5.14}$$

then the initial infection probabilities die out exponentially. If the infection strength $\tau = \beta/\delta$ is higher than τ_c, the infection probabilities will approach non-zero steady state values.

Therefore, a network with a larger maximum eigenvalue $\lambda_1(A)$ with a smaller threshold is more susceptible to probabilistic diffusion. Given a social or computer network, where the links represent who has the potential to infect whom, we can tell whether a virus will create an epidemic, or will quickly become extinct. This is a fundamental question in epidemiology and other fields. The diffusion threshold is the minimum level of virulence to prevent a virus from dying out quickly. In the virus propagation model, the effect of the underlying topology can be captured by just one parameter, the first eigenvalue $\lambda_1(A)$ of adjacency matrix A.

The difference in network topology generates different pathways to reach a steady state in diffusion. In general, the diffusion process in any network topology is characterized as phase transition, as shown in Figure 5.3. The diffusion process can be classified into three broad regimes: (i) the sub-diffusion regime, (ii) the critical regime, and (iii) the super-diffusion regime.

The sub-diffusion regime is the area where the infection strength $\tau = \beta/\delta$ is much smaller than the epidemic threshold τ_c. In this regime, the diffusion process is in an absorbing phase, and infected agents will disappear. The critical diffusion regime is the area where the infection strength is close to an epidemic threshold. This regime is very

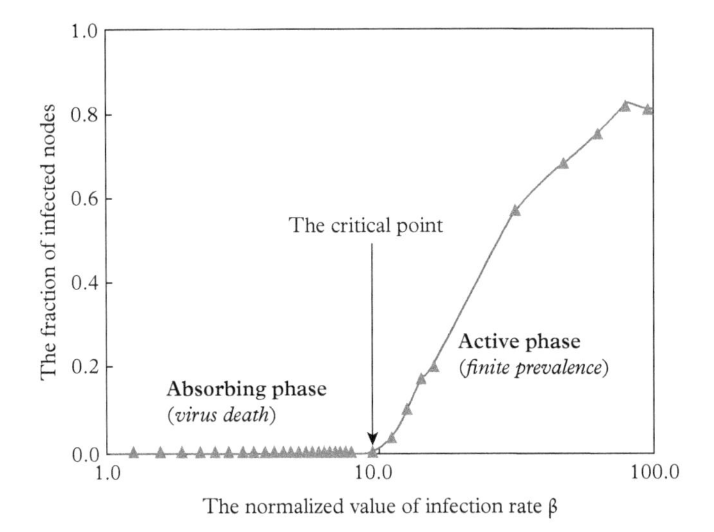

Figure 5.3 *The phase transition in diffusion dynamics. The y-axis shows the fraction of infected agents. At a subcritical regime, in which the infection strength, τ, is smaller than the diffusion threshold, τ_c, the fraction of infected agents converges to zero (absorbing phase). At the supercritical regime, in which the infection strength is greater than the diffusion threshold, the fraction of infected agents converges to a constant value (active phase). At the critical regime, in which the infection strength is close to the threshold, the diffusion process is unstable for two cases, in which (1) only a few agents are infected or (2) a large-number of agents may be infected depending on the initial conditions*

unstable, and two cases, one in which the diffusion outbreak is relatively small and a large-scale diffusion outbreak, occur depending upon the initial conditions. The super-diffusion regime is the area in which the infection strength is greater than an epidemic threshold. In this regime, a large-scale diffusion outbreak occurs, and a number of agents are infected.

Research on the relationship between network topology and the time evolution of diffusion processes may also be applied to other applications, such as the estimation of the underlying network topology from observed diffusion data. However, empirical data are usually unclear and essential for a deep understanding of the diffusion dynamics.

5.3 Maximizing and Minimizing Diffusion Thresholds

The pattern by which epidemics or information spreads from one person to another is largely determined by the properties of the pathogen carrying it or the information content. Furthermore, the interaction network structure within a population also determines to a great degree the manner by which the epidemics or information is likely to spread. This suggests that accurately modeling the interaction network is crucial in order to investigate the diffusion dynamics.

There are many cases where we want to devise efficient distributed algorithms, which are easy to implement, in order to help the dispersal of useful information. In designing a marketing strategy, the adoption of a certain product should be maximized. On the other hand, in epidemiology, the spread of infectious diseases should be minimized. Designing good intervention strategies to control harmful diffusion is also an important issue. It is also effective to modify the network topology to protect against negative diffusion or to promote positive diffusion at minimal cost. These design issues have also led to research on optimization problems in which the goals are to minimize or maximize the spread of some entity through a network.

In this section, we consider the network topologies that are best for minimizing or maximizing diffusion. A very simple way to minimize diffusion is to have a network of isolated agents. Therefore, we assume the network to be connected, without any isolated agents.

Let G be a connected network with N agents and L links, and let its associated adjacency matrix be A. The inverse of the maximum eigenvalue $1/\lambda_1(A)$ of adjacency matrix A becomes the minimum value, i.e., the threshold for diffusion in a network. The upper and lower bounds for the maximum eigenvalue $\lambda_1(A)$ of the adjacency matrix A are given as follows:

$$< k > \leq \lambda_1(A) \leq k_{max} \tag{5.15}$$

where $< k >$ denotes the average degree and k_{max} denotes the maximum degree of agents in G. The maximum eigenvalue $\lambda_1(A)$ is equal to or larger than the average degree $< k >$ and increases with the increase of the average degree $< k >$.

The equality on the left-hand side in eqn. (5.15) holds for a regular graph, in which all agents have the same number of links. Therefore, a regular network in which each agent has the same degree is the optimal for minimizing diffusion. On the other hand, the equality on the right-hand side holds for a complete graph in which all agents have $N-1$ links. Therefore, a complete network has the largest maximum eigenvalue of $\lambda_1(A) = N-1$, and this network is the optimal for maximizing diffusion.

Let us consider the influence of network topology on the spread of negative entities, such as digital viruses or rumors. Digital viruses live in cyberspace and use the Internet as their primary transport media and contacts among their victims as their propagation networks. Increasing threats from cybercrime justify research on the spread of viruses in networks. Negative information spread on new types of online social networks, such as Facebook and Twitter, spread in a manner similar to epidemic diffusion. While this analogy still needs to be verified and evaluated, epidemic theory is expected to be fruitful in assessing the rate of diffusion and how many users of social networks are reached by news.

If the network has an infinite number of agents and a certain average degree, the maximum eigenvalue also becomes infinite, which means that the comparison of the maximum eigenvalue has no meaning because no network becomes diffusive with no threshold. For example, diffusion can occur on even a homogeneous network, which is the least diffusive network, when networks of finite size are compared. Many networks, however, are designed or evolve considering the balance between the network performance and the size of the network. Therefore, we characterize the network topology having the largest maximum eigenvalue under the constraints of the network size, the numbers of agents and links.

Yuan (1988) obtains an upper bound of the maximum eigenvalue under the constraint of the number of agents N and links L as follows:

$$\lambda_1(A) \leq \sqrt{2L - N + 1}. \tag{5.16}$$

In general, the clique with the largest maximum eigenvalue promotes diffusion even if it has a relatively small diffusion probability. Using the relation in eqn. (5.16), we consider a heuristic method for designing a network with the largest maximum eigenvalue with the following two steps:

Step 1: Construct a partial complete graph (clique)

We construct a clique, a partial complete graph K_n, of n agents (core agents). Each agent in K_n is connected to every other agent, and the core agent has $n-1$ degrees. Since K_n has ${}_nC_2$ links and each of the remaining $N-n$ agents should have at least one link to other agents, the total number of links L is:

$$L = \frac{n(n-1)}{2} + N - n. \tag{5.17}$$

Then the clique size, the number of agents in K_n, is determined as:

$$n = \left\lfloor \frac{3 + \sqrt{9 + 8(L-N)}}{2} \right\rfloor \tag{5.18}$$

where $\lfloor x \rfloor$ is the floor function, which represents the maximum integer limited to x.

Step 2: Preferential or random attachment to a partial complete graph

The remaining $N-n$ agents are attached to the agents in clique K_n with preferential attachment or random attachment.

The networks designed with the above steps are referred to as the clique with preferential attachment (*CPA*) or the clique with random attachment (*CRA*). Some networks generated by the heuristic method are shown in Figure 5.4 and Figure 5.5. Once the network size, i.e., the numbers of agents and links, is specified, the theoretical upper limit of the maximum eigenvalue is given in eqn. (5.16). In general, the maximum eigenvalue of each network increases with the increase in the total number of links. Therefore, large diffusion may occur easily on any dense network with many links. On the other hand, it is hard to occur in a sparse network.

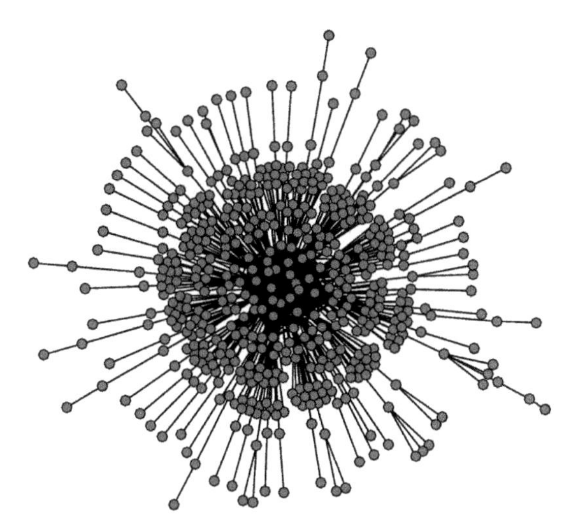

Figure 5.4 *The clique with preferential attachment (CPA). The total number of agents, N = 500 (number of core agents, n = 33), and the average degree is z = 4*

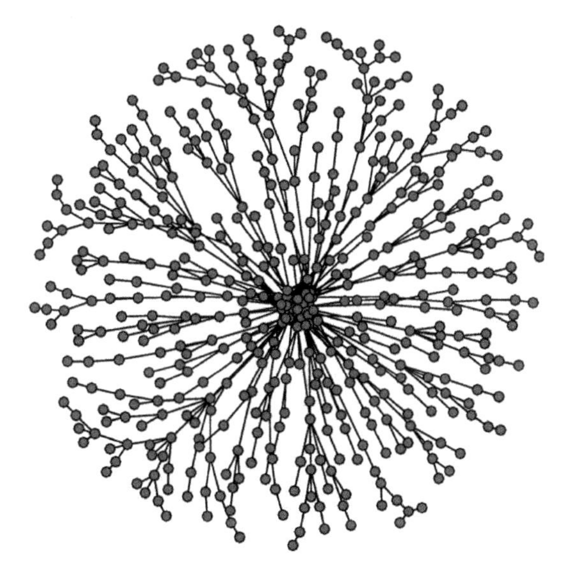

Figure 5.5 *The clique with random attachment (CRA)*

The maximum eigenvalue $\lambda_1(A)$ of adjacency matrix A provides useful information as to whether a network is diffusive, and Figure 5.6 compares the maximum eigenvalues of several networks that have the same number of agents and links. The networks generated with the *CPA* and the *CRA* have the largest maximum eigenvalue. Therefore, diffusion will take place easily in these networks with the lowest diffusion threshold. The maximum

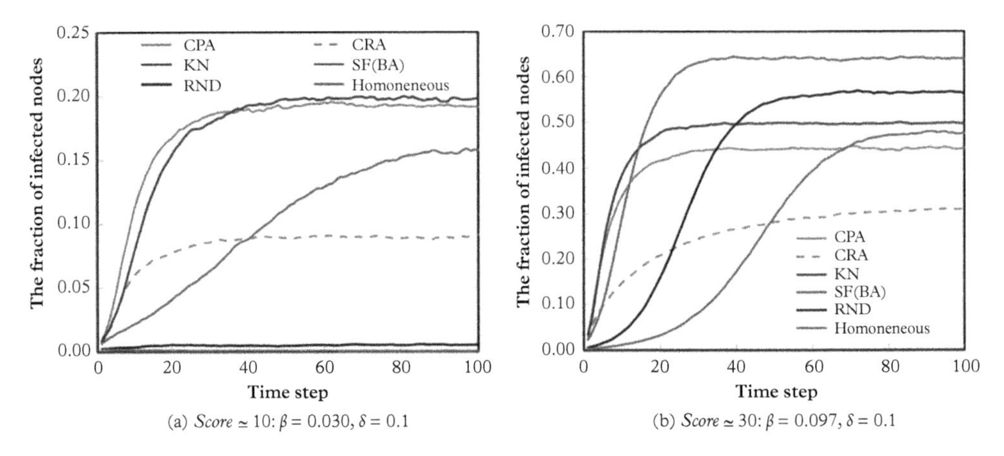

Figure 5.6 *Maximum eigenvalues as a function of average degree. Each network has $N = 500$ agents. The networks are the clique with preferential attachment (CPA), the clique with random attachment (CRA), the scale-free network by the Barabasi-Albert model (SF(BA)), the random network (RND), and the regular random network (RR)*

eigenvalue of the scale-free network with the Barabasi-Albert model (2000), *SF*(*BA*), is smaller than those of the *CPA* and the *CRA*, but is larger than that of the random network (*RND*). The random regular network (*RR*) has the smallest eigenvalue, and diffusion is most difficult in regular networks.

5.4 Comparison of Diffusion in Networks

In this section, we investigate how network topology affects diffusion processes by comparing diffusion dynamics in several networks. We also confirm that the diffusion threshold is given by the inverse of the maximum eigenvalue of the adjacency matrix of the network in eqn. (5.14) by simulation.

In order to compare the diffusion power of each network, we introduce the score, a scaling parameter of threshold, which is defined as follows:

$$score = \left(\frac{\beta}{\delta}\right) \lambda_1 (A_{CPA}) \tag{5.19}$$

where $\lambda_1(A_{CPA})$ is the maximum eigenvalue of the *CPA*, which has the largest maximum eigenvalue. The threshold of the *CPA* is given at a score of 1, and the diffusion on other networks starts at a score of greater than 1.

Initially, all agents, except the agent with the highest degree (*hub agent*), are set to be susceptible. The hub agent is infected, and we observe how this initially infected agent spreads to other agents through the network. Infected agents attempt to infect each of their neighbors with diffusion probability β at each time step. An attempt to infect an

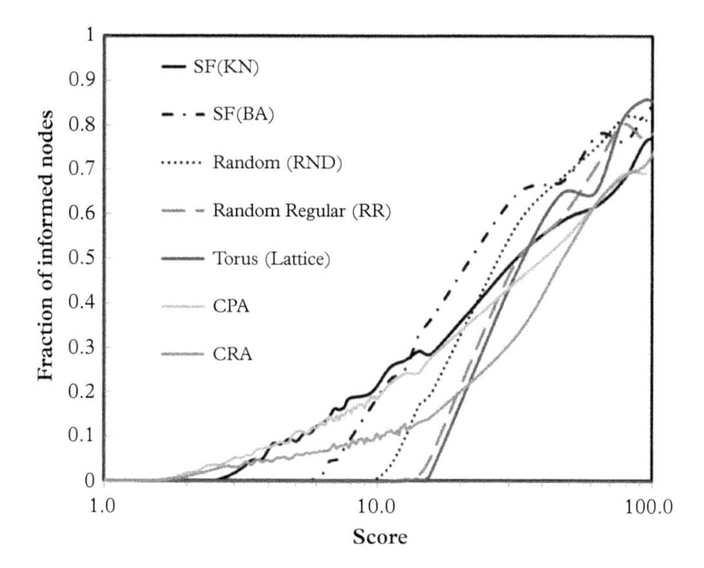

Figure 5.7 *The diffusion ratio on each network as the function of the score. All networks have N = 500 agents and the average degree of z = 4*

already infected agent has no effect. Each infected agent becomes susceptible again with curing probability δ. Each simulation will be continued until no agent is infected.

Figure 5.7 shows the diffusion ratio of each network, the ratio of infected agents in a population, with scores of 10 and 30, respectively, as time elapses. At a score of 10, the diffusion in the *CPA*, which has the largest maximum eigenvalue, starts very quickly, and the diffusion ratio converges at the highest level. The diffusion in the *CRA*, which has almost the same largest maximum eigenvalue as the *CPA*, also starts to diffuse quickly, but the diffusion saturates at a very low level. For the *SF*(*BA*), the diffusion starts relatively slowly but gradually diffuses widely and reaches almost the same level as the *CPA*. For the scale-free network with power index $\gamma = 2.1$(*KN*) starts to diffuse quickly and diffuses widely such as *CPA*. For the *RND*, the diffusion ratio is very low, and for the homogeneous networks such as *RR*, and torus (lattice) which have a smaller maximum eigenvalue, the diffusion rate is almost zero.

At a score of 30, the diffusion power is increased by three times compared to that at a score of 10, and higher diffusion ratios are observed for all networks. However, the diffusion speed differs depending on the network topology. Among the network topologies, the highest diffusion rate is observed in the *SF*(*BA*) and *KN*. For the *RND*, a high diffusion ratio is also observed. For the *RR*, the diffusion penetrates the network with a high diffusion ratio as time elapses, although the diffusion starts very slowly. The diffusion in the *CPA* starts quickly but saturates at the same level as for the *RR* and torus (lattice). The diffusion in the *CRA* also starts very quickly but saturates at the lowest level diffusion, which is approximately half that of the *RR*.

We run the simulation by increasing the score and the relative diffusion probability and plot the average diffusion ratio for each network over 100 runs in Figure 5.7. We can classify the diffusion dynamics depending on the threshold of each network in the following three regimes:

1. *Subcritical regime*: the relative diffusion probability is less than the threshold at which the diffusion ratio is almost zero and only a few agents are infected.
2. *Critical regime*: the relative diffusion probability is close to the threshold at which the diffusion starts and the diffusion ratio remains at a low level.
3. *Supercritical regime*: the relative diffusion probability is larger than the threshold at which a high diffusion rate is observed and many agents are infected.

The *CPA*, which has the largest maximum eigenvalue, starts to diffuse at the lowest score, i.e., *score* = 1. However, the increase in the diffusion ratio is relatively slow when the relative diffusion probability (score) is increased. On the other hand, the random regular network (*RR*) has the lowest eigenvalue and starts to diffuse at the highest score of approximately 20. However, the increase in the diffusion ratio is very fast when the score is increased and becomes higher than that of the *CRA*. *CRA* has the largest maximum eigenvalue (same as that of the *CPA*), and starts to diffuse at the lowest score of 1, but the diffusion ratio remains very low, even if the score is very high.

These simulation results imply that the diffusion threshold, the inverse of the maximum eigenvalue, is important for investigating the start of diffusion in networks. However, in order to investigate the diffusion process, especially in the supercritical regime, we need to know other properties of the network.

From eqn. (5.13), the influence of the eigenvector of the maximum eigenvalue is dominant, but the corresponding eigenvector (*principal eigenvector*) and other eigenvalues satisfying $\beta/\delta > 1/\lambda_i(A)$ also contribute to spreading diffusion. Expanding eqn. (5.1) for each agent i, the infection ratio of agent i at time t can be represented as follows:

$$v_i(t) = exp((\beta\lambda_1 - \delta)t)u_{1i}\|u_1\|v_i(0) + \ldots + exp((\beta\lambda_n - \delta)t)u_{ni}\|u_n\|v_i(0) \tag{5.20}$$

where the norm $\|u_k\|$ stands for the sum of all elements of the eigenvector corresponding to the k-th eigenvalue, which is defined as:

$$\|u_k\| = u_{k1} + u_{k2} + \cdots + u_{kn}. \tag{5.21}$$

We assume that the initial infection occurs sufficiently at random and every agent has an equivalent opportunity to be the initial infected agent. Each term in eqn. (5.20) indicates that the impact of the k-th eigenvalue λ_k on each agent i is governed by the product of the i-th component u_{ik} and the norm $\|u_k\|$ of the corresponding eigenvalue λ_k.

In addition, taking the average of the diffusion ratio of each agent i, the diffusion ratio over the network $y(t)$ can be calculated as:

$$
\begin{aligned}
y(t) &= \frac{1}{N}\sum\nolimits_{i=1}^{N} v_i(t) \\
&= \frac{1}{N}(exp((\beta\lambda_1 - \delta)t)\|u_1\|(u_{11} + u_{12} + \ldots + u_{1n})v_1(0) \\
&\quad + exp((\beta\lambda_2 - \delta)t)\|u_2\|(u_{21} + u_{22} + \ldots + u_{2n})v_2(0) + \cdots + \cdot \\
&\quad + exp((\beta\lambda_N - \delta)t)\|u_N\|(u_{N1} + u_{N2} + \ldots + u_{NN})v_N(0)) \\
&= \sum\nolimits_{i=1}^{N} exp((\beta\lambda_i - \delta)t\|u_i\|^2
\end{aligned}
\tag{5.22}
$$

Eqn. (5.22) can be also written as follows:

$$
y(t) = exp\left((\beta\lambda_1 - \delta)t\|u_1\|^2\left(1 + \sum_{k=2}^{N} exp\left(\beta\left(\lambda_k - \lambda_1\right)\right)\|u_k\|^2/\|u_1\|^2\right)\right). \tag{5.23}
$$

Eqn. (5.23) indicates that the diffusion power of the network is largely governed by λ_1, the largest maximum eigenvalue, and $\|u_1\|^2$, the square of the sum of the elements of the principal eigenvector. This equation also implies that, when $|\lambda_1 - \lambda_{k,(k\neq1)}|$ is large enough (because $\lambda_1 > \lambda_{k,(k\neq1)}$), and the value of:

$$
\|u_k\|^2/\|u_1\|^2 \equiv \rho \tag{5.24}
$$

is comparatively small, $y(t)$ can be considered to be approximately governed by the largest eigenvalue and its corresponding eigenvector. However, when $|\lambda_1 - \lambda_{k,(k\neq1)}|$ is small and ρ is large, the influence of the other eigenvalues are not negligible, so that the approximation method using only the largest eigenvalue and corresponding eigenvectors is not appropriate.

Billen et al. (2009) reported that spectrums are positively skewed as the number of triangles (i.e., clusters or three cycles) in the network increases. This implies that when the clustering coefficient of the network increases, the value of $|\lambda_1 - \lambda_{k,(k\neq1)}|$ increases, so that $y(t)$ tends to be governed by the largest eigenvalue. However, an empirical survey as well as numerical computation results indicate that the scale-free network, which has a high clustering coefficient, exhibits a small steady-state fraction of infections, $y(\infty)$, which implies that the value of ρ in eqn. (5.24) is more influential for deciding the dominance of each term in eqn. (5.22).

Figure 5.8 shows the distribution of ρ on the artificial networks, including the *SF*(*BA*), the *RND*, the *RR*, and the *CRA*. The *CRA* has a complete graph at the center of the network and additional links that are randomly attached to the core. This network-growing model is proposed to heuristically imitate the genetic optimization results to maximize the largest eigenvalue. As shown in Figure 5.8 in the *RND* and the *RR*, ρ corresponding to the largest eigenvalue is prominent and ρ corresponding to the second

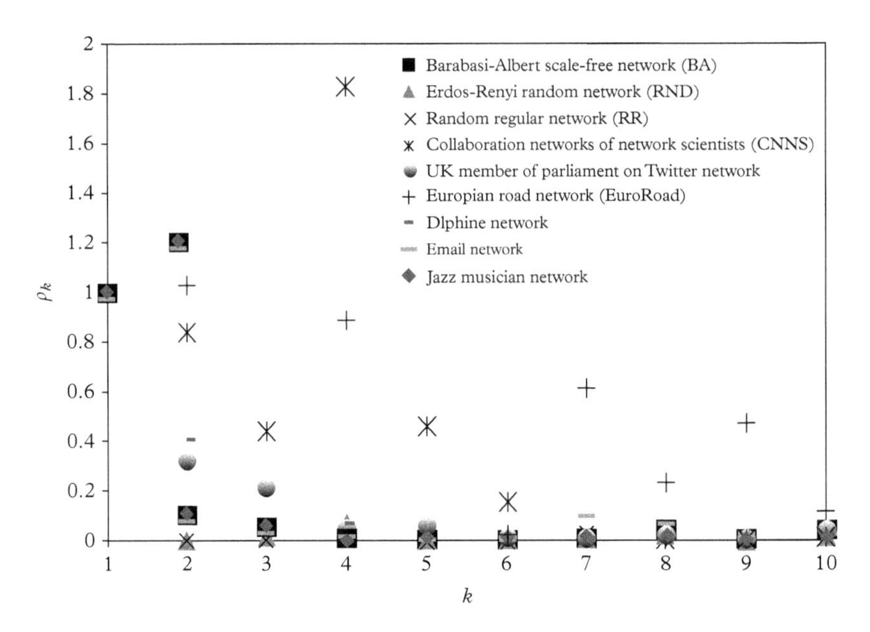

Figure 5.8 *Comparison of ρ defined in eqn. (5.24)*

largest eigenvalues is far smaller and almost negligible. In contrast, in the *BA* and the *CRA*, ρ corresponding to the second largest eigenvalues is comparatively large and is not negligible. In particular, in the *CRA*, ρ for the non-largest eigenvalues exhibits comparatively large values, which implies that the terms corresponding to the non-largest eigenvalues are not negligible and that the approximation method used to analyze the diffusion dynamics using only the largest eigenvalue cannot well represent the diffusion in the network.

Figure 5.9 shows the ten largest eigenvalues of the adjacency matrix of real networks: the co-author network of network scientists (*CNNS*), UK members of parliament on the Twitter network (*UKMPTN*), the Euro road network (*EuroRoad*), and the dolphin (*Dolphin*) network [open data set], in which the influence from the non-largest eigenvalues are important. These results indicate that, when analyzing the diffusion dynamics of networks, we need to consider the effects of the maximum eigenvalues, as well as those of other eigenvalues. Moreover, approximation methods that use only the largest eigenvalue, e.g., eqn. (5.22), are not appropriate for some networks, which exhibit the spectral property whereby ρ for the non-largest eigenvalues are comparatively large.

Both the *CPA* and the *CRA* have the largest maximum eigenvalue but their second largest maximum eigenvalues are smaller than that of *SF*(*BA*), which reflects that, in these networks, only the maximum eigenvalue contributes to diffusion, and in *SF*(*BA*), other eigenvalues $\lambda_i(A)$, $i \neq 1$, also contribute to diffusion if they satisfy $\beta/\delta > 1/\lambda_i$. The magnitude eigenvalues of *RND* and *RR* do not decrease sharply such as those of the *CPA* and the *CRA*, which indicates that high diffusion rates are observed in the regime of the

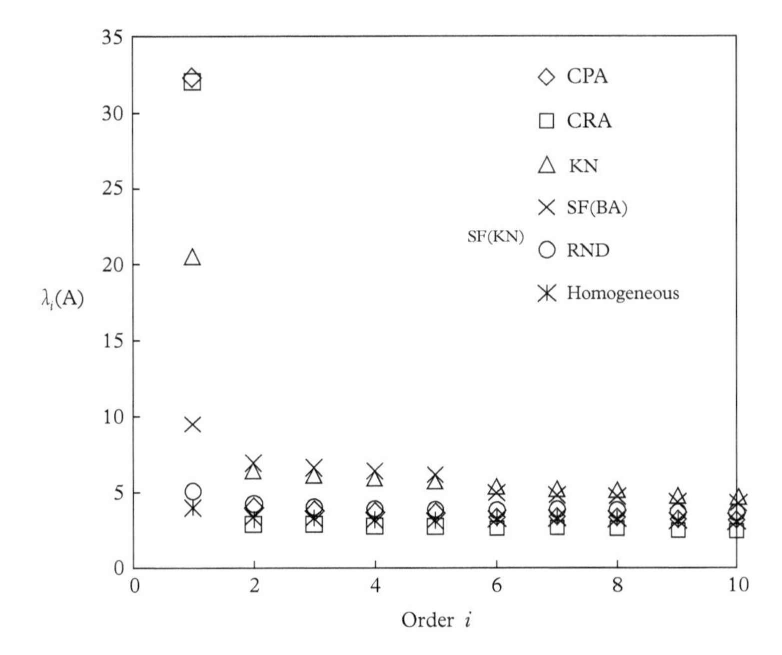

Figure 5.9 *The i-th maximum eigenvalue of the adjacency matrix of networks*

high relative diffusion rate (high score) in Figure 5.7. In any other case, the magnitude relationship of the maximum eigenvalue of networks does not necessarily represent the network diffusion power and the remainder of the eigenvalues, and their corresponding eigenvectors also contribute to diffusion.

5.5 Key Agents in Diffusion

In the previous sections, we explored the topological conditions under which a large-scale diffusion is possible. Network-wide diffusion is initially triggered by a small number of seed agents, called initiators. The impact of individual agents in the global performance of the system inevitably depends on both the network topology and the specificities of the dynamics. For diffusive processes, we quantify how networks could be controlled efficiently by manipulating appropriate agents.

In epidemic processes, for instance, targeting agents for effective vaccination strategies is an important policy issue. When the *SIR* or *SIS* model is applied to a contact network, the diffusion threshold, i.e., the inverse of the maximum eigenvalue, of the network plays an important role in the dispersal process. However, this considers only an aggregate representation of the network structure, which can result in inaccurate analysis.

Social influence is also the ability of an agent to manipulate the propagation process by inducing other agents to adopt or reject the transmission. We seek important agents who play key roles in diffusion. The relation between the influence of the seed agents that trigger information diffusion and the attitude of the remaining members toward the information is important. In order to characterize the influence of seed agents, we examine their position within the network, based on the promise that agents that are more central will reach many other agents.

In this section, we provide a framework for defining *centrality measures*, which represent roles in diffusion dynamics. In particular, we are interested in knowing, (a) how each agent is vulnerable to diffusion, (b) the role of each agent in amplifying diffusion, and (c) how to manipulate key agents in the network in order to promote or protect against diffusion at minimal cost.

Many network centrality measures have been proposed to describe an agent's importance based on its position in a network. *Degree centrality (DC)* is based on the number of connections of each agent. The *betweenness centrality (BC)* measures the number of shortest paths through a certain agent. *Alpha-centrality (AC)* is an index for measuring the long-term influence in the networks which is defined as follows:

$$C_\alpha = (I - \alpha A)^{-1} u \tag{5.25}$$

where C_α is a vector in which each element corresponds to the value of the alpha centrality $C_\alpha(i)$ of agent i, $\boldsymbol{u} = (1, 1, \ldots, 1)^T$ is a vector in which all elements are one, I is a unit matrix, and $0 < \alpha < 1/\lambda_1(A)$ is an arbitrary parameter that controls the weight of influences among agents.

The relationship between centrality and network dynamics is important. We need to determine influence measures that explicitly represent how strongly the behavior of each agent affects collective behavior. However, identifying key agents or agents in collective phenomena remains a challenge (Klemm, 2013). The questions as to how to single out the role of individual agents in collective behavior remain, as does determining the best way to quantify their importance. The key issue is that the contribution of an agent to the collective behavior is not uniquely determined by the structure of the network but rather is a result of the interplay between network dynamics and network structure.

In diffusion, the choice of the spreading parameter, being the average number of agents infected when choosing the seed agent uniformly, controls the average diffusion ratio. The diffusion process is classified into the three regimes: (1) the *subcritical regime*, in which the relative diffusion probability β/δ is smaller than the diffusion threshold and only a few agents are infected, (2) the *critical regime*, and edge of diffusion, in which the spreading parameter is close to the epidemic threshold, and (3) the *supercritical regime*, in which the relative diffusion probability is larger than the diffusion threshold and many agents are infected.

Key agents in diffusion differ depending on the regime of the spreading parameter.

(1) Subcritical regime: $\beta/\delta < \tau_c = 1/\lambda_1(A)$

For the subcritical regime, in which the relative diffusion probability β/δ is very low, the network is in an absorbing configuration, and no agent is infected. In this case, we focus on the outbreak size, which is defined as the number of agents having been infected at least once before reaching an absorbing configuration.

In the discrete-time version of the *SIS* model in eqn. (5.9), we define the transition matrix M as follows:

$$M = (1 - \delta)I + \beta A. \tag{5.26}$$

If we represent the set of diffusion probabilities of each agent at time t by the vector $p(t)$, the diffusion ratio of each agent at time $t + 1$ is given as:

$$p(t+1) = Mp(t) = M^t p(0) \tag{5.27}$$

where vector $p(0)$ is the initial value.

We can define the vulnerability index as the expected diffusion ratio per agent during its entire diffusion process when each agent is infected in turn and all other agents are susceptible at time 0. We define the *vulnerability index* $v_i (1 \leq i \leq N)$ of each agent i as:

$$v = \sum_{t=0}^{T-1} p(t)/T = (I + M + M^2 + M^3 + \ldots + M^t + \ldots)p(0)/T \doteq (I - M)^{-1} p(0)/T. \tag{5.28}$$

By setting the infection strength by the ratio of the infection rate to the cure rate, which is identical to parameter $\alpha \equiv \beta/\delta$, and assuming that each agent becomes an initiator with equal probability, then eqn. (5.28) can be written as:

$$v = (I - ((\beta/\delta)A)^{-1} p(0)/\delta T) \propto (I - \alpha A)^{-1} u \tag{5.29}$$

where $u = (1, \ldots, 1)^T$ is the unit vector.

We can also define the *amplification index* representing the spreading efficiency of each agent, which indicates the average outbreak size when initiating the dynamics with each agent in turn, and all others susceptible as:

$$w^T = u^T (I - M)^{-1} \propto u^T (I - \alpha A)^{-1}. \tag{5.30}$$

If the network is undirected, the adjacency matrix A is symmetric. In this case, the vulnerability index and amplification index become equivalent, i.e.,

$$w^T = v. \tag{5.31}$$

Therefore, in the subcritical regime, the vulnerability index and amplification index are obtained as *alpha-centrality (AC)* in eqn. (5.29) by setting α in eqn. (5.28) as $\alpha \equiv \beta/\delta$.

One of the properties of *AC* is that, when $\alpha \approx 0$, the degree of each agent is dominant in *AC*, i.e., the agents with higher alpha centralities correspond to agents with centralities of higher degree. Therefore, when the infection strength $\alpha \equiv \beta/\delta$ is very close to zero, *AC* in eqn. (5.29) is approximated as:

$$(I - \alpha A)^{-1} u = \alpha A u = \alpha D. \tag{5.32}$$

Therefore, the vulnerability index (and amplification index) and the *degree centrality (DC)* are almost the same when the infection strength is very low.

(2) Critical regime: $\beta/\delta = \tau_c = 1/\lambda_1(A)$

For the critical regime, in which the effective diffusion rate $\alpha = \beta/\delta$ is close to the diffusion threshold $1/\lambda_1(A)$ given below, eqn. (5.30) converges to the eigenvector centrality, i.e.,

$$lim_{\alpha \to 1/\lambda_1(A)} (I - \alpha A)^{-1} u = e, \tag{5.33}$$

where

$$Ae = \lambda_1(A)e. \tag{5.34}$$

The principal eigenvector corresponding to the largest eigenvalue satisfying eqn. (5.34) is defined as the *eigenvector centrality (EVC)*. In this regime, in which the infection strength is close to the diffusion threshold from below, the diffusion centrality is measured by the *EVC*, and agents with higher *EVC* correspond to the agents with higher eigenvector centralities when $\alpha \cong 1/\lambda_1(A)$.

However, if the infection strength $\alpha = \beta/\delta$ is larger than the diffusion threshold, $1/\lambda_1(A)$, eqn. (5.25) diverges, which means that the *AC* cannot be applied to the regime over the diffusion threshold. In this regime, we use the continuous version of the *SIS* model in eqn. (5.9). In the region in which the diffusion ratio of each agent $v_i(t)$ is small, the nonlinear term in eqn. (5.9) can be negligible and can be approximated as follows:

$$\frac{dv_i(t)}{dt} = \beta \sum_{j=1}^{N} a_{ij} v_j(t) - v_i(t)\delta. \tag{5.35}$$

Eqn. (5.35) can be also written as follows:

$$V(t) = exp((-\delta I)t) exp((\beta A)t) V(0). \tag{5.36}$$

We define the relative diffusion rate of agent i as:

$$w_i = \frac{v_i}{\sum_{i=1}^{N} v_i}, \quad \sum_{i=1}^{N} w_i = 1. \tag{5.37}$$

Rewriting the dynamics of eqn. (5.35) in terms of the relative diffusion rate in eqn. (5.37) we have:

$$\frac{dw_i(t)}{dt} = \sum_{j=1}^{N} a_{ij} w_j(t) - w_i(t) \sum_{i,j=1}^{N} a_{ij} w_j(t). \tag{5.38}$$

We now consider the principal eigenvector e of the largest eigenvalue of βA, which is identical to that of A. The right-hand side of eqn. (5.38) becomes:

$$\sum_{j=1}^{N} a_{ij} e_j - e_i \sum_{i,j=1}^{N} a_{ij} e_j. \tag{5.39}$$

The principal eigenvector e of the largest eigenvalue of A is a stationary solution of eqn. (5.38). Therefore, in this regime where the infection strength is close to the diffusion threshold from above, the diffusion centrality is also measured by the *EVC*.

(3) Supercritical regime: $\beta/\delta > \tau_c = 1/\lambda_1(A)$

In the supercritical regime, the infection ratio of each agent $v_i(t)$ is large and the nonlinear term eqn. (5.9) is not negligible. Mieghem et al. (2009) showed that the N-intertwined model in eqn. (5.9) converges to the steady-state, which is obtained by setting $dv_i(t)/dt = 0$, and we have:

$$v_i(\infty) = 1 - \frac{1}{1 + \tau \sum_{j=1}^{N} a_{ij} v_j(\infty)}. \tag{5.40}$$

The value in the steady state is denoted as $v_i(\infty)$ and can be approximated as:

$$1 - \frac{1}{1 + \frac{k_i}{k_{min}}(\tau k_{min} - 1)} \le v_i(\infty) \le 1 - \frac{1}{1 + \tau k_i}. \tag{5.41}$$

Therefore, in the supercritical regime, the diffusion centrality is well approximated by the *degree centrality (DC)*.

In summary, the choice of the infection strength β/δ controls the diffusion power of the network and is evaluated by the average number of agents infected when choosing the seed agent uniformly. In the subcritical regime, in which the infection strength is smaller than the diffusion threshold, the *alpha-centrality (AC)* in eqn. (5.29) encodes the interplay between topology and diffusion dynamics. In the critical regime, in which the effective diffusion ratio is close to threshold from below or above and at the edge of diffusion, the *eigenvector centrality (EVC)* encodes the interplay between topology and dynamics. In the supercritical regime, in which the effective diffusion ratio is far above the diffusion threshold and many agents are infected, the *degree centrality (DC)* encodes the interplay between topology and diffusion dynamics.

Next, we investigate how well the diffusion centrality index may forecast the actual *SIS* spreading dynamics, which is measured as the spreading vulnerability and amplification of each agent that we define as the expected fraction of agents reached by an epidemic outbreak initiated with each agent being infected (seed agent) in turn, and all others agents susceptible. Table 5.1 shows the predictive power of dynamic influence for spreading efficiency as a function of the infection probability in larger real-world networks.

Table 5.1 shows that *AC* is a good predictor in the subcritical regime. The predictive power is quantified by the rank order correlation. The *EVC* is a good predictor of spreading efficiency in the critical regime, where $\beta/\beta_c \cong 1$. In the critical regime between these extremes, infectious seed agents are perturbations that trigger relaxation dynamics at all scales. This is reflected in a dynamic dominated by a marginal linear mode and a variety of possible final states. Dynamic predictions at criticality require a global view of the network structure and the final state is determined by the conservation law associated with the principal eigenvector.

For infection strengths that are far greater than or less than the critical value, however, the degree centrality or the degree of each agent is a better predictor of spreading efficiency. In the subcritical regime, spreading is sparse and typically confined to the neighborhood of the seed agent, while in the supercritical regime, the epidemics spread to the entire network. In this regime, hub agents with large connectivities enable large-scale diffusion to occur. Indeed, high-connectivity agents are also vulnerable to frequent infection.

Table 5.1 *Comparison of the predictive power of each measurement*

Network	Centrality	Rank correlation coefficient for specific values of Score						
		0.5	0.6	0.7	0.8	0.9	1	2
	DC	0.3843	0.5110	0.6483	0.6942	0.8172	0.8500	0.9574
RND	EVC	0.3727	0.4706	0.6300	0.7204	0.8600	0.9352	0.9170
N = 498	AC	0.3976	0.5171	0.6655	0.7397	0.8798	0.9325	0.9838
	IDEC	0.3990	0.5232	0.6661	0.7342	0.8660	0.9110	0.9954
	DC	0.1824	0.2770	0.2663	0.3187	0.3458	0.3480	0.4466
BA	EVC	0.3028	0.4111	0.5833	0.7159	0.8671	0.9428	0.8746
N = 500	AC	0.2899	0.4149	0.5846	0.7125	0.8705	0.9359	0.8856
	IDEC	0.2883	0.4349	0.5858	0.7033	0.8600	0.9234	0.9244
	DC	0.4643	0.5156	0.6034	0.5976	0.6214	0.6209	0.7988
CNNS	EVC	0.2666	0.2144	0.3855	0.5373	0.5412	0.6524	0.3897
N = 379	AC	0.4727	0.5394	0.6818	0.7780	0.8563	0.8450	0.9560
	IDEC	0.4795	0.5269	0.6525	0.6884	0.7500	0.7511	0.9467
	DC	0.3796	0.5015	0.5933	0.6458	0.6578	0.6124	0.8653
UKMPTN	EVC	0.3040	0.3411	0.4010	0.5654	0.6627	0.7313	0.5455
N = 419	AC	0.3669	0.5484	0.6785	0.7861	0.8506	0.8539	0.9663
	IDEC	0.3828	0.5468	0.6676	0.7581	0.8011	0.7929	0.9877

We provide consistent evidence that such a category of agents plays a crucial role in diffusion dynamics. We establish how these agents, who do not occupy key topological positions, cause a large multiplicative effect that results in a high spreading efficiency.

Diffusion centrality measures are also useful to devise good intervention strategies to control diffusion processes. This problem is crucial when dealing with harmful information such as human diseases or computer viruses. We refer to these processes as harmful diffusions. In epidemic diffusion, we should remark that the most efficient spreaders are not necessarily the same as those targeted by efficient vaccination strategies in order to contain epidemics. We distinguish between centralized and decentralized intervention strategies. In centralized intervention strategies, there is a controller who has a limited amount of intervention resources. We need to study the problem of allocating these limited resources among the network agents so that the spread of the diffusion process is minimized. In decentralized intervention strategies, each individual in the network makes his/her own decision to protect himself/herself. In such settings, we are interested in questions such as whether there is a stable set of intervention strategies or the cost of decentralized solutions compared with an optimal centralized solution. At the network level, the aim of vaccination is to increase the diffusion threshold τ_c in order to render the spreading dynamics subcritical. Lastly, we augment our research on intervention strategies with the consideration of the behavioral change at the individual level, which would lead to a new kind of network dynamics. However, the extent of the perversity and its dependence on network structure as well as the precise nature of the behavior change remains largely unknown.

6 Agent-Based Cascade Dynamics

The spread of new ideas, behaviors, or technologies was extensively studied using epidemic models in Chapter 5. However, propagation can also be complex, and a certain proportion of an agent's neighbors must have adopted a behavior before the agent will decide to adopt it as well. As a result, there are many social phenomena that differ from diffusion-like epidemics. Epidemic models are limited in that they treat the relationships between all pairs of agents as equal, not accounting for the potential effects of individual heterogeneity on the decision to adopt a behavior or idea or on interpersonal relationships. In this chapter, we consider a model of diffusion where an agent's behavior is the result of his/her deliberate choice. Each agent has a threshold that determines the proportion of neighbor agents that must be activated. Such a model can also be applied to the study of the diffusion of innovation, cascading failure, or a chain of bankruptcy.

6.1 Social Contagion and Cascade Phenomena

A huge number of new products, movies, and books are produced each year and some of these emerge out of obscurity to become big hits. New practices are introduced and they become either popular or remain obscure, and established practices can persist over years. The notion of how something becomes popular is very relevant to the concept of *social contagion*. When we observe a society over years, we see a recurring pattern according to which new social phenomena are constantly emerging and disappearing. The study of social contagion includes all social phenomena that can spread in a society. The basic mechanisms of social contagion, in particular, how social phenomena start in the absence of explicit communication among people, have been studied in many related scientific disciplines.

One mechanism of social contagion is *social influence*. For instance, the spread of new practices through a society depends to a large extent on the fact that people influence each other. We can observe many situations where people influence each other. As individuals see an increasing number of people doing the same thing, they generally become more likely to do it as well. What people do affects what other people do, and all of these situations have common features in which individual behavior is influenced by behaviors of other people.

Agent-Based Modeling and Network Dynamics. First Edition. Akira Namatame and Shu-Heng Chen.

The French social psychologist Le Bon (1895) popularized the idea of social contagion as a specific form of social influence. He analyzed the specific unconscious process by which information or beliefs are spread throughout a social group, taking the form of mass contagion. He asserted that we admire and aspire to originality, and claim to be different; however, we are far less original than we think. We borrow our ideas and practices promiscuously, and we imitate the behaviors of others. Most of us have a fundamental inclination to behave as we see others behaving, and we could hypothesize that people imitate others simply because of an underlying human tendency to conform.

This is an important observation, but it leaves some crucial questions unresolved. In particular, by taking imitation as a given, we miss the opportunity to ask why people influence each other. Although this is a difficult question, we may identify reasons for which individuals with no *a priori* desire to conform to what others are doing imitate the behavior of others. People usually care more about aligning their behavior with the behavior of friends, colleagues, or close neighbors. Another reason is based on the fact that the behaviors of other people convey useful information and there is a benefit to be gained from aligning with them. There is a growing literature on social contagion, including articles on how individuals are influenced by their friends and colleagues. Examples include studies on the spread of new technologies (Arthur, 1989), herding behavior in stock markets (Banerjee, 1992), and the adoption of conventions or social norms (Ellison, 1993; Morris, 2000).

In some cases, the success of a new product in becoming socially contagious can be attributed to a good advertisement. It can also be accounted for by word-of-mouth recommendations among people that promote individual purchasing decisions. The studies on innovation spread have focused on settings in which individual decisions about adoption are driven primarily by cost-benefit calculations. The social contagion observed in many social media, such as Facebook and Twitter, depends on the incentives of people who have already adopted. When people are connected via such social media, it becomes possible for them to influence each other; then, their incentives to adopt are increased, and the so-called *cascade phenomena* then follow.

In social contagion, many things spread from one person to another as in epidemic diffusion. However, there are fundamental differences between social contagion and epidemic diffusion in terms of their underlying mechanisms. Social contagion tends to involve decision-making on the part of the affected individuals, whereas epidemic diffusion is based on the probability of contact with other individuals who carry such a disease-causing pathogen. This difference highlights the process by which one person influences another. Social contagion models focus particularly on the cases where the underlying decision processes of the individuals become important. In the case of social contagion, people are making decisions to adopt an innovation or opinion. Therefore, social contagion models focus on relating the underlying decision-making at the individual level to the larger external effects at the collective level.

Schelling (1978) formulated the *threshold model*, which is applicable to a variety of cascade phenomena. The variants of this model also describe a large number of cascade phenomena that become *self-sustaining* once the level of activities passes some minimum level defined as a *threshold*. What is common to threshold models is that the agent's behavior depends on the number of other agents behaving in a particular way. If a sufficient number are doing the same thing, each agent also adopts it, and this behavioral tendency is modeled with an agent's specific *threshold rule*.

Granovetter (1978) described the adoption process of innovation or behaviors based on an agent's threshold rule. In this case, a threshold rule is derived from the utility-maximizing behavior of each agent, which also depends on the behaviors of other agents. If more agents adopt a particular behavior, it becomes more attractive to each agent and then the agent moves toward adopting the same behavior. Granovetter suggested that cascades involving social movements, attitudes, and ideas are likely to be transmitted in this manner. He also gave some other interesting examples, such as entrapment into criminal behavior or mutually reinforcing choices related to the adoption of birth control practices, in particular by groups of young people.

Many contagion processes exhibit an *S-shaped pattern* (Rogers et al., 2005) that we can understand intuitively. Initially, there are a small number of adopters of the new product largely independent of the social context. They simply want to try a product regardless of what other people do. This slowly gets the adoption process started, and consequently, the social contagion process kicks in. The initial adoption leads other individuals to start adopting. This increased adoption means that more individuals have friends or acquaintances who have adopted, which leads to more adoption, and so forth. The more acquaintances or colleagues who choose a product, the more desirable it becomes to individuals who have not adopted it, and this *reinforcement mechanism* from others assures the increase in the rate of adoption. Eventually, the contagion process begins to be self-sustaining, until the people who have not adopted the product are those for whom the product entails high costs or few benefits.

Rogers (2003) empirically studied a set of recurring reasons why an innovation can fail to spread through a population even when it has a significant relative advantage compared to existing products or practices. In particular, he clarified that the success of an innovation depends on the so-called *tipping points*. For instance, when a new innovation arrives from outside a community, it tends to be difficult for it to penetrate the community where traditional product practices prevail until a sufficient number of people have adopted it. The amount of information available to consumers is increasing as more social media becomes available and global markets become larger and more complex than ever before. As a result, consumers must spend much time thinking and hesitating before making their adoptions.

Spielman (2005) defined an innovation system as a network of agents, together with the institutions or organizations that condition their behavior and performance with respect to generating, exchanging, and utilizing knowledge. An innovation system reflects one aspect of value chain analysis in that it brings agents together in the application of knowledge within a value chain. This definition highlights the need for a holistic view of the nature and structure of interactions among agents linked to one another as an agent network. The adoption of an innovation by one agent in a network can have a positive or negative influence on the behavior of other agents, a fact that results in an outcome that is often unintentional or unpredictable.

The cascade model in networks also differs from diffusion in networks in some important aspects. When a few active agents trigger the activation of their neighbors, hub agents with many connections may efficiently allow drastic diffusion of, for example, disease or rumors. Propagation through networks can also be simple, meaning that contact with one infected agent is sufficient for transmission. Many diseases spread in this manner. However, not all propagation constitutes the simple activation of agents after exposure to their active neighbors, and there is another type of diffusion that can be modeled using threshold-based behavioral rules. Unlike an epidemic diffusion model, where contagion events between pairs of individuals are independent, the cascade model introduces dependencies; that is, the effect that a single infected agent has on a given agent depends critically on the states of the other connected agents. This is a natural condition to impose for interdependent decisions, and such dependency is described as a threshold rule.

An idea or innovation will appear and can then either die out quickly or make significant inroads into a society. In order to understand the extent to which such ideas are adopted, it is important to understand how the adoption dynamics are likely to unfold within the underlying agent network. While the research on social contagion recognizes the importance of interactions among agents, it rarely explains or measures unintended impacts, either positive or negative, on agents within a network. More to the point, current methodologies provide insufficient explication of *network externalities*, and this suggests that alternative tools are required to provide an improved understanding of the roles of networks. However, a full characterization of the roles of networks is not possible without considering the interplay between agent behavior and the networks through which agents are connected.

A new social movement starts with a small set of initiators who adopt it initially. There exists a subset of agents (*seed set*) who have already adopted the innovation, and then the new social movement spreads radially to other agents who are connected through the agent network. We can analyze the social contagion process by building a model that combines agent incentives and popularity in a network. The contagion model is extended over a graph representing a network of agents. Consider the decision of an agent to buy or not buy a new product, where the benefits accruing to that agent from using the product increase with the number of his/her neighbors who also use the product. Agents will differ in a variety of respects that affect their decisions to purchase the product. Each agent is modeled to adopt the product if the fraction of the other agents connected through a network who have adopted it is above an agent's specific threshold. From the network perspective, agents might differ in terms of the number of their friends or colleagues, thus making the product more valuable to certain agents. Therefore, we also consider how the structure of the underlying agent network can affect the agents' decision-making process.

Watts (2002) presented a general cascade model. He considered networked agents whose choices are determined by those of the connected other agents according to a simple threshold rule and presented a possible explanation of the cascade phenomenon in terms of the threshold and the network topology. His work focused on the existence of some critical points and the characterization of their relationship to the topology of the agent network. When the network is sparse and interpersonal influences are sufficiently sparse, the propagation of cascades is limited by the connectivity of the network. When the network is sufficiently dense, cascade propagation is also limited by the stability of the individual agents. In the middle regime of network density, the chance of a large cascade occurring increases. This regime also displays a more dramatic property, exhibiting a very small probability that a large-scale cascade will occur. This feature would make cascades exceptionally hard to anticipate.

In the following sections, we discuss *cascade dynamics* and explore how networks serve to aggregate threshold-based agents' behaviors and produce collective outcomes of interest. Cascade dynamics are similar to diffusion dynamics and insights from the study of diffusion dynamics, such as epidemics, apply to cascade dynamics as well. Both help us consider the processes by which many things spread on networks. We consider the way agents can explore chains of social contact patterns at the network level that help facilitate or inhibit social contagion. We investigate in particular the specific topological patterns via a cluster of vulnerable agents in order to clarify the pathway to a *global cascade*.

6.2 Threshold Model

We can observe many situations where agents impinge on other agents. Agents are influenced by the decisions, actions, and advice of others when making a wide variety of decisions. Similarly, the actions that agents take affect the actions of other agents. Therefore, understanding how and when *social influence* arises and how individual decisions aggregate in the presence of such mutual influence is one of the central issues in the study of *collective action*.

Granovetter (1978) proposed analyzing social phenomena as resulting from agents' *deliberate action*, which is nevertheless *embedded* in social structures. He considered that *social relations* are crucial for understanding agent action. On the other hand, social relations are the result of aggregating purposive individual actions, and explanations should also account for the emergence of structures of social relations. He illustrated this notion using the hypothetical example of a crowd poised on the brink of a riot. Because all involved are uncertain about the costs and benefits associated with rioting, each agent of the crowd is influenced by his/her peers, such that each can be characterized as behaving according to the threshold rule. An agent will join a riot only when sufficiently many others do; otherwise, he/she will refrain.

The reasons for agent behavior changes lie solely in the social content; however, it is likely that social content also depends on actual behavior in relevant agent interactions. As a solution, we can analyze social phenomena as resulting from agent behavior, which is nevertheless embedded in social structures. It is a continuation of the study of agent behavior in the sense that we study goal-directed agent behavior as embedded in systems of social relations. If social relations are crucial for understanding agent behavior, and if they are the result of purposive agent behavior, this two-sided *causality* should be considered. To establish this *micro–macro* link, the collective effects on agent behavior must be modeled. We focus on the stylized facts of macroscopic emergent phenomena that are the results of the *bi-directional interactions*.

Granovetter's threshold model is a novel method for analyzing the outcomes of collective action when agents are faced with a choice to adopt some new state (*active*), such as a behavior, belief, or an innovation, or to remain in their existing state (*inactive*). His model is motivated by considering a population of agents, each of whom must decide between two alternative actions and whose decisions depend explicitly on the actions of other members of the population. In decision-making scenarios, an agent's payoff is an explicit function of the actions of others. In these decision problems, therefore, regardless of the details, an agent has an incentive to pay attention to the decisions of other agents. In economic terms, this entire class of problems is known generically as *binary decisions with externalities*. As simplistic as it appears, a binary decision framework is relevant to surprisingly complex problems. Both the detailed mechanisms involved in binary decision problems and also the origins of the externalities can vary widely across specific problems. Nevertheless, the decision itself can be considered a function solely of the relative number of other agents who are observed to choose one alternative rather than the other.

Granovetter argued that the threshold model could serve to bridge the gap between micro- and macro-level models. To establish this *micro–macro link*, the collective effect on each agent decision should be modeled. Agents' microscopic behaviors reflecting their micro-motives combined with the behaviors of others often produce unanticipated outcomes. Consequently, small changes in individual preferences may lead to large and unpredictable macro-level changes.

In Chapter 3, we formulated the binary choice problem with *positive externality*. Each agent chooses either $A(x_i = 1)$ or $B(x_i = 0)$, which are introduced to the market. Defining the relative merit (subject value) of B over A as the threshold $\phi = b/(a + b)$, the threshold-based decision rule of agent i is given (the same as eqn. (3.5)):

$$x_i = \begin{cases} 1 \ \ if \ \ p \geq \phi \ (Choose\ A) \\ 0 \ \ if \ \ p < \phi \ (Choose\ B) \end{cases}, \tag{6.1}$$

where p is the fraction of agents choosing A. The relevant agent decision function in eqn. (6.1) exhibits a strong *threshold* nature. Agents may display *inertia* in switching states, but once their personal threshold has been reached, the action of even a single neighbor can tip an agent from one state to another. Each agent in the population has a threshold according to some probability distribution $f(\phi)$, which may also capture the relevant *psychological attributes* of agents with respect to the particular decision at hand. The distribution of threshold $f(\phi)$ represents both the average tendencies and the heterogeneity present in the population. Lowering or raising the mean of the threshold distribution $f(\phi)$ corresponds to lowering or raising the general susceptibility of the collective action of the population.

The contagion process is referred to as *progressive* if, when an agent has switched from one state B to another state A, he/she remains at A in all subsequent time steps. The other type of process is *non-progressive* in which, as time progresses, agents can switch from state B to A or from A to B, depending on the states of their neighbors. In the case of a progressive process, there can be a cascading sequence of agents switching to A starting from B. In the case of a non-progressive process, a network-wide equilibrium is reached in the limit such that most agents adopt A or B, or it may involve coexistence, with the agents being partitioned into a set adopting A and a set adopting B.

Collective behavior depends on heterogeneity in agents. Equilibrium collective behavior of interacting agents is characterized by the threshold distribution over the population. We denote the *cumulative distribution function* of the threshold, which is defined as:

$$F(\phi) = \int_0^{\phi} f(x)dx. \tag{6.2}$$

The function $F(\phi)$ is the fraction of the agents whose threshold falls below ϕ. Examples of the threshold distributions are shown in Figure 6.1. The vertical axis represents the ratio of agents having a particular threshold. Figure 6.1 (a) shows a normal distribution with the peak at $\phi = 0.5$; the majority of the agents have an intermediate

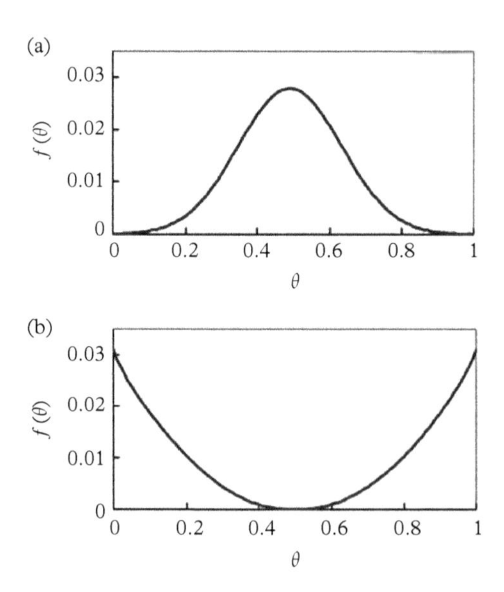

Figure 6.1 *The threshold density function of the agent population captures the variability of individual characteristics. The vertical axis represents the ratio of agents having a particular threshold. (a) The majority of the agents in the population has intermediate thresholds* $\phi = 0.5$. *(b) The agent population is split into two groups having extreme thresholds that are close to* $\phi = 0$ *and* $\phi = 1$

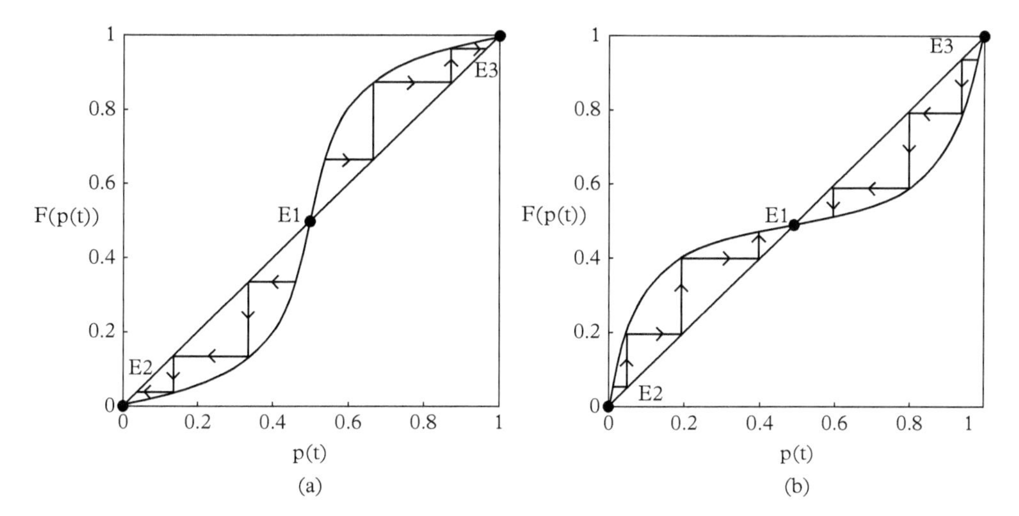

Figure 6.2 *Phase portraits of the dynamics in eqn. (6.1). These figures show the accumulative threshold distributions of $F(p)$, and the convergences of the collective choice with the threshold densities in Figure 6.1(a) and (b)*

threshold around $\phi = 0.5$, and therefore, their preference for one of the two alternatives, A or B, is ambiguous. Figure 6.1(b) shows that the agent population splits into two groups having extreme thresholds that are close to $\phi = 0$ and $\phi = 1$. Their cumulative distribution functions are shown in Figure 6.2 (Namatame and Iwanaga, 2002).

Let us consider a population in which some fraction $p(t)$ in the agent population is assumed to choose A (is active) and the remainder fraction of the agents $1 - p(t)$ chooses B (is inactive) at each time step t. Each agent compares $p(t)$, the current active fraction of the agents in A during time step t, with their threshold ϕ, and shifts to A if $p(t) > \phi$, and otherwise remains at B. Then, the cascading sequence of agents switching to A is described as follows. Starting from any initial condition, $p(0)$, the fraction of active agents in A at the beginning because of some exogenous process, the fraction of active agents at the next time step is that which can be found by drawing a vertical line on Figure 6.2(a) to intersect at $p(0)$. This fraction now becomes the input for the next iteration of the map, which is achieved graphically by drawing a horizontal line from the map to the diagonal in Figure 6.2(a). The process now repeats, thus generating values for $p(1) = F(p(0)), p(2) = F(p(1)), \ldots, p(t) = F(p(t-1))$, and the proportion of active agents in A in sequence is given as:

$$p(t+1) = F(p(t)) \tag{6.3}$$

and so on until some equilibrium value of eqn. (6.3) is reached, after which no further changes occur.

We can predict the ultimate proportion of active agents choosing A. Given the threshold density $f(\phi)$, the problem is one of finding equilibrium, which is obtained using the fixed point of eqn. (6.3), which satisfies:

$$p^* = F(p^*). \tag{6.4}$$

We consider two threshold distributions in Figure 6.1. The phase portrait in Figure 6.2(a) shows the convergence of the dynamics of eqn. (6.4) with the threshold density of Figure 6.1(a). The fixed point $F(p)$ is obtained as the point where it intersects with the 45-degree line. There are three equilibria: the right-extremity, the left-extremity, and the middle. Among these, equilibria E_1 at the right-extremity and E_3 at the left-extremity are stable, and the equilibrium E_2 in the middle is unstable. In this example, two nearly identical initial conditions result in exactly opposite outcomes. Suppose the initial proportion of active agents in A at the beginning is given at the point that is slightly less than the proportion at E_2. The collective choice converges to E_1, where all agents are inactive in B. On the other hand, if the given initial proportion of active agents in A is slightly higher than E_2, the collective choice converges to E_3, and in this case all agents are active in A. The conclusions reached using threshold models are strikingly different from those reached using epidemic ones. A small change in the initial condition near the equilibrium acts as a kind of switch, which can result in polar opposite outcomes, whereas away from equilibrium even large changes in the initial condition will converge on the same outcome.

We now consider another threshold distribution, as shown in Figure 6.1(b). With this threshold distribution, some fractions of the agents ($\phi \cong 0$) are active in A independently of the behavior of other agents. Similarly, some fraction of the agents ($\phi \cong 1$) are inactive in B regardless of others. These *hardcore* agents consider only what they actually want to do personally. The cumulative threshold distribution is given in Figure 6.2(b). The fixed point of $F(p)$ is obtained as the point where it intersects with the 45-degree line, and there is a stable equilibrium, E_1, in the middle. The phase portrait in Figure 6.2(b) illustrates the convergence of the dynamics to E_1, where half of the agents are active in A and the second half are inactive in B, and the agent population is segregated into two extreme groups, starting from any initial point.

The threshold-based behavioral rule in eqn. (6.1) represents the rational choice of the binary decisions with *positive externality*; that is, when other agents' choice of A makes an agent more likely to choose A. We now consider the opposite type, the binary decision with *negative externality*. When the influence is negative, the threshold rule indicates the binary decisions for which the corresponding decision externality is negative, that is, when others' choice of A makes an agent less likely to choose A, which generates dynamics that are qualitatively different from those generated when the influence is positive.

An agent chooses either $A(x_i = 1)$ or $B(x_i = 0)$. The relative merit (subject value) of B over A is defined as the threshold $\phi = b/(a+b)$, and the rational choice of the binary decision with *negative externality* is described as the threshold rule (the same as eqn. (3.7)):

$$x_i = \begin{cases} 1 \;\; if \;\; p \leq \phi \;(Choose\ A) \\ 0 \;\; if \;\; p > \phi \;(Choose\ B) \end{cases}, \tag{6.5}$$

where p is the fraction of the agents choosing A.

Each agent in the population has a threshold according to some threshold distribution $f(\phi)$. Let us consider a population of which some fraction $p(t)$ is assumed to have been activated and the remainder of the population $1-p(t)$ is inactive at t. The fraction of the active agents $p(t)$ choosing A at time $t+1$ can be described simply in terms of the active fraction at the previous time step and the mapping:

$$p(t+1) = 1 - F(p(t)). \tag{6.6}$$

The function $1 - F(p(t))$ is defined as the *complement cumulative threshold distribution*, which can be derived easily by observing that at any point in time t, $p(t+1)$ is just the fraction of the population whose thresholds are above $p(t)$.

We consider a collective choice with the threshold density shown in Figure 6.1(a). The complement cumulative distribution function is shown in Figure 6.3(a). The fixed point of $1-F(p(t))$ is obtained as the point where it intersects with the 45-degree line. There is only one equilibrium E_2 in the middle. However, this equilibrium is unstable. The phase portrait in Figure 6.3(a) shows that the collective choice does not converge to the middle and it becomes a cyclic behavior between two extreme cases: all agents choose A or B. For the threshold density shown in Figure 6.1(b), the complement cumulative threshold distribution is given in Figure 6.3(b). In this case, there is also one equilibrium E_1 in the middle, which is a stable point. The phase portrait in Figure 6.3(b) shows the convergence of the dynamics. Starting from any initial point, the collective choice converges to E_1, where half of the agents choose A and the other half choose B; the agent population is split into two opposite groups. More precisely, the above results mean that when agents are sufficiently heterogeneous in terms of threshold, the collective choice is close to the case where two agents have the opposite preference. On the other hand, a small diversity means a small difference in threshold, and the collective choice becomes cyclic.

The threshold model yields an important insight. Collective outcomes are not easily intuited from the individual attributes, that is, thresholds. Consequently, small changes

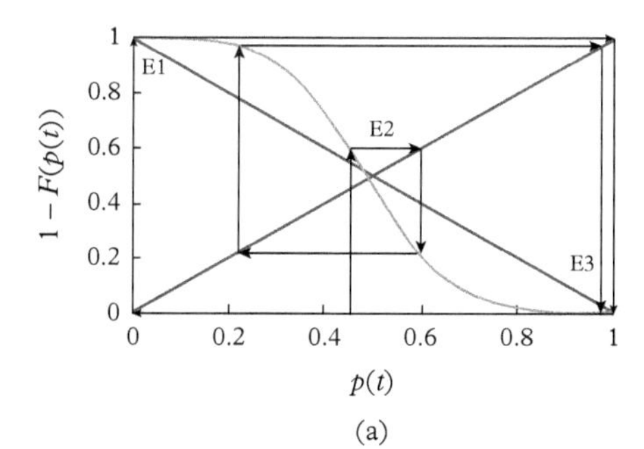

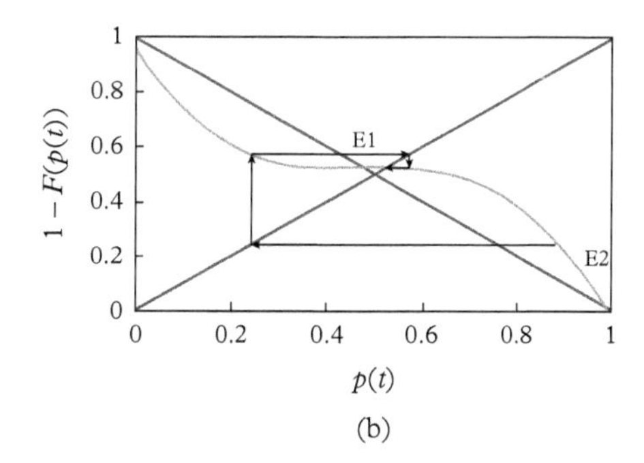

Figure 6.3 *Complementary cumulative distribution functions of thresholds, $1-F(p(t))$. (a) The cyclic behavior between $p = 0$ and $p = 1$, when the threshold density is given as shown in Figure 6.1(a); (b) the convergence to $p = 0.5$, when the threshold density is given as shown in Figure 6.1(b)*

in agent threshold may lead to large and unpredictable collective level changes. The purpose as well as the contingent behaviors of agents produce coherent collective behavior or *cyclic behavior*, and the resulting dynamics can be quite complex. The unexpected occurrence consists precisely of the emergence of a macrostructure from the bottom up, the origin of which is the local threshold rules that outwardly appear quite remote from the collective outcomes they generate. In short, it is not the emergent macroscopic object per se that is unexpected, but rather the *accumulative effect* of the local adaptive decisions that generate complex behaviors at the macroscopic level. Therefore, knowing the preferences, motives, or beliefs of agents at the microscopic level can provide only a necessary and not a sufficient condition for the explanation of the aggregated outcome.

The lesson to be learned is that the dynamics of adaptive agents can be exquisitely sensitive to its composition of micro-motives of agents. The agent's threshold is distributed across the collective according to some threshold distribution, and there is a *critical point* that alters the collective outcome. Therefore, interacting situations where an agents' behavior depends on those of other agents are those that usually do not permit any simple summation or extrapolation to the aggregates. We cannot simply jump to a conclusion about aggregate behavior from what we know or can guess about agent preferences. To make this connection, we have to look at the dynamics that relate the micro-motives of agents and the aggregated macro-behavior.

Specifically, a top-down decision-making approach may be required to deal with the practical aspects of the realistic behavior of agents, rather than the present bottom-up mechanisms. Such an approach refers to the *catastrophe theory*, which can be used to determine the set of conditions wherein the agents would finally choose one among the competitive options. This approach can adequately explain and classify abrupt conflict phenomena when a dynamical system reaches or exceeds a bifurcation point. These phenomena are characterized by sudden and dramatic shifts in collective dynamics arising from small changes in the distribution of the agents' thresholds. After the *bifurcation*, defining the multiple dynamical states in which the agents' collective choices are no longer superimposed and the collective system can reach a stable equilibrium or possibly enter into unstable and chaotic conditions can be useful.

6.3 Cascade Dynamics in Networks

How do behaviors, products, ideas, and failures spread through networks? These issues can all be addressed using the conceptual framework of a threshold model. Furthermore, the dynamics of networks is influenced by the specific relationships between individuals within that community. Which network structures favor the rapid spread of new ideas, behaviors, or products? In this section, we address this question.

The diffusion process enhances an innovation via the feedback of information about the innovation's utility across various users; this feedback can then be used to improve the innovation. This aspect is similar to the micro–macro loop, which is an essential part of emergence. Most human social activities are substantially free of centralized management, and although people may be concerned about the end result of the aggregate, individuals' decisions and behaviors are typically motivated by self-interests. To make the connection between microscopic behavior and macroscopic patterns of interest, we should look at the system of interactions among agents. This can be described as the network topology. Network topologies determine a basic and important form of social interactions among agents, and microscopic behaviors of agents largely determine the diffusion patterns observed at the macroscopic level.

In the previous section, we focused on the threshold model according to which agents adopt a new state (*active*) based on the perceived fraction of others who have already adopted the same active state. We have chosen to study *all-to-all* influence networks, in which all agents are influenced equally by the states of all others. Many agent interactions occur at a local, rather than a global, level. People are often not as concerned about the whole population's decisions as about the decisions made by friends and colleagues. For instance, they may adopt political views that are aligned with those of their friends. When we perform this type of local interaction, we focus on the structure of the network modeled as a graph $G = (V, E)$ with a set of nodes V and edges E, and examine how agents are influenced by their particular network neighbors.

In this section, we consider the threshold model in a network framework where, in contrast to the all-to-all influence model, agents are assumed to be influenced directly only by a small subset of agents to whom they are connected. We explore in detail how the properties of the threshold rule and corresponding influence network can impact the cascade dynamics. In particular, we determine the likelihood that large cascades of influence can originate from small initiators, the ability of prominent agents to trigger such cascades, and the importance of influence network structure in triggering and propagating large cascades.

Watts (2002) proposed that some generic features of contagion phenomena can be explained in terms of the connectivity of the network by which influence is transmitted between agents. Specifically, Watts' model addressed the set of qualitative observations that a very large cascade (*global cascade*) can be triggered by some initiators (*seeds*) that are very small relative to the agent network size; however, this rarely occurs. Watts presented a possible explanation of this phenomenon in terms of interacting agents whose actions are determined by those of their neighbors. The network topology is also essential for characterizing the cascade dynamics. The cascade process can have a critical value at both the agent and network level. At the agent level, the critical value is the proportion of neighbors who must have adopted in order for an agent to adopt, which is defined as the threshold. At the network level, the critical value is determined by the connectivity of agents and average degree of the network.

We model the cascade dynamics and identify the critical values at which the contagion phenomenon becomes self-sustaining. We consider a collection of networked N agents. The degree distribution, the agent network connectivity, is denoted by the probability distribution, $P(k)$. Let us suppose that each agent has the same threshold ϕ between $[0, 1]$, and that this represents the weighted fraction of an agent's neighbors that must become active in order that agent will become active. That is, each agent i faces a situation where he/she must decide whether he/she will be active (adopt a new innovation), $x_i = 1$, or adhere to his/her present position (not adopt) and to be inactive, $x_i = 0$.

Agents shift their states based on the fraction of adjacent agents who are active at the previous time step. If the fraction of neighboring active agents is larger than the threshold ϕ, the inactive agent i becomes active, $x_i = 1$, at the next time step. The state-transition of agent i is described by the following *deterministic threshold-rule*:

$$x_i(t+1) = \begin{cases} 1 \ \ if \ \ \dfrac{\sum_{j \in N_i} x_j(t)}{k_i} \geq \phi \\ 0 \ \ if \ \ \dfrac{\sum_{j \in N_i} x_j(t)}{k_i} < \phi \end{cases}, \tag{6.7}$$

where N_i denotes a set of connected agents of agent i, and k_i is the size of N_i, or the degree of agent i.

In Section 6.2, we considered a *non-progressive* cascade process in which, as time progresses, agents can switch from state $x = 0$ to $x = 1$ or from $x = 1$ to $x = 0$, depending on the states of their neighbors. In this section, we consider a progressive cascade process in which, when an agent has switched from one state $x = 0$ to $x = 1$, he/she remains at $x = 1$ in all subsequent time steps. Given an initial set of active agents, *initiators*, with all other agents inactive, the progressive cascade process unfolds deterministically in discrete steps: in step t, all agents that are active in step $t - 1$ remain active, and we activate any agent the fraction of whose active neighbors is at least his/her threshold, ϕ. We call this chain reaction of switches to $x = 1$, the *cascade of adoptions* of $x = 1$. Then, we may have two fundamental possibilities:

1. The cascade runs for a while but stops while there are still many inactive agents at $x = 0$. In this case, the majority of the population is composed of inactive agents, and active agents at $x = 1$ are very few (*local cascade*).
2. There is a large-scale cascade in which most agents in the network switch from $x = 0$ to $x = 1$. In this case, the majority of the population is composed of active agents at $x = 1$ (*global cascade*).

We define the *network threshold* for distinguishing the second from the first possibility. We assume that all agents are homogeneous, having the same threshold ϕ. We start with an infinitesimally small fraction of active agents at $x = 1$, as the initial adopters, while all other agents are inactive at $x = 0$. All inactive agents then repeatedly evaluate and decide whether to switch from $x = 0$ to $x = 1$ using the deterministic threshold rule given in eqn. (6.7). If, for a given specific threshold ϕ^* between $[0, 1]$, the resulting cascade of adoptions of $x = 1$ eventually causes many agents to switch from $x = 0$ to $x = 1$, then for any threshold ϕ smaller than ϕ^* the set of initial seeds causes a global cascade at the threshold ϕ^*, which is defined as the *network threshold.*

The existence of a network threshold corresponds to a *phase transition* from local to global cascades. We now show how the network threshold is related to the network topology. We define a *vulnerable agent* who can be activated by just one active neighboring agent. In this case, an agent with the degree k becomes vulnerable if the two parameters satisfy:

$$k < 1/\phi. \tag{6.8}$$

The manner in which the cascade progresses over the network is determined by the degree of the connectivity between vulnerable agents. From eqn. (6.8), the lower connectivity (degree) of an agent, the higher vulnerability to cascade. If an agent with low degree k has a high threshold ϕ, a cascade in networks does not spread and the network in question becomes robust to cascade.

The cascade process can have a critical value at both the agent and the network level. At the agent level, the critical value is the proportion of an agent's neighbors who must be active in order for the agent to be active, which is defined as the *behavioral threshold*. An agent behavioral rule based on the threshold has important implications for the network dynamics. At the network level, the critical value of a threshold is determined by the degree of connectivity among agents.

Therefore, the network topology is essential for characterizing the network threshold that in turn characterizes the cascade dynamics. Let us consider two networks, a *star network* and a *wheel network*, as shown in Figure 6.4. Both networks have a hub agent in the center, and the threshold of each agent is $\phi = 0.5$. In a star network, as shown in Figure 6.4(a), there is no vulnerable agent, and a global cascade rarely occurs unless a hub agent is selected as an initiator. Therefore, a hub agent plays the role of a kind of obstacle or firewall for global cascade. If a hub agent changes to $x = 1$, all periphery agents shift to $x = 1$ and a global cascade occurs. In order to change the state of a hub agent, a large fraction of agents in the periphery should shift to $x = 1$. If the initiator to be changed to $x = 1$ is randomly chosen among all agents, the probability of a hub agent being chosen as an initiator is very low, and therefore, global cascades rarely occur. However, when a hub agent is selected as an initiator, a global cascade occurs. This network property that a global cascade may occur is rarely realized; however, when it does occur, the cascade is huge and is frequently referred to as *robust yet fragile*. This property is commonly observed in many networked systems. That is, they may appear stable for long periods and withstand many external shocks and therefore they are robust, but they may suddenly and apparently inexplicably exhibit a large-scale fragility (Carlson and Doyle, 2002).

In the wheel network shown in Figure 6.4(b), all the periphery agents are vulnerable and they form a large cluster. They are easy to shift to $x = 1$ if only one neighboring agent shifts to $x = 1$, and if one of the periphery agents is selected as an initiator and shifts to $x = 1$, the sequence of shifts continues in a cluster of periphery agents. Therefore, if a single periphery agent shifts to $x = 1$, a succession of the same behavior follows. When a substantial fraction of periphery agents shifts, a hub agent in the center, who is the most reluctant to shift, also shifts to $x = 1$. In this case, each vulnerable agent in the periphery cannot directly influence a hub agent to shift, but the accumulated power of vulnerable agents has a strong influence on a hub agent who cannot easily be shifted. After a hub

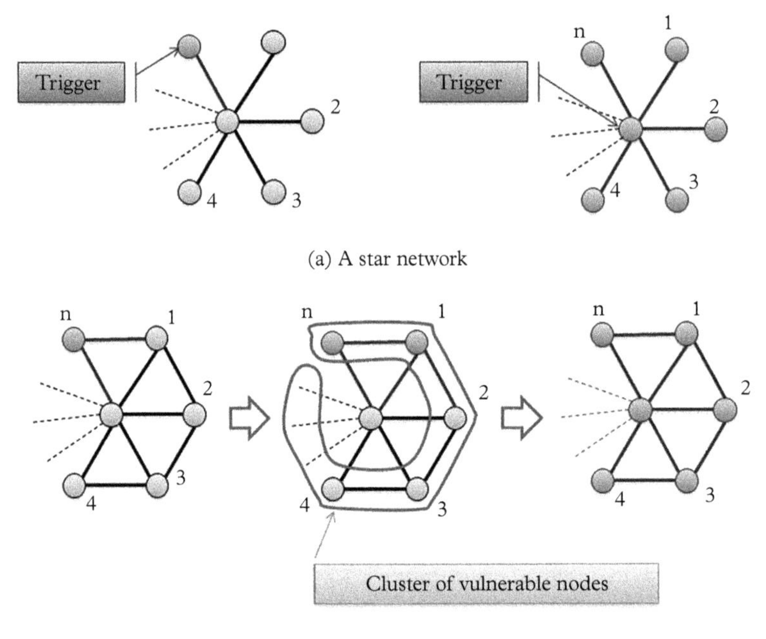

Figure 6.4 *The effects of network topology on cascade dynamics: (a) a star network, (b) a wheel network. All agents have the same threshold, $\phi = 0.5$. (a) In the star network, there is no vulnerable agent, and a cascade occurs only if a hub agent is selected as an initiator. (b) In the wheel network, all periphery agents are vulnerable and they form a large cluster. If one of the periphery agents is selected as an initiator, the sequence of shifts continues in a cluster of all periphery agents and then the hub agent in the center also changes*

agent shifts, he/she will influence the rest of the agents to shift. In this respect, the cluster of vulnerable agents works as a *catalyst* of global cascade.

The network threshold is hard to solve explicitly. The *mean-field approach* is widely used in many areas of science, since it describes successfully the main qualitative features of stochastic dynamics. The mean-field approach approximates the random discrete time dynamics with some assumptions. Lopetz-Pintado (2006) explored the condition for a global cascade in a general network framework using the mean-field approach with some assumptions. First, the agent population is sufficiently large, and the connectivity of linked agents is stochastically independent and there should be an absence of network clustering. The agent network is also connected and there are no isolated agents.

The agent network is represented by the degree distribution $P(k)$. The probability of an agent that is connected to an active agent at state $x = 1$ having k degree (links) is given by:

$$P(\textit{linked node has k links}) = kP_k/z, \tag{6.9}$$

where z is the average degree, i.e.,

$$z \equiv \sum_{k=1}^{\infty} kP(k). \tag{6.10}$$

The probability β_k that a linked agent with degree k is a vulnerable agent, is:

$$\beta_k = \int_{\phi=0}^{1/k} f(\phi)d\phi, \tag{6.11}$$

where $f(\phi)$ is the threshold distribution.

If a linked agent is vulnerable, he/she may produce $k-1$ new active agents, and if a linked agent is not vulnerable, he/she produces no active agents. The expected number of active agents produced by one active agent is:

$$\sum_{k=1}^{\infty}((k-1)\beta_k \frac{kP(k)}{z} + (1-\beta_k)\frac{kP(k)}{z}) = \sum_{k=1}^{\infty}(k-1)\beta_k kP(k)/z. \tag{6.12}$$

If the expected number of active agents produced by one active agent is at least one, then a cascade is initiated, and the condition of global cascade is described by:

$$\sum_{k=1}^{\infty}(k-1)\beta_k kP(k)/z \geq 1. \tag{6.13}$$

When we obtain the degree distributions $P(k)$ and threshold distributions $f(\phi)$, we can obtain the cascade condition by solving eqn. (6.13).

We denote the proportion of active agents with k degrees (links) that are active ($x = 1$) at time t by $\rho_k(t)$. We call a link an *active link* if it is connected to an *activated agent* at state $x = 1$. To allow the cascade to progress, at least one active link is needed. Let $\phi(t)$ denote the probability that any given link points to an active agent, which is obtained as:

$$\theta(t) = \sum_{k} \frac{kP(k)}{z}\rho_k(t). \tag{6.14}$$

It should be noted that $\theta(t)$ is used as a mean-field parameter, since it is assumed to be the same for all agents independently of their connectivity in the network.

We now obtain the probability that an agent with degree k is connected to exactly k_1 active agents, which follows a binomial distribution given as:

$${}_kC_{k_1} = \theta^{k_1}(1-\theta)^{k-k_1}. \tag{6.15}$$

We denote the probability of an agent having k degree and, if k_1 from among k their neighbors are active, changing to $x = 1$ as $P(x = 1 \mid k_1, k)$. Similarly, the probability of an agent having k degree and, if k_1 from among k its neighbors are inactive, remaining at $x = 0$ is denoted by $P(x = 0 \mid k_1, k)$. Given the probability that any given link points to an active agent, $\theta(t)$, an agent with degree k becomes active ($x = 1$) at a rate:

$$rate(1 \mid k, \theta(t)) = \sum_{k_1=0} P(x = 1 \mid k_1, k)_k C_{k_1} \theta^{k_1} (1-\theta)^{k-k_1} \rho_k(t) \tag{6.16}$$

and becomes inactive ($x = 0$) at a rate:

$$rate(0 \mid k, \theta(t)) = \sum_{k_1=0} P(x = 0 \mid k_1, k)_k C_{k_1} \theta^{k_1} (1-\theta)^{k-k_1} \rho_k(t). \tag{6.17}$$

The rate change of the fraction of active agents $\rho_k(t)$ is then given by:

$$\begin{aligned} &\frac{d\rho_k(t)}{dt} \\ &= (1-\rho_k(t)) \sum\nolimits_{k_1=0}^{k} P(x = 1 \mid k_1, k)_k C_{k_1} \theta^{k_1} (1-\theta)^{k-k_1}. \\ &-\rho_k(t) \sum\nolimits_{k_1=0}^{k} P(x = 0 \mid k_1, k)_k C_{k_1} \theta^{k_1} (1-\theta)^{k-k_1} \end{aligned} \tag{6.18}$$

At equilibrium, i.e., $d\rho_k(t)/dt = 0$, we have:

$$\rho_k(t) = \sum_{k_1^k=0} P(x = 1 \mid k_1, k)_k C_{k_1} \theta^{k_1} (1-\theta)^{k-k_1}. \tag{6.19}$$

Upon replacing eqn. (6.19) in eqn. (6.14), we obtain the recurrence formula:

$$\theta = H(\theta), \tag{6.20}$$

where:

$$H(\theta) = \sum_{k} \frac{kP(k)}{z} \sum_{k_1=0}^{k} P(x = 1 \mid k_1, k)_k C_{k_1} \theta^{k_1} (1-\theta)^{k-k_1}. \tag{6.21}$$

If the equation $\theta = H(\theta)$ has a stable equilibrium at some point $\theta > 0$, there exists a stable equilibrium with a positive fraction of active agents, and in this case a global cascade occurs. We can easily observe that $H(0) = 0$ and $H(1) = 1$. These situations correspond to the values, $\rho_k = 0$ and $\rho_k = 1$, respectively. The existence of a positive fixed point $\theta > 0$ is guaranteed if $H'(0) > 1$. We are interested in determining which threshold values ϕ make the state $\theta = 0$ an unstable fixed point. This condition is equivalent to computing the values of ϕ satisfying:

$$\frac{dH(\theta)}{dt}\Big|_{k} = \frac{1}{z}\sum_{k} kp(k)\left(\sum_{k_1=kq+1}^{k}(k_1\theta^{k_1}(1-\theta)^{k-k_1} - (k-k_1)\theta^{k_1}(1-\theta)^{k-k_1-1})\right). \tag{6.22}$$

If the condition eqn. (6.8) holds, we have:

$$\frac{dH(\theta)}{dt}\Big|_{q=0} = 0.$$

In this case, we have:

$$\frac{dH(\theta)}{dt}\Big|_{q=0} = \frac{1}{z}\sum_{k=1}^{\lfloor 1/\phi \rfloor} k^2 P(k). \tag{6.23}$$

We say that a cascade occurs if, starting at an initial state with a very small fraction of active agents ($x = 1$), the cascade dynamics converges to a stable state with a positive fraction of active agents at $\theta > 0$. If there exists a *network threshold* ϕ^*, the global cascade occurs in the region of threshold $\phi < \phi^*$. We can implicitly determine such a network threshold by solving:

$$\phi^* = \underset{\phi}{argmin}\frac{1}{z}\sum_{k=1}^{\lfloor 1/\phi \rfloor} k^2 P(k). \tag{6.24}$$

The network threshold ϕ^* in eqn. (6.24) defines the lowest upper bound for the second moment of the degree distribution summed up from $k = 1$ to the inverse of the agent threshold $\lfloor 1/\phi \rfloor$.

From eqn. (6.8) we can infer that the higher the average connectivity of the network, the lower the threshold for global cascade. By solving eqn. (6.24), we obtain the cascade condition by assuming all agents have the same threshold ϕ. In this case, we can describe the cascade condition graphically as the function of the threshold ϕ and the average degree z. Global cascades can occur only within a certain region of the parameter space of (z, ϕ), which is defined as the *cascade window*. Whereas global cascades occur inside this region, outside this region, cascades are typically small. In the mean-field dynamics described above and a network with connectivity distribution $P(k)$, some networks have a wider and others a smaller cascade window, depending on their degree distributions.

Figure 6.6 shows the cascade windows of the following typical networks with the number of agents, $N = 500$, and the average degree $z = 8$ (the total links are $L = 2000$).

Random network: RND

A random network is constructed by the Erdos-Renyi random model. Each link is included in the graph with probability p independent of every other link. A graph generated by the Erdos-Renyi model, which is represented as $G(N, p)$, has binomial distribution.

Exponential network: EXP

The degree distribution is set as $p(k) = ce^{-k/z}$, where c is some constant.

Scale-free network: SF(BA)

The degree distribution of a scale-free network of the Barabasi-Albert model is $p(k) = ck^{-3}$, where c is some constant.

Scale-free network: KN

The degree distribution of a scale-free network is set to $p(k) = ck^{\gamma}$, with the power coefficient $\gamma = 2 + 1/(z - 1)$.

The cascade window of each network has a particular shape, as illustrated in Figure 6.5. If the average degree and threshold are inside the cascade window, a single active agent can initiate a global cascade. However, if they are outside the cascade window, the initial trigger agent has a weak influence and a cascade does not occur.

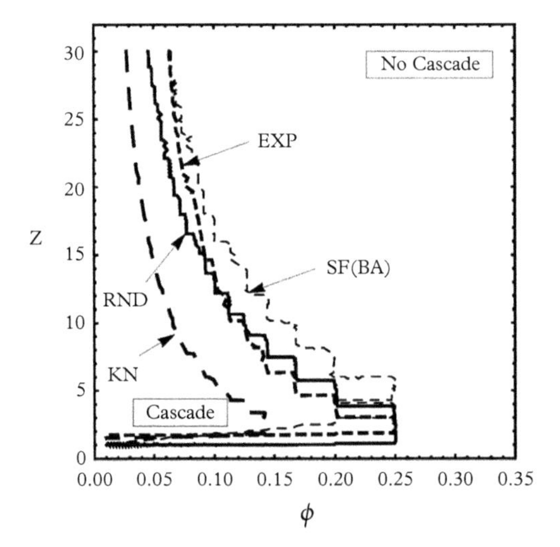

Figure 6.5 *The cascade window as a function of the threshold value ϕ of agents and the average degree z of networks: a random network (RND), an exponential network (EXP), a scale-free network of the Barabasi-Albert model, (SF(BA)) with the power coefficient $\gamma = 3$, and a scale-free network (KN) with the power coefficient $\gamma = 2.1$*

A comparison of the four networks shows that *KN* has the smallest cascade window and *SF(BA)* has the largest cascade window, while the cascade window of *EXP* and *RND* is of intermediate size. Therefore, a large-scale cascade likely occurs on *SF(BA)* and a cascade on *KN* is mostly limited.

The manner in which a cascade in these networks spreads is qualitatively distinct from that in which it spreads in all-to-all networks, described in Section 6.2. The extra condition of the cascade window constitutes a major difference between the network interaction model and the all-to-all interaction model. A large cluster of vulnerable agents must exist to allow the activation and spread of cascade. When one of the vulnerable agents is activated, the rest of the agents in the cluster begin to follow in short order, whereupon non-vulnerable other agents are also activated. In the more realistic modeling with sparse interaction networks, the success of a global cascade depends on not only the agent threshold, but also the connectivity of the agents. In brief, a cascade in its early stages can spread only via vulnerable agents who can be activated by only a single active neighbor. In order for a widespread global cascade to occur, the network must contain a large cluster of vulnerable agents.

In cascade dynamics, two regimes are identified in which the network is susceptible to very large cascades; however, they occur very rarely. When cascade propagation is limited by the connectivity of the network, a power law distribution of cascade sizes can be observed (Watts, 2002). However, when the network is highly connected, cascade propagation is limited instead by the local stability of the agents themselves, and the size distribution of cascades is bimodal, implying a more extreme kind of instability, which is correspondingly more difficult to anticipate. It is found that in the first regime, where the distribution of network neighbors is highly skewed, the most connected agents are far more likely than average agents to trigger cascades, but that this is not so in the second regime.

The origin of large but rare cascades that are triggered by small initial active agents is a phenomenon that manifests itself diversely as cultural fads, collective action, the diffusion of norms and innovations, and cascading failures in socio-technical networks. This section presented a possible explanation of this phenomenon in terms of networked agents whose decisions are determined by the actions of their neighbors according to a simple threshold rule. When the network is highly connected, cascade propagation is limited by the local stability of the agents themselves. When the global cascade represents a bad phenomenon, such as the sequence of bankruptcy, it will be difficult to prevent or mitigate the global cascade by the usual cost and benefit thinking scheme, since the expected loss is sufficiently large. Many systems are designed to maximize their utility and minimize their cost, including the cost of security. As a result, in the worst case, we may have to accept the entire loss caused by the occurrence of a bad cascade.

6.4 Optimizing Cascades in Networks

Cascades are a network phenomenon in which small local changes can result in wide spread events, such as fads, innovations, or power failures. There are two kinds of cascade, *good cascades* and *bad cascades*, depending on important consequences of the transmission of influence across a network. For the spread of useful information, new ideas, behaviors, or innovation, we may want to maximize a good cascade. On the other hand, we need to minimize a bad cascade, such as the spread of a rumor, social unrest, power failure, or bankruptcy.

If agents are represented by nodes in a network, where a link indicates that one agent influences another, then some topologies of this network make it more likely that an innovation will be widely adopted. Moreover, in some networks, if only a small proportion of agents adopt an innovation, the innovation will spread quickly to a large component of the network. In this case, methods for determining which initial changes result in the largest cascades are of interest in the research field of marketing and other areas. Given such agent networks, a natural question emerges: if we wish to maximize or minimize the size of a cascade, which subset of agents should be targeted for initiators? What kinds of network topology maximize or minimize the size of a cascade as compared with other potential network configurations? The issues of triggering maximum possible cascade or preventing cascade at the minimum level are as yet poorly understood, but clearly lead to new research questions. According to the results presented in Section 6.2, changing the connectivity of the agent network and the agent threshold can have important implications for controlling the mutual influence that propagates throughout an agent network.

In this section, we seek the optimal network that maximizes or minimizes the size of a cascade and investigate the specific topological patterns via a cluster of vulnerable agents to clarify the pathway of the spread of the cascade. This problem is equivalent to seeking the network topology that provides a maximum or minimum cascade window. We can characterize the success of cascade using only two parameters: the agent threshold ϕ and the average degree z representing the connectivity of the agents. A large-scale of cascades of influence can occur only within a certain region of the space of the two parameters (ϕ, z) defined as the cascade window. If the pair of the average degree and threshold is inside the cascade window, an initial trigger agent can cause a global cascade; however, if it is located outside the cascade window, the trigger agent has little influence on the spread over the network. Therefore a network topology with a smaller cascade window is more reluctant to cascade, and one with a larger cascade window is more likely to trigger a global cascade.

We consider a heuristic network design model for obtaining an optimal network with the minimum or the maximum cascade window. According to the definition of the threshold model in eqn. (6.1), a hub agent with many links is the most reluctant to change, and many active agents are needed to cause a hub agent to change, since a hub agent has the role of preventing a global cascade if a large cluster of hub agents is formed in the agent network. In contrast, periphery agents with very few links change most easily under the influence of a few active agents. Therefore, forming a large cluster of periphery (vulnerable) agents becomes the pathway to global cascade.

A basic idea of the heuristic model is to create a large cluster of hub agents or a large cluster of periphery agents by controlling some parameters. In this heuristic design model, at each time, one new node (agent) with m links and c new links is generated and these are connected to hub nodes or vulnerable nodes with some probability. Initially, a small connected network with a few hub nodes and some periphery nodes is prepared as a primary network.

With probability $1-p$, a new node with m links is connected to m existing hub nodes. Furthermore, c new links are used to connect among the hub nodes. The probability of a hub node to which a new node is linked or a newly generated link is connected being chosen is given as:

$$P(k) = \frac{(k)^{\beta}}{\sum (k)^{\beta}}. \tag{6.25}$$

Therefore, existing hub nodes with many links are more likely to be linked to a new node. The exponent β controls the selection of a hub node.

With remaining probability p, a new node is connected to m periphery nodes with a small number of links. Furthermore, newly generated c links are used to connect among periphery nodes. In this case, the probability of a periphery node to which a new node is linked or a newly generated link is connected being chosen is:

$$P(k) = \frac{(1/k)^{\beta}}{\sum (1/k)^{\beta}}. \tag{6.26}$$

Therefore, an existing periphery node with a small number of links is more likely to be linked to a new node.

If the controlling parameter p is close to 0, most of the newly generated nodes and links are connected to existing hub nodes. In this case, the network grows to a *core-periphery network* with two types of nodes: a core node and a periphery node. The core (hub) nodes are connected to both core nodes and periphery nodes, but a periphery node is connected only to a core node. On the other hand, if p is close to 1, new nodes are connected mainly to periphery nodes and new links are also used to connect periphery nodes, and the network grows as a random network. By controlling the parameter p, the probability of selecting a cluster of hub nodes in order to add new nodes and links, we can control the growths of the cluster of hub nodes (robust to change) or the cluster of periphery nodes (vulnerable to change). This heuristic method is useful for finding the network with the maximum cascade window or that with the minimum cascade window. The average degree z of the generated network is given as $z \simeq 2(c+m)$, where m is the number of links of a new node and c is the number of newly generated links generated at each time period. For instance, if we set the parameters, $p = 0$, $\beta = 1$, $c = m = 1$, a scale-free network with power index $\gamma = 2.1$ is generated.

We summarize the heuristic network design model in the following steps (see Figure 6.6).

Step 1: Construct a primary connected network with a few hub nodes.

Step 2: Choose Step 3 with probability $1-p$ or Step 4 with probability p.

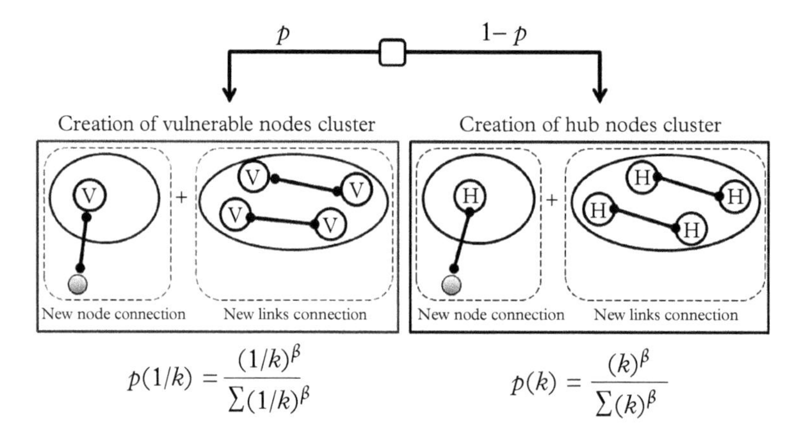

$$p(1/k) = \frac{(1/k)^{\beta}}{\sum(1/k)^{\beta}} \qquad p(k) = \frac{(k)^{\beta}}{\sum(k)^{\beta}}$$

Figure 6.6 *Heuristic network generation method. The controlling parameter p represents the probability of using a newly introduced node and links to form a cluster of hub nodes. The encircled characters V and H represent a periphery node and a hub node, respectively*

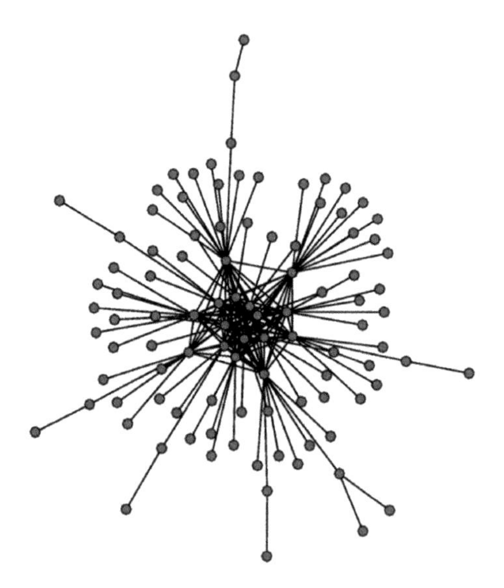

Figure 6.7 *Network topology of a core-periphery network with preferential attachment (CPA). This network is generated by the heuristic model by setting the parameters: $p = 0$, $\beta = 2$, $c = m = 2$. $N = 100$ nodes and $z = 4$. The number of nodes in the center core is 16*

Step 3: Add a new node with m links and new c links to hub nodes with the probability represented in eqn. (6.25), and go to Step 5.

Step 4: Add a new node with m links and new c links to periphery nodes with the probability represented in eqn. (6.26), and go to Step 5.

Step 5: Go to Step 2 if a stop criterion is not met, which is defined by the number of generated nodes.

Figure 6.7 shows the network topology generated by the heuristic model when the parameters are set to: $p = 0$, $\beta = 2$, $c = m = 2$. The generated network has $N = 500$ nodes and we define it as a *clique network with preferential attachment (CPA)*. The cascade window of *CPA* is shown Figure 6.8 in comparison with the cascade windows of a random network (*RND*) and a scale-free network (*KN*) with power index $\gamma = 2.1$. *CPA* has the smallest cascade window, and it mostly prevents global cascade.

Figure 6.9 shows the network topology generated by the heuristic model when the parameters are set to: $p = 0.3$, $\beta = 2$, $c = m = 2$. The generated network has $N = 500$ nodes and we call it as a *concentric ring network (CRN)*. The cascade window of *CRN* is shown Figure 6.10 in comparison with the cascade windows of the scale-free network of

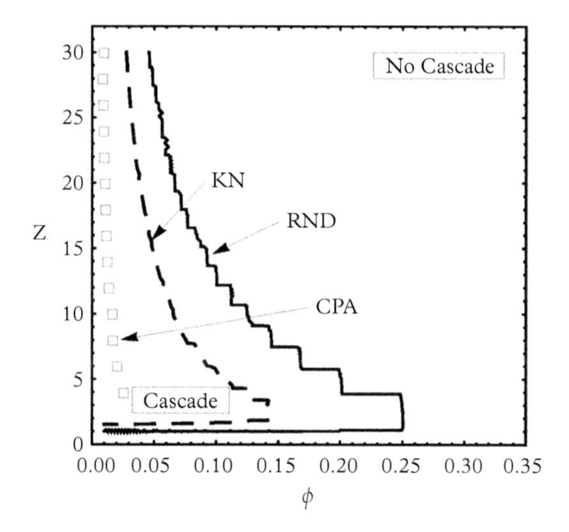

Figure 6.8 *Comparison of cascade windows. The dashed line encloses the region of the (ϕ, z) in which global cascades occur for the same parameter settings in the full dynamical model for $N = 500$ (averaged over 100 random single-node perturbations). A clique network with preferential attachment (CPA), a random network (RND), and a scale-free network (KN) with power index $\gamma = 2.1$*

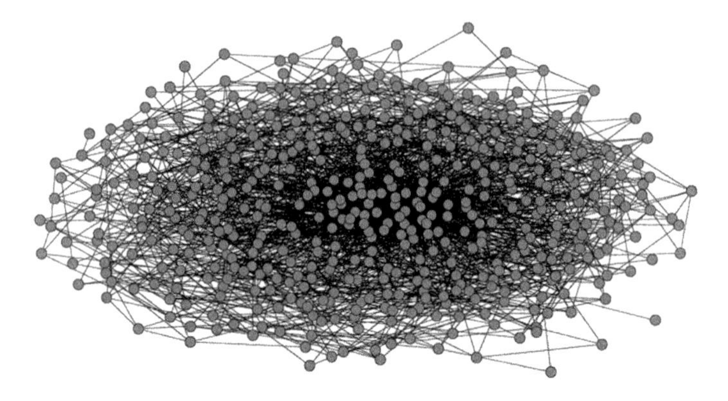

Figure 6.9 *Network topology of a concentric ring network (CRN) with the number of nodes $N = 500$ and the average degree $z = 10$, which has the largest cascade window. This network is generated by the heuristic model by setting the parameters: $p = 0.3$, $\beta = 2$, $c = 4$, $m = 1$*

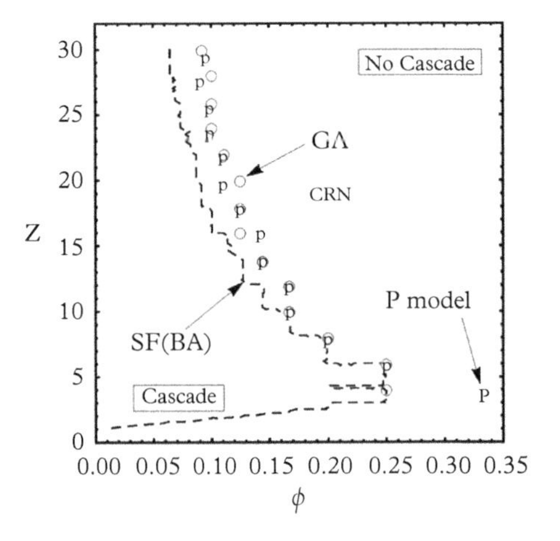

Figure 6.10 *Comparison of cascade windows. A centric ring network (CRN) and a scale-free network of the Barabasi-Albert model (SF(BA)) are shown for comparison. The solid circles outline the region in which global cascades occur for the same parameter settings in the full dynamical model for $N = 1000$ (averaged over 100 random single-node perturbations)*

the Barabasi-Albert model (*SF(BA)*). *CRN* has the largest cascade window and it mostly promotes global cascade.

In general, both *CPA* and *CRN* have a few core agents and many vulnerable agents, but a cascade is minimized in *CPA* and maximized in *CRN*. Figure 6.12 shows the

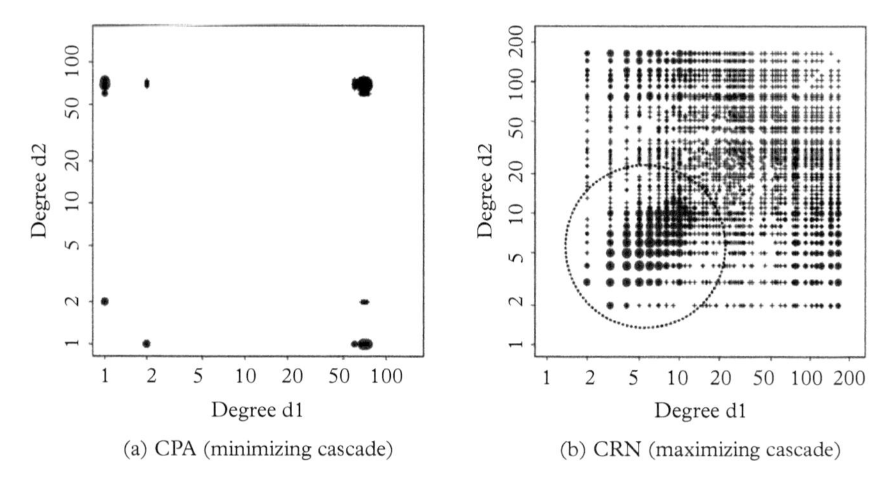

Figure 6.11 *The map of degree–degree relationship between nodes. The average degree of each network is z = 10. The diameter of each dot on the map is proportional to the logarithm of the frequency of the degree–degree relationship. (a) A clique with preferential attachment (CPA) for minimizing cascade. (b) A centric ring network (CRN) for maximizing cascade*

map of degree–degree relationships between agents. The diameter of each dot on the map is proportional to the logarithm of the frequency of the degree–degree relationship. *CPA* has one big cluster of core agents (clique). The peripheral agents have no direct connection to each other and they are connected only to core agents. Therefore, a cascade cannot spread via peripheral agents. *CRN*, on the other hand, forms a multi-level agent network, as shown in Figure 6.10, and the hub agents are also well connected with many periphery agents at lower hierarchy levels. Hub agents located on higher hierarchies have many links, depending on their hierarchical positions. In this type of multi-level wheel network, many vulnerable agents in the periphery cannot directly influence hub agents to change, but the accumulated power of the vulnerable agents has a strong influence when they form a large cluster.

We can observe that a global cascade occurs if the agent threshold ϕ and the average degree z are inside the cascade windows shown in Figure 6.9 or Figure 6.11. At the border of the cascade window, the cascade occurrence becomes obscure. The average is usually a statistic index that is useful for understanding the properties of the phenomena concerned. The results show that the global cascade can occur even if the average cascade size is almost 0. This gap between the frequency of global cascade and the average size of cascade may lead to an underestimation of the probability of events driven by cascade dynamics. The cascade dynamics around the boundary of the cascade window are stochastic and we need to observe a few occurrences of global cascade among many trials initiated by a randomly chosen trigger agent.

Simulation studies were conducted to observe the gateway for a global cascade and compare the influence of network topologies on cascade dynamics. In Figures 6.12 to 6.16, we show the frequencies of global cascade on several networks. We calculated

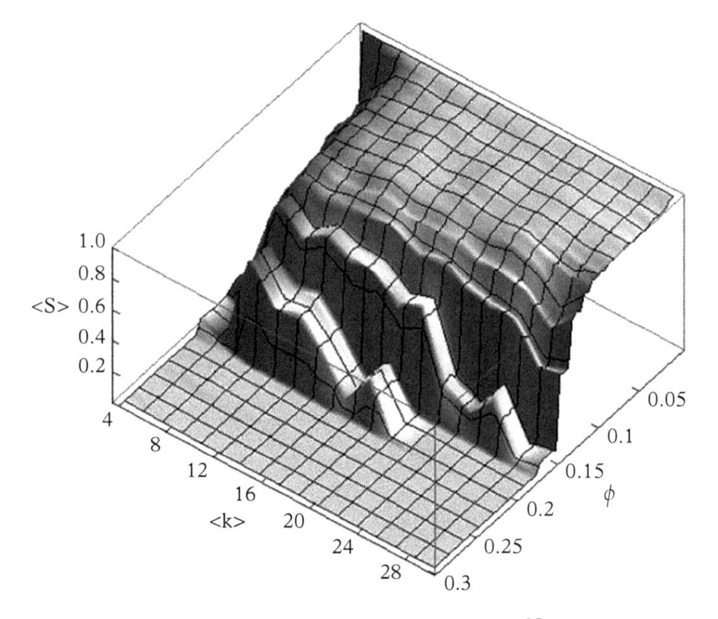

Figure 6.12 *The average cascade size,* $< S > = \sum_{i=1}^{N} x_i/N$, *the average ratio of the active agents among 1000 simulations, on a centric ring network (CRN), which yields the largest cascade window. The agent threshold is set between* $\phi \in [0, 0.3]$ *and the average degree is set between* $z \in [4, 30]$. *One agent is selected randomly as the initial active agent and all other agents are inactive for each simulation. The agent network size is* $N = 500$

the cascade size, $< S > = \sum_{i=1}^{N} x_i/N$, as the average fraction of the active agents in a reiterative simulation repeated over 1000 times in which the two parameters, the agent threshold ϕ and the average degree $z = < k >$, were varied. In each simulation, one agent was selected as a trigger agent.

The simulation results for *CRN* are shown in Figure 6.12. It is very interesting that the average cascade size is relatively small at a small average degree z, even if the network has the largest cascade window, as shown in Figure 6.10. We can observe that a global cascade occurs in the wide area of the combination of agent threshold ϕ and average degree z, as compared with the other networks.

The simulation results of *CPA* with the smallest cascade window are shown in Figure 6.13. We can observe that a global cascade occurs only in the limited area of the combination of agent threshold ϕ and the average degree z, as compared with the other networks. However, the average cascade sizes are small. For instance, *CPA* with a smaller average degree is a star-like network structure and a global cascade occurs only if the hub agent is selected as a trigger agent. The specific structure of *CPA* reduces the frequency of global cascade. As a result, the average cascade size becomes small when an initial trigger agent is chosen randomly.

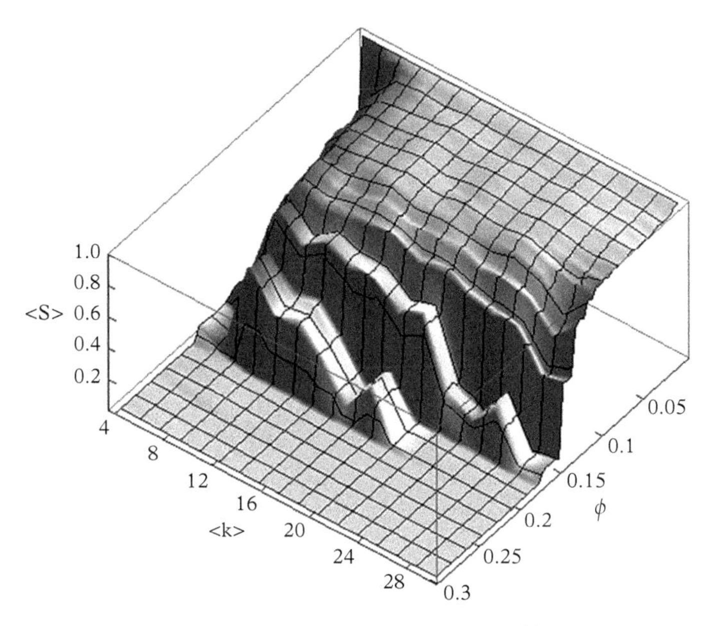

Figure 6.13 *The average cascade size, $< S > = \sum_{i=1}^{N} x_i/N$, the average ratio of the active agents among 1000 simulations, on core-periphery with preferential attachment (CPA). The agent threshold is set between $\phi \in [0, 0.3]$ and the average degree is set between $z \in [4, 30]$*

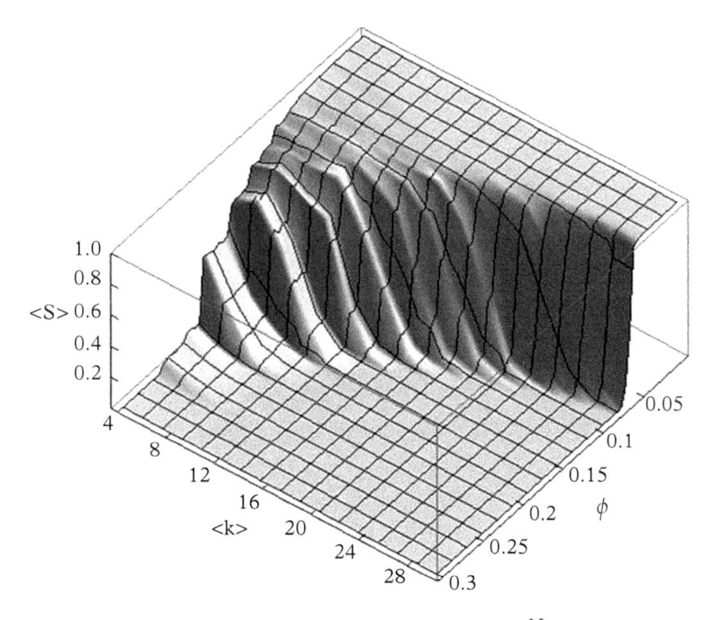

Figure 6.14 *The average cascade size, $< S > = \sum_{i=1}^{N} x_i/N$, the average ratio of the active agents among 1000 simulations, on a scale-free network of the Barabasi-Albert model, SF(BA). The agent threshold is set between $\phi \in [0, 0.3]$ and the average degree is set between $z \in [4, 30]$*

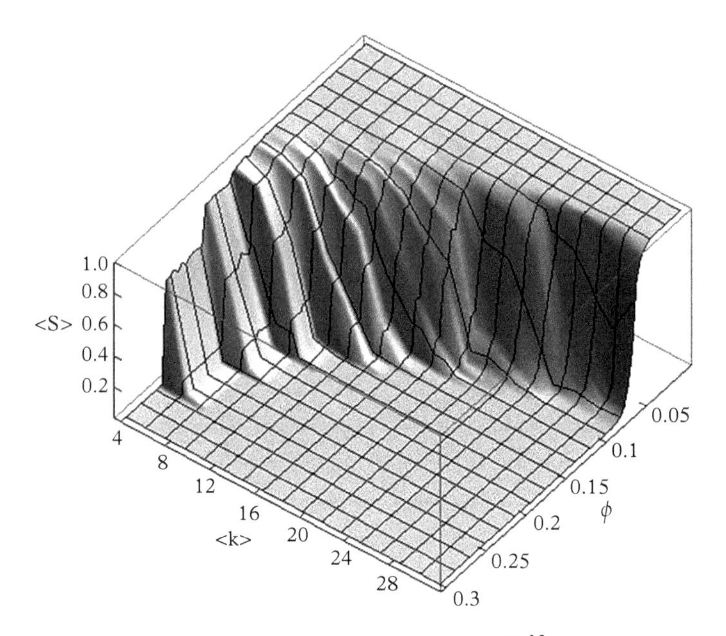

Figure 6.15 *The average cascade size,* $< S > = \sum_{i=1}^{N} x_i/N$*, the average ratio of the active agents among 1000 simulations, on an exponential network (EXP). The agent threshold is set between* $\phi \in [0, 0.3]$ *and the average degree is set between* $z \in [4, 30]$

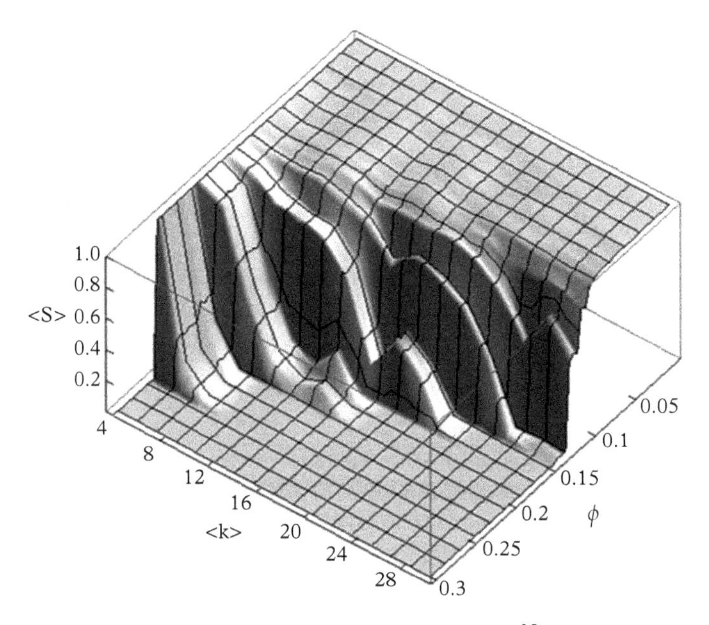

Figure 6.16 *The average cascade size,* $< S > = \sum_{i=1}^{N} x_i/N$*, the average ratio of the active agents among 1000 simulations, on a random network (RND). The agent threshold is set between* $\phi \in [0, 0.3]$ *and the average degree is set between* $z \in [4, 30]$

For networks other than *CRN*, there are roughly two common features. The average cascade size jumps from almost 0 to 1 as a step function around the border of the cascade window. As the results presented in this section make clear, changing the agent threshold and agent connectivity can have important implications both for the scale of cascades that may propagate throughout a network and the manner in which these cascades may be seeded.

6.5 Cascade Dynamics with Stochastic Rule

In Section 6.4, we discussed in detail how the properties of the agent network topologies combined with deterministic threshold rules can impact cascade dynamics. In particular, we determined the likelihood that a global cascade of influence originates from a very small number of initiators, and then, characterized network structures that favor or inhibit the spread of a cascade. In this section, we consider the cascade dynamics with the *stochastic threshold model*. We extensively study how this simple extension to a stochastic cascade process can affect the occurrence of global cascades.

An agent is influenced by the behaviors of other agents with whom they are in contact. To study such situations, an understanding of the agent network structure at hand and its interplay with agent decisions is required. Game theory treats a complementary point of view of contagion by considering scenarios where the agents' behavior is the result of a rational choice among competing alternatives (Young, 1998). Results of studies of the dynamics of coordination games in which game theory was applied provide a simple condition for a new action or innovation to become widespread in the network. A rational choice of each agent is characterized by a threshold that determines the fraction of neighbors that must be activated.

In Chapter 3, we considered scenarios where the agents' behavior is the result of a strategic choice between two competing alternatives, A or B. For instance, there are two competing technologies A and B, and agents must use the same technology in order to communicate. By assuming that all agents are identically situated in the sense that every agent's outcome, regardless of the choice made, depends on the number of agents who choose A or B, these interactive situations can be modeled by decomposing them into the underlying 2×2 games with payoffs, as shown in Table 3.1. The whole outcome depends on the threshold rule and the network structure. By an agent network or graph $G = (V, E)$ we mean a simple, undirected, connected graph with the total number of agents (nodes) $|V| = N$ and the total number of links (edges) $|E| = L$. The agent network G governs who talks to whom and there is a link between a pair of agents (i, j). They play a coordination game with two strategies, A or B: if both agents i and j choose A, then they each receive a payoff of $1 - \phi$; if they each choose B, then they each receive a payoff of ϕ. However, if they choose opposite technologies, then they each receive a payoff of 0, reflecting the lack of interoperability. The payoff parameter ϕ takes any value between [0,1]. For any agent, the payoff for a choice of A or B depends on how many other agents who are connected choose A or B, and the optimal choice rule of each agent is given by:

$$If \begin{cases} p \geq \theta \ , \ choose \ A(x=1) \\ p < \theta \ , \ choose \ B(x=0) \end{cases}, \tag{6.27}$$

where p represents the ratio of neighbor (connected) agents who choose A.

The agents with the deterministic threshold rule in eqn. (6.27) can change their states in the following deterministic way. If for some threshold ϕ, a fraction of neighbor agents adopts, the agent's state changes to adopt as well. Suppose all agents initially choose B. Thereafter, a small number of agents begin adopting A instead of B. Then, we apply the deterministic threshold rule in eqn. (6.27), that is, switch to A if the fraction of neighbors who have already adopted A is larger than the threshold ϕ, to each agent in the network to update her state. Agents display inertia in switching choices, but when their personal threshold has been reached, the action of even a single neighbor can tip them to switch.

Cascading in behavior means a succession of the same behavior. In this section, we consider a non-progressive cascade. There can be a cascading sequence of agents switching to A, such that a network-wide equilibrium is reached in the limit with almost all agents adopting A or B, or coexistence may be involved, such that the agents are partitioned into a set adopting A and a set adopting B. If many agents choose B, no agent is motivated to choose A unless a sufficient number of other agents do switch beyond the critical point $p = \phi$, the fraction that chooses A. If only a few agents choose A, they will subsequently switch back to B. However, if many agents follow and choose A, then agents who choose B will soon switch to A. It is sufficient merely that agents make the right choice at the beginning for the selection of A to be achieved with complete efficiency, if $\phi < 0.5$.

The attractors of the collective dynamics sometimes correspond to a local optimal of the payoff landscape. An effective method for extracting a collective system from a local optimal is to add a degree of indeterminism. This method can be described as the injection of noise or random perturbation into the system, which causes it to deviate from its preferred trajectory. The role of *noise* or *mistake* as an internal perturbation is to push the collective system upward, towards a higher potential of the payoff landscape. This may be sufficient to allow the collective system to escape from a local optimal, after which it will again start to climb up toward a better state.

Kandori and Mailath (1993) introduced the *uniform error model* in coordination games and showed that a mistake in implementing strategy brings an important result. When each agent i updates her strategy, she chooses a strategy that maximizes her expected payoff with probability 1ϵ and chooses a strategy uniformly at random with probability ϵ. Kandori et al. verified the influence that the act of making such a mistake exerted on all agents. When agents occasionally make small mistakes, they collectively evolve into full coordination, choosing the same strategy A, if $\phi < 0.5$.

We now consider the uniform error version of the deterministic threshold rule in eqn. (6.27) as:

$$If \begin{cases} p \geq \theta \begin{cases} choose\ A\ with\ probability\ 1-\epsilon \\ choose\ A\ or\ B\ with\ probability\ \epsilon \end{cases} \\ p < \theta \begin{cases} choose\ B\ with\ probability\ 1-\epsilon \\ choose\ A\ or\ B\ with\ probability\ \epsilon \end{cases} \end{cases}, \tag{6.28}$$

where $\epsilon > 0$ is a small error probability. This is defined as the *stochastic threshold rule.*

The difference between the stochastic threshold model and the deterministic threshold model is that the former agents are not necessarily active in A, even if the fraction of their active neighbors in A is above their threshold, and with some small probability ϵ, an agent remains inactive in B, possibly because of a reluctance to change; or an agent becomes active in A with some small probability, even if the fraction of her active neighbors in A is below her threshold, possibly because of an error. These minor mistakes, however, generate significant variability in the outcome of the collective choice process. The concept of path dependence originated as an idea that a small initial advantage with a few minor random shocks along the way could alter the course of history (David, 1985). In this sense, the stochastic threshold model may capture the notion of *path dependence.*

An alternative approach to the stochastic threshold model represented in eqn. (6.28) is the *log-linear model* suggested by Blume (1993). Given a real number $\beta > 0$, agent i chooses A with probability:

$$p = \frac{exp(\beta q(1-\theta))}{exp(\beta q(1-\theta)) + exp(\beta(1-q)\theta)} = \frac{1}{1 + exp(-\beta(q-\theta))} \tag{6.29}$$

where q represents the ratio of the neighbors who choose A. The parameter β measures the rationality of the agent and the larger β is, the more likely it is that the agent chooses an optimal strategy given the choices of his/her neighbors. If we set $\beta(\beta \to \infty)$, the rule in eqn. (6.29) is close to the deterministic rule in eqn. (6.27).

There are two basic cases in equilibrium. The first case is where all agents coordinate in choosing A and the second case is where all agents coordinate in choosing B. The simulation results show agents with the stochastic threshold rule succeed in coordinating and choosing A, and this perfect coordination does not depend on the network topology that determines the interaction of agents.

We consider five network topologies: (i) a regular ring lattice network where all agents are connected to their nearest neighbors regularly; (ii) a random regular graph where all agents are connected to the same number of neighbors who are randomly chosen; (iii) a random network; and (iv) a scale-free network of the Barabasi-Albert model *SF(BA)*, and (v) a scale-free network with power index $\gamma = 2.1$ (*KN*). The size of each network is $N = 3000$ agents and the average degree of each network is $z = 4$. All agents in each network have a threshold $\phi = 0.25$. We conducted 1000 simulation runs, and for each

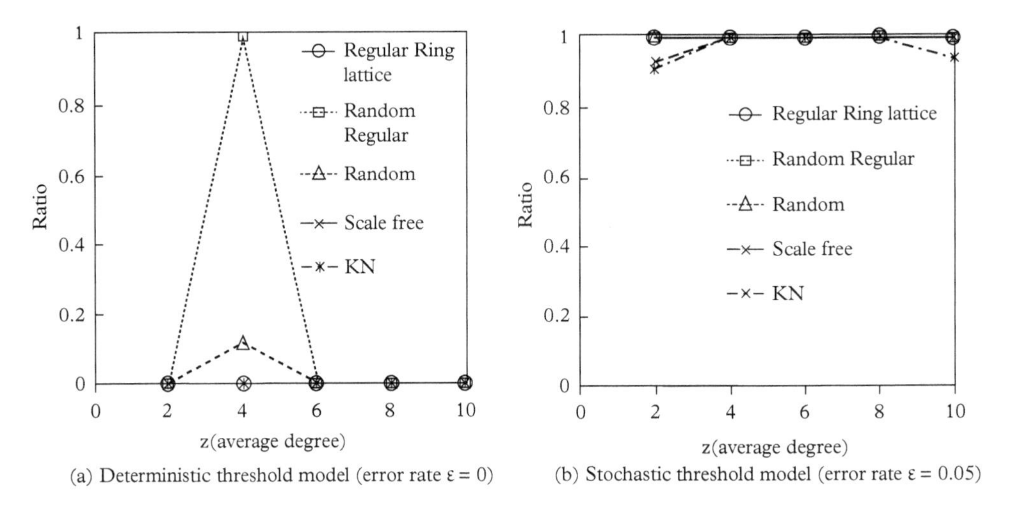

Figure 6.17 *The average fraction of agents who adopted A on each agent network in 1000 simulation runs. Each network has N = 3000 agents and the average degree is z = 4. Threshold of each agent is $\phi = 0.25$: (a) Each agent has deterministic threshold rule ($\beta \to \infty$ in eqn. (6.29)), (b) each agent has stochastic threshold rule ($\beta = 18$ in eqn. (6.29))*

simulation run, randomly selected three agents (0.1% of the whole population) as trigger agents to adopt A. In each simulation, the total time steps are bounded up to 200.

Figure 6.17(a) shows the occurrence of global cascade where more than 99% of agents among $N = 3000$ change from B to A with $\beta \to \infty$ in eqn. (6.29) and all agents behave under the deterministic threshold rule. In this case, all agents succeed in choosing A, without depending on the network topology and the average degree. Each network has a specific cascade window, and a global cascade occurs only in some interval of the average z between 2 and 6. For instance, in a regular ring lattice network where the average degree is $z = 4$, the fraction of the agents adopting A is close to 1. The fraction of the agents who adopt A in a regular random network is almost zero for any average degree, because each simulation time is bounded up to 200 steps, and then, more time steps may be required for some agents to adopt. The fraction of agents who adopt A becomes almost 0 on all networks when the average degree is more than $z = 6$.

Figure 6.17(b) shows the occurrence of global cascade in the case where agents use the stochastic threshold rule in eqn. (6.29) with $\beta = 18$. The occurrence of a global cascade is almost certain in any network topology. The cascade window restricting the spread of A in Figure 6.17(a) disappears, and almost all agents succeed in adopting A. A stochastic threshold-based choice is a method for introducing the adoption of A into every agent network and every region of average degree. Only a few agents are required to trigger the cascade of the adoption of A.

A minor mistake in each agent's choice under the stochastic threshold model may alter the cascade dynamics significantly. The stronger the mistake, the greater the probability that the cascade dynamics will be able to escape the relatively shallow valleys and thus reach a potentially better outcome. However, cascade dynamics with some mistakes will never be able to become truly stable in a global optimum since whatever level it reaches it will still be perturbed and pushed into another state.

Different network structures and different update rules at the agent level may yield different results. Nevertheless, this example illustrates a general cascade phenomenon that we conjecture holds across a range of situations. The results based on the stochastic threshold model differ strongly from those derived using the deterministic threshold model. The cascade widow will disappear in any network topology and a new behavior or innovation, for instance, will become widespread in any network topology. In the stochastic threshold model, the network topology or the connectivity among agents plays an ambiguous role and it allows a global cascade to spread on any network topology, and it affects only the speed at which the cascade penetrates (Young, 2009, 2011).

7

Agent-Based Influence Dynamics

Social influence refers to the behavioral change of individuals affected by others in a society. Social influence is an intuitive and well-accepted phenomenon in social networks. The strength of social influence depends on many factors, such as the strength of relationships between people, each individual in the networks, the network distance between users, temporal effects, and characteristics of networks. In this chapter we focus on the computational aspect of social influence analysis and describe the measures and algorithms related to it. More specifically, we discuss the qualitative or quantitative measurement of influence levels in an agent network. Using agent-based modeling and network analysis, we shed light on these issues and investigate better connections that foster mutual influence in an egalitarian society.

7.1 Mutual Influence and Consensus

People are influenced by others in many situations, such as when they make decisions about what opinions they should hold, which products they should buy, what political positions they should support, and many other things. We should go beyond this observation and consider some of the reasons why mutual influence occurs. There are many settings in which it may in fact be rational for an agent to imitate the choices of other agents, even if the agent's own information suggests an alternative choice. Agents are influenced by the decisions, actions, or advice of others when making a wide variety of decisions, both consciously and unconsciously. Understanding how and when social influence arises, and how agent decisions aggregate in the presence of such influence are central issues in many disciplines. The *influence network* describes the network of who influences whom and how it can impact the dynamics of collective decisions.

Social influence is defined as a change in an individual's thoughts, feelings, attitudes, or behaviors that results from the presence of others. Social influence is also defined as a change in an individual's actions that results from interaction with another individual or a group (Moussald et al., 2013). Social influence is one of major concerns involved in the issue of opinion formation and also in a variety of other social phenomena, such as learning from others. Social influence is distinct from *conformity* or power. Conformity occurs when an individual expresses a particular opinion or behavior in order to fit in

Agent-Based Modeling and Network Dynamics. First Edition. Akira Namatame and Shu-Heng Chen.

to a given situation or to meet the expectations of a given other, although the individual does not necessarily hold that opinion or believe that the behavior is appropriate. Power is the ability to force or coerce an individual to act in a certain way by controlling the individual's attitude.

Social influence is also the process by which individuals make real changes to their feelings and behaviors as a result of interaction with others who are perceived to be similar, attractive, or experts. People adjust their beliefs with respect to others to whom they feel similar in accordance with psychological principles such as balance. Individuals are also influenced by the majority. If a large portion of an individual's referent social group holds a particular attitude, it is likely that the individual will adopt it as well.

The most studied mechanism of social contagion is social influence. In this case, the phenomenon of social contagion as a specific form of social influence received attention from social science researchers as far back as the nineteenth century, when the French social psychologist Le Bon (1895) helped popularize the idea of social contagion. He analyzed an unconscious process by which information or beliefs are spread throughout a social group, taking the form of contagion. Many applications, such as those for marketing, advertisement, and recommendations, have been built on the basis of the implicit notion of social influence between people. Today, with the exponential growth of online social media, such as Facebook and Twitter, social influence can for the first time be measured over a large-scale social network. We frequently observe many situations where opinions, behavior, and many other things spread from person to person through a social network in which people influence their friends. Another important element of social influence is influence networks. However, in spite of recent interest in the topic, it is still the case that little is known about the structure of large-scale social networks, and in what manner social influence propagates on them.

In organizations, networking enables every individual from the top to the lowest members in the organizational hierarchy to share information and gain increased situational awareness. Sharing information allows individuals to feel capable of making their own decisions with increased confidence. Rank within a hierarchical organization becomes less relevant as a source of power. Given this kind of an egalitarian society, people need to decide whom to trust to provide them with the information that they want. Interaction between individuals does not involve merely sharing information. Its function is to construct a shared reality consisting of agreed-upon opinions. Individuals have their own opinions, beliefs, or preferences. These endogenous opinions and preferences can be compared with and adjusted to each other. In principle, it is possible to describe the conditions of communication that allow a correct description of preference intensities. In the communication process, individuals must influence one another to arrive at a common interpretation of their shared reality.

The pioneering work on the study of influence in networks is largely due to DeGroot (1974). Agents have beliefs about some common question of interest. At each time period, agents communicate with their neighbors in the agent network and update their beliefs. The updating process is simple: an agent's new belief is the average of her neighbors' beliefs in the previous period. Over time, diverse beliefs among agents converge to a consensus if the network is connected and satisfies a weak aperiodicity condition. This allows us to make a subtle distinction between *consensus* and *agreement*. Consensus occurs between individuals when they have the same intensity of preference. All members of a group are then equally likely to accept or reject a statement of desire or preference. Agreement signifies a state of the world where all individuals have the same preference ranking, but not the same intensity of preferences. Hence, they agree on the preference ranking, although they do not have consensus about their preferences.

Golub and Jackson (2010) focused on situations where there is some true state of nature that agents are trying to learn and to which each agent's initial belief is equal. An outside observer who could aggregate all of the initial beliefs could develop an estimate of the true state that would be arbitrarily accurate in a large society. Their question is: in which networks will agents using the simple updating process all converge to an accurate estimate of the true state? Given that communication frequently involves repeated transfers of information among large numbers of agents in networks, fully rational learning becomes infeasible. Nonetheless, it is possible that agents using fairly simple updating rules will arrive at the same outcomes as those achieved through fully rational learning. Golub and Jackson identified networks in which agents converge to fully rational beliefs despite using simple and decentralized updating rules and networks in which beliefs fail to converge to the rational limit under the same updating.

For example, if all agents listen to only one particular agent, then their beliefs converge, but they converge on the basis of that agent's initial information and thus the beliefs are not accurate, in the sense that they have a significant probability of deviating substantially from the truth. In contrast, if all agents place equal weight on all agents in their communication, then they immediately converge to an average of all the beliefs and then, by a law of large numbers, agents in a large society all hold beliefs close to the true value of the variable. Golub and Jackson called a networked society that converges to this accurate limit *wise*. Their main result is a characterization of *wisdom* in terms of influence weights: a society is wise if and only if the influence of the most influential agent vanishes as the society grows. Building on this characterization, they focused on the relationship between social network structure and wisdom. They also provided some examples of network structures that prevent a society from being wise.

There are two competing hypotheses about the manner in which network structure affects influence among agents. The *strength of weak ties* hypothesis predicts that networks with many *long ties* will spread a social behavior farther and more quickly than a network in which ties are highly clustered. This hypothesis treats the spread of behavior as a simple contagion, such as disease or information: a single contact with an infected agent is usually sufficient to transmit the behavior. The power of long ties is that they reduce the redundancy of the diffusion process by connecting people whose friends do not know each other, thereby allowing a behavior to rapidly spread to new areas of the network.

The second hypothesis states that, unlike disease, social behavior is a *complex contagion*: people usually require contact with multiple sources of infection before being convinced to adopt a behavior. This hypothesis predicts that because clustered networks have more redundant ties, which provide *social reinforcement* for adoption, they will better promote the diffusion of behaviors across large populations. Despite the scientific and practical importance of understanding the spread of behavior through social networks, an empirical test of these predictions has not been possible, because it requires the ability to independently vary the structure of a social network.

Social networks are primary conduits of information, opinions, and behaviors. They carry news about products, jobs, and various social programs, drive political opinions and attitudes toward other groups, and influence decisions. In view of this, it is important to understand how individual beliefs and behaviors evolve over time, how this depends on the network structure, and whether or not the resulting outcomes are efficient. Centola (2010) investigated the effects of network structure on diffusion experimentally by studying the spread of health behavior through artificially structured online communities. Participants made decisions about whether or not to adopt a health behavior based on the adoption patterns of their health buddies. Arriving participants were randomly assigned to one of two experimental conditions: a clustered-lattice network and a random network. In the clustered-lattice network condition, there was a high level of clustering created by redundant ties that linked each agent's neighbors to one another. His results showed that the likelihood of agent adoption was much greater when agents received social reinforcement from multiple neighbors in the social network. This produced the striking finding that the behavior spread farther and faster across a clustered-lattice network than across a corresponding random network.

7.2 Opinion Dynamics and Influence Network

Social influence is the process by which individuals adapt their opinion, revise their beliefs, or change their behavior as a result of interactions with others. Social psychologists have long been interested in social influence processes, and whether these processes will lead over time to the convergence or divergence of attitudes or opinions (Friedkin, 2012). In a strongly interconnected society, social influence plays a prominent role in many self-organized phenomena, such as herding in cultural markets, the spread of ideas and innovations, and the amplification of fears during epidemics.

Adopting a network perspective on social influence, this section develops mutual influence models among agents. Given the agent network structure, we assume that each agent gradually revises his/her opinion or attitude toward those of his/her contacts, depending on the relative strength of the social tie to each contact. The influence network emphasizes the collective dynamics involved in the formation of agents' opinions or attitudes. This section focuses on the conditions under which opinions converge in the long run.

Consider a group of agents among whom some process of opinion formation takes place. In general, an agent neither simply shares nor strictly disregards the opinion of any other agent, but takes into account the opinions of others to a certain extent when forming his/her own opinion. This can be modeled by different weights that any of the agents places on the opinions of all the other agents. Assume that there are N agents, which we shall sometimes refer to as *nodes*. Let $G = (V, E)$ be the network with the agent set V and the set of edges E. Thus, E is a collection of ordered pairs of edges between two agents i,j, where agent i is connected to agent j. G is assumed to be the weighted network. Each edge i,j has a weight $w_{ij} = w_{ji} > 0$, which we shall interpret as a measure of the *mutual influence* that agent i and agent j exert on one another. We assume that $w_{ij} = w_{ji} = 0$, whenever i,j is not an edge. The influence network is completely specified by a symmetric $N \times N$ matrix of weights $W = (w_{ij})$, which is defined as the *direct influence matrix*.

We also permit agents to influence themselves; in this case, own influence is reflected by $w_{ii} > 0$. If agent i holds an initial opinion $x_i(0)$, which is a scalar between 0 and 1, we arrange these N initial opinions as an $N \times 1$ column vector $x(0)$. Agent i's opinion in the next period is assumed to be a weighted average of the initial opinions held by agent i's contacts:

$$x_i(1) = \sum_{j=1}^{N} w_{ij} x_j(0). \tag{7.1}$$

Having set the vector of initial opinions $x(0)$ and the direct influence matrix W, we can obtain the long-run opinions through iteration. For instance, each agent's new opinion can be determined by computing a weighted average of the previous opinions using weights given in each row of W. In general, letting $x(t)$ denote the vector of opinions of N agents in period t, the opinion-revising dynamic process is given by the discrete-time dynamics:

$$x(t+1) = Wx(t). \tag{7.2}$$

In the case where the dynamical changes of the opinions of all agents in eqn. (7.2) converge:

$$lim_{t\to\infty} x(t) = lim_{t\to\infty} W^t x(0) = W^* x(0). \tag{7.3}$$

Following Friedkin and Johnsen (1999), influence structure from and to agent i is shown in Figure 7.1 and it is useful to distinguish *direct influence* from *total influence*. Agent j is said to influence agent i *directly* if $w_{ij} > 0$ and agent j is said to influence agent i *indirectly* if $w_{ij} = 0$, but there exists some t such that the (i,j)-th element of W^t is positive. Direct influence is reflected by eqn. (7.2), and the direct influence matrix W maps the previous period of opinions into the next period of opinions. In contrast, the mutual influence in the long run or *total influence* is reflected by:

$$lim_{t\to\infty} W^t = W^*. \tag{7.4}$$

The total influence matrix W^* maps initial opinions $x(0)$ into $W^* x(0)$, which constitutes the long-run opinions.

Each element w_{ij} of W indicates that agent i is influenced by agent j or represents the weight or trust that agent i places in the current opinion of agent j in forming his/her opinion. Therefore, the direct influence matrix W can be interpreted from the point

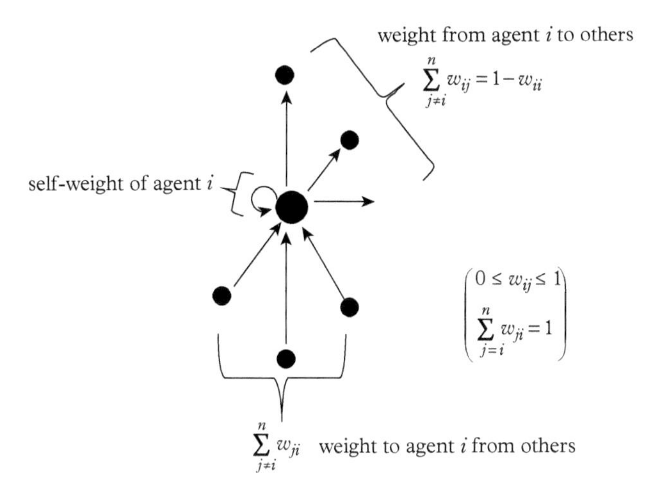

Figure 7.1 *The influence structure from and to agent i*

of view of the agent who shows respect or from the point of view of the person who influences those who ask for advice. If we denote the i-th row vectors of W as $u_i = \{w_{i1}, w_{i2}, \ldots, w_{iN}\}$, it constitutes the vector of weights that agent i gives to everyone in the network, and thus, it consists of the influence of the network on agent i. Similarly, if we denote the i-th column vectors of W as $v_i^T = \{w_{1i}, w_{2i}, \ldots, w_{Ni}\}$, it constitutes the vector of weights given to agent i by every other agent in the network and thus it consists of the influence that agent i has on all agents in the network. If the weight matrix W gives the transition probabilities that an individual will adopt a different view in the new period then an agent will maximize the informational content of his/her deliberation by taking the weighted average over the other agents' opinions.

If every agent influences every other agent, the influence matrix structure is said to be *irreducible*. In order to guarantee the convergence of eqn. (7.2), we assume that the direct influence matrix W is a row or column stochastic matrix.

Case 1: W is a row-stochastic matrix: $(\sum_{j=1}^{N} w_{ij} = 1, 1 \leq i \leq N)$.
If the direct influence matrix W is a row-stochastic matrix (the sum of the elements of each row vector is one), there exists the left eigenvector $\pi = \{\pi_1, \pi_2, \ldots, \pi_N\}$, which satisfies:

$$\pi = \pi W. \tag{7.5}$$

In this case, the power of the influence matrix W^t converges as:

$$lim_{t \to \infty} W^t = \begin{pmatrix} \pi_1, \pi_2, \ldots, \pi_N \\ \pi_1, \pi_2, \ldots, \pi_N \\ \\ \pi_1, \pi_2, \ldots, \pi_N \end{pmatrix}, \tag{7.6}$$

i.e., every row of the total influence matrix W^* equals π, the left eigenvector of W. From eqn. (7.2), the opinions of all agents in the long run converge to a common value, the weighted initial opinions of all agents:

$$x_1(t) = x_2(t) = \ldots = x_i(t) = \ldots = x_N(t) = \sum_{i=1}^{N} \pi_i x_i(0) \quad (\textit{for } t \to \infty). \tag{7.7}$$

Therefore, if the influence matrix is a row-stochastic matrix, agent i influences all other agents to the same degree, π_i, $(1 \leq i \leq N)$.

Case 2: W is a column-stochastic matrix: $(\sum_{i=1}^{N} w_{ij} = 1,\ 1 \leq j \leq N)$.
If the influence matrix W is a column-stochastic matrix (the sum of the elements of each column vector is one), there exists the right eigenvector $\lambda = \{\lambda_1, \lambda_2, \ldots, \lambda_N\}$ satisfying:

$$\lambda = W\lambda. \tag{7.8}$$

The power of the direct influence matrix W^t converges:

$$lim_{t\to\infty} W^t = \begin{pmatrix} \lambda_1, \ \lambda_1, \ldots, \ \lambda_1 \\ \lambda_2, \ \lambda_2, \ldots, \ \lambda_2 \\ \\ \\ \lambda_N, \ \lambda_N, \ldots, \ \lambda_N \end{pmatrix}. \tag{7.9}$$

From eqn. (7.2) and eqn. (7.3), the opinion of each agent converges to:

$$x_i(t) = \lambda_i \sum_{i=1}^{N} x_i(0), \quad 1 \leq i \leq N \quad (\textit{for } t \to \infty). \tag{7.10}$$

In this case, the long-run opinions of each agent differ from each other.

Case 3: W is a double stochastic matrix: $(\sum_{j=1}^{N} w_{ij} = 1,\ \sum_{i=1}^{N} w_{ij} = 1,\ 1 \leq i,j \leq N)$.
If the direct influence matrix W is a *double stochastic matrix*, i.e., it is both a row- and column-stochastic matrix, the power of the direct influence matrix W^t converges to:

$$lim_{t\to\infty} W^t = \begin{pmatrix} 1/N, \ 1/N, \ldots, \ 1/N \\ 1/N, \ 1/N, \ldots, \ 1/N \\ \\ \\ 1/N, \ 1/N, \ldots, \ 1/N \end{pmatrix}. \tag{7.11}$$

From eqn. (7.2) and eqn. (7.3), the opinions of all agents converge to the average of the initial opinions as:

$$x_i(t) = \sum_{i=1}^{N} x_i(0)/N, \quad 1 \leq i \leq N \ \ (\textit{for } t \to \infty). \tag{7.12}$$

In this case, each agent gives and receives the same level of influence from all other agents.

By setting the vector of initial opinions, we can obtain the long-run opinions through iteration. We can determine the long-run opinion of each agent by computing a weighted average or the average of the initial opinions using the weights given in the total influence (W^*) matrix. Intuitively, these weights do not reflect the proximate direct influences on agents, but rather the total influences. That is, every agent's long-run opinion is determined by the same weighted average of the initial opinions. To illustrate, consider the following three examples drawn from DeGroot's paper (1974).

Example 7.1

Consider a trivial interaction matrix between two agents in:

$$W = \begin{bmatrix} 1 & 0 \\ 0 & 1 \end{bmatrix} \quad W^* = \begin{bmatrix} 1 & 0 \\ 0 & 1 \end{bmatrix}.$$

The two agents do not communicate with each other. Then, W^t retains the same values and it converges. However, $\pi W = \pi$ is true for any distribution π and the stationary distribution is not unique. The opinion dynamics in this example is such that each agent always retains his/her initial opinion.

Similarly, there are two groups of agents who communicate only with the agents of the same group, and do not communicate with agents in the other group:

$$W = \begin{bmatrix} W_A & 0 \\ 0 & W_B \end{bmatrix}.$$

The opinion dynamics in this example is such that each agent's opinion in the same group is mixed only within the same group.

Example 7.2

Let us consider the interaction matrix of two agents:

$$W = \begin{bmatrix} 0 & 1 \\ 1 & 0 \end{bmatrix}.$$

continued

Example 7.2 *continued*

The powers of W periodically change, and we have:

$$W^2 = \begin{bmatrix} 1 & 0 \\ 0 & 1 \end{bmatrix} \quad W^3 = \begin{bmatrix} 0 & 1 \\ 1 & 0 \end{bmatrix} \quad W^4 = \begin{bmatrix} 1 & 0 \\ 0 & 1 \end{bmatrix}.$$

The stationary distribution is obtained by solving $\pi W = \pi$, which yields $\pi = (1/2, 1/2)$. However, the opinion dynamics does not converge to this stationary distribution starting from any initial condition. The reason that W^t does not converge to the stationary distribution is that the state changes periodically and the opinions of both agents oscillate.

Example 7.3

Motivated by the above example, the following definitions lead to the classification of the interaction matrix being aperiodic:

$$W = \begin{bmatrix} 1/2 & 1/2 & 0 \\ 1/4 & 3/4 & 0 \\ 1/3 & 1/3 & 1/3 \end{bmatrix}.$$

Agent 1 assigns equal weight to his/her own opinion and that of agent 2, but she does not assign weight to agent 3. Agent 2 assigns a greater weight to his/her own opinion and the remaining weight to agent 1, but does not assign weight to agent 3. Agent 3 assigns equal weight to the three opinions, including his/her own opinion.

W^t converges to the stationary distribution:

$$W^* = \begin{bmatrix} 1/3 & 2/3 & 0 \\ 1/3 & 2/3 & 0 \\ 1/3 & 2/3 & 0 \end{bmatrix}.$$

The opinion of each agent converges to the same value:

$$lim_{t \to \infty} x_i(t) = (1/3)x_1(0) + (2/3)x_2(0), \quad 1 \le i \le 3, \tag{7.13}$$

but the opinion of agent 3 is not reflected in the final opinion.

Example 7.4

Consider a set of the networked agents, where each agent i has the degree k_i. Agent i assigns the equal weight $1/k_i$ to his/her connected agents. We define interaction matrix $A = \{a_{ij}\}$, where each element a_{ij} has the binary value $a_{ij}(= a_{ji}) = 1$, if agent i is connected to agent j; otherwise, $a_{ij}(= a_{ji}) = 0$. We define the direct influence matrix W as $W = D^{-1}A$, where $D = (k_i)$ is the $N \times N$ diagonal matrix with each diagonal element k_i. In this case, the left eigenvector satisfying $\pi W = \pi$ is obtained as:

$$\pi_i = k_i / \sum_{i=1}^{N} k_i, \quad 1 \le i \le N. \tag{7.14}$$

The opinion of each agent converges to the same value:

$$lim_{t \to \infty} x_i(t) = \sum_{i=1}^{N} \pi_i x_i(0) = \sum_{i=1}^{N} (k_i/K) x_i(0), \; K = \sum_{i=1}^{N} k_i, \quad 1 \le i \le N. \tag{7.15}$$

As an example, we consider the star network shown in Figure 7.2. The degree of the center agent is $k_i = N - 1$ and that of all other agents $k_i = 1$, $2 \le i \le N$. Each element of the left eigenvector is given as:

$$\begin{aligned} \pi_1 &= k_1/K = 1/2 \\ \pi_i &= k_i/K = 1/2(N-1), \quad 2 \le i \le N. \end{aligned} \tag{7.16}$$

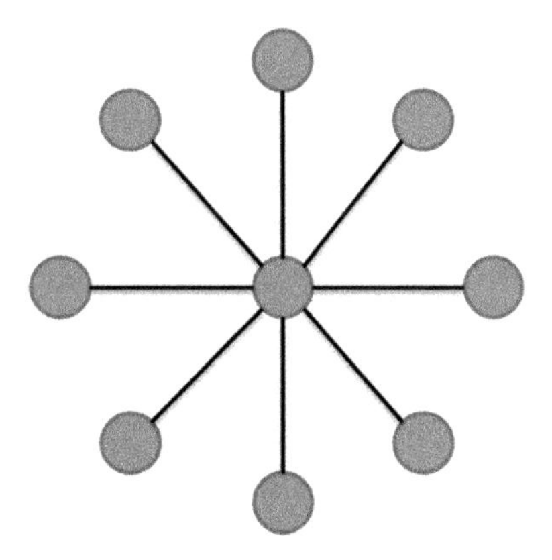

Figure 7.2 *A star network with N agents and the hub agent 1 in the center has the degree N – 1, and all other periphery agents have the degree 1*

continued

Example 7.4 *continued*

The long-run opinions of all agents are the same:

$$lim_{t\to\infty}x_i(t) = (1/2)x_1(0) + \sum_{i=2}^{N} x_i(0)/2(N-1), \quad 1 \leq i \leq N. \tag{7.17}$$

In this case, the initial opinion of the hub agent in the center is strongly reflected in the long-run opinions of all agents.

Example 7.5

Consider again a set of networked agents. In this example, each agent i receives the equal weight $1/k_i$ from his/her connected agents. In this case, the direct influence matrix W is defined as $W = AD^{-1}$. The right eigenvector satisfying $W\lambda = \lambda$ is also obtained as:

$$\lambda_i = k_i / \sum_{i=1}^{N} k_i, \quad 1 \leq i \leq N. \tag{7.18}$$

In this case, the opinion of each agent i converges to a different value:

$$lim_{t\to\infty}x_i(t) = \lambda_i \sum_{i=1}^{N} x_i(0), \quad 1 \leq i \leq N. \tag{7.19}$$

If we consider the star network shown in Figure 7.2 with the direct influence matrix $W = AD^{-1}$, each element of the right eigenvector is also given as:

$$\begin{aligned} \lambda_1 &= k_1/K = 1/2 \\ \lambda_i &= k_i/K = 1/2(N-1), \ 2 \leq i \leq N. \end{aligned} \tag{7.20}$$

The long-run opinion of each agent i converges to a different value:

$$\begin{aligned} lim_{t\to\infty}x_1(t) &= \sum\nolimits_{i=1}^{N} x_i(0)/2 \\ lim_{t\to\infty}x_i(t) &= \sum\nolimits_{i=1}^{N} x_i(0)/2(N-1), \ 2 \leq i \leq N. \end{aligned} \tag{7.21}$$

Therefore, the opinion value of the hub agent in the center is very high, but that of all other agents is low.

Friedkin and Johnsen (1999) proposed an attractive model quantifying how each agent is open to the influence of the other agents rather than anchored to his/her initial opinion. More precisely, the opinion dynamics with the persistence of initial opinions is given by:

$$x(t+1) = \alpha Wx(t) + (1-\alpha)x(0), \tag{7.22}$$

where $1-\alpha$ is the level of the persistence of initial opinion and α is a coefficient of social influence between 0 and 1. In the basic model in eqn. (7.2), the coefficient α equals 1. Friedkin and Johnsen generalized the model so that the initial opinion of each agent may have a permanent effect on his/her long-run opinion. It is straightforward to solve for the long-run opinion vector, $x(\infty)$. By induction, we obtain the general formula:

$$x(t+1) = \left(\alpha^{t+1} W^{t+1} + (1-\alpha)\sum_{\pi=0}^{t} \alpha^{\pi} W^{\pi}\right) x(0). \tag{7.23}$$

Therefore, the long-run opinion of all agents can be obtained as:

$$x(\infty) = \alpha Wx(\infty) + (1-\alpha)x(0). \tag{7.24}$$

We can rewrite eqn. (7.24) as:

$$x(\infty) = (1-\alpha)(I-\alpha W)^{-1}x(0). \tag{7.25}$$

Thus, $(1-\alpha)(I-\alpha W)^{-1}$ is the total influence matrix for the opinion dynamics with the persistence of initial opinions in eqn. (7.22). In the basic model with $\alpha = 1$, we saw that all opinions converge; however, this is no longer true in the generalized model with $\alpha < 1$.

7.3 Consensus Formation in Networks

Consensus means reaching an agreement regarding a certain item of interest that depends on the states of all agents. The history of consensus problems in computer science is long and forms the foundation of the field of distributed computing (Boyd et al., 2006; Kar et al., 2008; Olshevsky and Tsitsiklis, 2011). The formal study of consensus problems in groups of experts originated in the field of statistics (DeGroot, 1974). The original idea was an aggregation of information with uncertainty obtained from multiple experts or sensors.

In this section, we discuss consensus dynamics in agent networks. Agents constantly interact with each other in different manners and for different purposes. Somehow these agent interactions produce some coherence at the collective level. It is important for agents to act in coordination with the other agents. To achieve coordination, individual agents do not need to share complete information. The agents involved may have a very limited view of some part of the whole world, but their opinions are coordinated to a large degree and produce a collective outcome. This relies on a small set of micro-prerequisites leading to full emergence of stationary outcomes. The outcome may refer to consensus, polarization, or fragmentation of opinions, according to the structure of the agent network.

A consensus algorithm is an interaction rule that specifies the information exchange between agents and all their neighbors on the network. In order to achieve a certain convergence and robustness in dynamical environment change, an appropriate distributed algorithm must be designed. A consensus algorithm is an interactive method that satisfies this need by providing the group with a common coordination variable. It is useful to observe the similarity between consensus algorithms, where the opinions of all agents converge to a prescribed function, typically the average of the initial opinions, and agreement algorithms, which are used for coordinating the states of all agents where the states of all agents converge to a common value. The consensus problem requires a distributed algorithm that the agents can use to reach agreement on a common value, starting from an initial situation where the agents have different values. Distributed algorithms that solve the consensus problem provide the means by which networks of agents can be coordinated, although each agent may have access to different local information.

Distributed algorithms for solving consensus problems are often useful as subroutines in more complicated network coordination problems and they have historically appeared in many diverse areas. For instance, the design of sensor networks capable of reaching a consensus on a globally optimal decision without requiring a fusion center is a problem that has received considerable attention. The theoretical framework for posing and solving consensus problems for networked systems was well surveyed by Olfati-Saber et al. (2007). Hines and Talukdar (2007) provided general strategies for developing distributed controls of power networks. The autonomous agent network that they used combined distributed control with a repetitive procedure that combined the advantages of both long-term planning and reactive control. At the beginning of each repetition, the state of the system to be controlled is measured. For this reason, the control of many complex systems is distributed to many autonomous agents. These agents usually operate with only local information and follow simple risk sharing rules. Because in general an agent has incomplete information, it can, at best, make locally correct decisions, which can be globally wrong. This is the general challenge of designing autonomous agent networks: to design the agents such that locally correct decisions are simultaneously globally correct.

The level of stability of a coordinated state is a measure of the system's ability to yield a coherent response and to distribute information efficiently among agents. The conditions for achieving consensus ultimately depend on the network, which specifies the manner in which agents interact. The network topology has a major impact on the convergence of distributed consensus algorithms, namely, the distributed consensus algorithm converges much faster in a certain network topology than in other topologies of connectivity patterns.

The analysis of consensus problems relies heavily on matrix theory and spectral graph theory. The interaction topology of an agent network is represented using the adjacency matrix $A = (a_{ij})$ of the underlying undirected graph $G = (V, E)$ with the set of

N agents, V, and the set of edges, E. The neighbors of agent i to whom he/she is connected are denoted as N_i. Each agent i has a real-valued initial opinion (or state) $x_i(0)$. The consensus problem asks for a distributed algorithm that the agents can use to compute the average of their opinions. Each agent knows only the latest opinions of the connected agents. The opinion of agent i at time $t + 1$ is updated based on the proportional differences to his/her neighbor agents, defined as:

$$x_i(t+1) = w_{ii}x_i(t) + \sum_{j \in N(i)} w_{ij}x_j(t) = x_i(t) + \sum_{j \in N(i)} w_{ij}(x_j(t) - x_i(t)), \tag{7.26}$$

where $w_{ii} + \sum_{j \in N(i)} w_{ij} = 1$.

Eqn. (7.26) has the preserving property and the sum of the opinion values is the same over time:

$$\sum_{j=1}^{N} x_j(t+1) = \sum_{j=1}^{N} x_j(t), \quad t = 0, 1, 2, \ldots. \tag{7.27}$$

We specify the weights of agent i on other agents including himself or herself as:

$$w_{ii} = 1 - \alpha k_i, \quad w_{ij} = \alpha a_{ij}, \quad j \neq i, \tag{7.28}$$

where k_i is the cardinality of N_i, the number of neighbors of agent i (or the degree of agent i). Each agent i places a high priority and the same level of reliability on each neighbor. In this case, eqn. (7.26) has the form:

$$\begin{aligned} x_i(t+1) &= (1-\alpha)x_i(t) + \alpha \textstyle\sum_{j \in N(i)} a_{ij}(x_j(t) - x_i(t)) \\ &= (1 - \alpha k_i)x_i(t) + \alpha \textstyle\sum_{j \in N(i)} x_j(t). \end{aligned} \tag{7.29}$$

Essentially, the opinion adjustment proceeds as follows: each agent i compares his/her current opinion with each of his/her neighbors' opinions. This process is repeated until all agents detect their opinions to be locally balanced.

In matrix form, we can describe eqn. (7.29) as:

$$x(t+1) = ((I - \alpha(D - A))x(t), \tag{7.30}$$

where $D = (k_i)$ is the diagonal matrix with each diagonal element, $k_i, i = 1, 2, \ldots, N$. In this formulation, the weight matrix W in eqn. (7.2) is given as:

$$W = I - \alpha(D - A). \tag{7.31}$$

From the formulation in eqn. (7.2), we obtain a solution to the averaging consensus problem of eqn. (7.30) if and only if each element of the left eigenvector π with eigenvalue equal to 1 satisfies $\pi = 1/N$. This requirement, i.e., $\pi = (1/N, \ldots, 1/N)$ is the left eigenvector of W in eqn. (7.31) with an eigenvalue equal to 1, translates to the following property: W needs to be doubly stochastic. If the underlying graph $G = (V, E)$ is an undirected network, the adjacency matrix $A = (a_{ij})$ is symmetric. Therefore, the weight matrix W in eqn. (7.31) is also symmetric, i.e., $W = W^T$, and W is a doubly stochastic matrix. In this case, the power of W in the long-run converges to:

$$lim_{t\to\infty} W^t = W^*. \tag{7.32}$$

Under the distributed algorithm in eqn. (7.30), the opinion adjustment is *egalitarian*, i.e., each agent obtains in equilibrium a common value amounting to the average of all initial opinions:

$$x_1(t) = x_2(t) = \ldots = x_i(t) = \ldots = x_N(t) = \sum_{i=1}^{N} x_i(0)/N \ \ (\textit{for } t \to \infty). \tag{7.33}$$

Now, we consider another opinion update scheme. We now define the weights placed by agent i on other agents including himself/herself as:

$$w_{ii} = 1 - \alpha, \quad w_{ij} = \alpha a_{ij}/k_i, \ \ j \neq i. \tag{7.34}$$

Agent i places a high priority on his/her own opinion and places a low priority, but the same level of reliability, on the other connected agents. The opinion of agent i at time $t + 1$ is updated based on the average opinion of his/her neighbors, which is defined as:

$$x_i(t+1) = (1 - \alpha)x_i(t) + \alpha \sum_{j\in N(i)} x_j(t)/k_i. \tag{7.35}$$

In matrix form, we can describe this as:

$$x(t+1) = ((1 - \alpha)I + \alpha D^{-1}A)x(t). \tag{7.36}$$

In this formulation, the weight matrix W in eqn. (7.2) is given as:

$$W = (1 - \alpha)I + \alpha D^{-1}A. \tag{7.37}$$

This weight matrix is a stochastic matrix (all of its row-sums are 1), but it is not symmetric. In this case, the power of W in the long-run converges to the steady-state matrix with each row vector π, which is the left-eigenvector of W. Each element of π is:

$$\pi_i = k_i / \sum_{i=1}^{N} k_i, \ \ 1 \leq i \leq N. \tag{7.38}$$

Under the distributed consensus algorithm in eqn. (7.37), each agent obtains in equilibrium a common value, which is the weighted average of the initial opinions, i.e.,

$$x_1(t) = x_2(t) = \ldots = x_i(t) = \ldots = x_N(t) = \sum_{i=1}^{N} \left(k_i \Big/ \sum_{i=1}^{N} k_i \right) x_i(0) \quad (t \to \infty). \tag{7.39}$$

We refer to the distributed consensus algorithm in eqn. (7.30) as the *average-consensus algorithm*, and that in eqn. (7.36) as the *weighted average consensus algorithm*. We can use consensus dynamics to solve complex problems collectively. Human interactions give rise to the formation of different kinds of opinions in a society. The study of opinion dynamics and consensus formations sheds light on some issues common to many disciplines. Consensus dynamics and associated social structure lead to collective decision-making. Opinion formation is a kind of process of collective intelligence evolving from the integrative tendencies of social influence with the disintegrative effects of individualization.

The *wisdom of crowds* refers to the phenomenon in which the collective knowledge of a community is greater than the knowledge of any one individual. According to this notion, consensus problems are closely related to a variety of issues, such as collective behavior in nature, the interaction among agents in robot control, and the building of efficient wireless sensor networks. However, the design of appropriate networks is complex and may pose a multi-constraint and multi-criterion optimization problem. Surowiecki (2004) explored an idea that has profound implications: a large collection of people are smarter than an elite few, regardless of their superior intelligence and ability to solve problems, foster innovation, reach wise decisions, or even predict the future. He explained that the wisdom of crowds emerges only under the right conditions: diversity, independence, decentralization, and aggregation. His intuitive notion, which counters the notion of the madness of the crowd as traditionally understood, suggests new insights into the issue of the manner in which our complex social and economic activities should be organized.

In contrast, Lorentz et al. (2011) demonstrated by experimental evidence that even mild social influence can undermine the wisdom of the crowd effect in simple estimation tasks. In their experiment, subjects could reconsider their response to factual questions after having received average or full information of the responses of other subjects. The authors compared the subjects' convergence of estimates and improvements in accuracy over five consecutive estimation periods. They included a control condition, in which no information about others' responses was provided. Although the groups were initially wise, knowledge about the estimates of others narrowed the diversity of their opinions to such an extent that it undermined the wisdom of crowd effect. In particular, the social influence effect diminished the diversity of the crowd without ameliorating its collective error.

Sunstein and Kuran (2007) defined *availability cascade* as a self-reinforcing process of collective belief formation by which an expressed perception triggers a chain reaction that gives the perception increasing plausibility through its rising availability in public discourse. The driving mechanism involves a combination of informational and reputational motives. Individuals endorse the perception partly by learning from the apparent beliefs of others and partly by distorting their public responses in the interest of maintaining social acceptance. Availability campaigns, for instance, may yield social benefits, but sometimes they do harm, which suggests a need for safeguards. Therefore, we may need some methodologies to alleviate potential hazards of cascade dynamics through social media.

The consensus problem is also related to *synchronization*. Synchronization is the most prominent example of coherent behavior, which is strongly affected by the network structure. Examples of synchronization can be found in a wide range of phenomena, such as the firing of neurons, laser cascades, chemical reactions, and opinion formation. The range of stability of a synchronized state is a measure of the system's ability to yield a coherent response and to distribute information efficiently among agents, while a loss of stability fosters pattern formation. However, in many situations the formation of a coherent state is not pleasant and should be mitigated. For example, the onset of synchronization can be the root of traffic congestion in networks and the collapse of constructions. Louzada et al. (2012) proposed the use of contrarians to suppress *undesired synchronization*. They performed a comparative study of different strategies, requiring either local or total knowledge, and showed that the most efficient one requires solely local information. Their results also revealed that, even when the distribution of neighboring interactions is narrow, significant improvement is observed when contrarians sit at the highly connected elements.

7.4 Optimal Networks for Fast and Slow Consensus

The analysis of consensus problems relies heavily on the properties of the graph modeling of the interaction among agents. The conditions for achieving a consensus ultimately depend on the spectral conditions of an agent network. The network topology has a major impact on the convergence of distributed consensus algorithms, namely, the distributed consensus algorithm converges much faster in a certain network topology than in other topology connectivity patterns.

We investigate the effect of the network structure on the speed of consensus formation. The findings can help to identify specific topological characteristics, such as density and the critical agents that affect the consensus dynamics. One basic question is how to design the weight matrix W defined by eqn. (7.31) so that the averaging consensus algorithm converges very quickly. The agent network topology is represented by the $N \times N$ adjacency matrix $A = \{a_{ij}\}$, where each element a_{ij} has the binary value. Consider the average consensus dynamics of N agents, which is the continuous-time counterpart of the discrete-time model in eqn. (7.31):

$$\dot{x}_i(t) = \sum_{j \in N(i)} a_{ij}(x_j(t) - x_i(t)), \tag{7.40}$$

where $x(t)$ represents a column vector consisting of $x_i(t), 1 \leq i \leq N$, which represents the state (opinion) of each agent i. We also denote the neighbors of agent i by $N(i)$. Here, reaching a consensus means asymptotically converging to the same state by means of an agreement characterized by eqn. (7.40):

$$x_1(t) = x_2(t) = \ldots = x_i(t) = \ldots = x_N(t) = c, \ (for\ t \to \infty). \tag{7.41}$$

The level of stability of a coordinated state is a measure of the system's ability to yield a coherent response and to distribute information efficiently among agents. Assuming that the adjacency matrix A is symmetric ($a_{ij} = a_{ji}$ for all i, j), the continuous dynamics in eqn. (7.40) converge to the average of the initial states of all agents:

$$c = \frac{1}{N}\sum_{i=1}^{N} x_i(0). \tag{7.42}$$

The dynamics of eqn. (7.40) can be expressed as:

$$\dot{x}(t) = -Lx(t), \tag{7.43}$$

where L is the *graph Laplacian* defined as:

$$L = D - A, \tag{7.44}$$

where $D = (k_i)$ is the diagonal matrix with elements $k_i = \sum_{j \neq j} a_{ij}$.

Similarly, the continuous-time counter part of the discrete-time weighted average consensus dynamics of eqn. (7.37) is:

$$\dot{x}(t) = -D^{-1}Lx(t). \tag{7.45}$$

According to the definition of *graph Laplacian* in eqn. (7.44), all row-sums of L are zeros, and therefore, it always has a zero eigenvalue, which corresponds to the unit-eigenvector $1 = (1, 1, \ldots, 1)^T$. If the agent network is connected without isolated agents, an equilibrium of the dynamic system in eqn. (7.43), which is a state where all agents agree, is globally exponentially stable. This implies that, irrespective of the initial value of the state of each agent, all agents reach the consensus value, which is equal to the average of the initial state values. In general, finding the average value of the initial conditions of the agents is more complicated, and the implication for the calculation of the average is important. For example, if a network has $N = 10^8$ agents and each agent can communicate only with $log_{10}N = 8$ neighbors, the average value is reached starting from any initial condition. The consensus algorithm provides a systematic mechanism for computing the average quickly in such a large network.

The solution of eqn. (7.43) can be obtained analytically as:

$$x(t) = exp(-Lt)x(0). \tag{7.46}$$

This equation can be rewritten using the set of eigenvalues of the Laplacian matrix L as:

$$x(t) = z \begin{pmatrix} e^{-\lambda_1(L)t} & & & \\ & e^{-\lambda_2(L)t} & & \\ & & \cdots & \\ & & & e^{-\lambda_N(L)t} \end{pmatrix} z^T x(0), \tag{7.47}$$

where $\lambda_i(L), 1 \leq i \leq N$, represents the i-th minimum eigenvalue of the graph Laplacian L and z is the set of the corresponding eigenvector. When time t becomes sufficiently large, eqn. (7.46) converges to:

$$x(\infty) = z \begin{pmatrix} 1 & & & \\ & 0 & & \\ & & \cdots & \\ & & & 0 \end{pmatrix} z^T x(0) = \left(\frac{1}{N}\right) 11^T x(0). \tag{7.48}$$

It should be noted that the eigenvector corresponding to eigenvalue 0 of the matrix L is normalized as $1/\sqrt{N}(1, 1, \ldots, 1)^T$. The speed of convergence of eqn. (7.47) is determined mainly by the second minimum eigenvalue $\lambda_2(L)$, which is also defined as *algebraic connectivity*. Therefore, the network with a larger algebraic connectivity $\lambda_2(L)$ achieves consensus more quickly.

The discrete-time counterpart of the continuous-time consensus dynamics in eqn. (7.31) is:

$$x(t+1) = (I - \alpha L)x(t) = Wx(t), \tag{7.49}$$

where α is some small constant. The dynamics in eqn. (7.49) converge to:

$$lim_{t\to\infty} W^t = \left(\frac{1}{N}\right) 11^T. \tag{7.50}$$

The necessary and sufficient condition of the convergence to eqn. (7.50) is:

$$\rho(W - (1/N)11^T) < 1, \tag{7.51}$$

where $\rho(W - (1/N)11^T)$ represents the spectrum radius of the matrix $(W - (1/N)11^T)$, which is defined as:

$$\rho(W - (1/N)11^T) = max\{1 - \lambda_2(L), \alpha\lambda_N(L) - 1\}. \tag{7.52}$$

For a faster consensus, the spectrum radius ρ should be smaller, and is minimized at $\alpha = \alpha^*$, which is given as:

$$\alpha^* = \frac{2}{\lambda_2(L) + \lambda_N(L)}. \tag{7.53}$$

The spectrum radius ρ^* of the matrix $(W - (1/N)11^T)$ at $\alpha = \alpha^*$ is:

$$\rho^* = \frac{\lambda_N(L) - \lambda_2(L)}{\lambda_2(L) + \lambda_N(L)} = \frac{1 - R}{1 + R}, \tag{7.54}$$

where $R = \lambda_2(L)/\lambda_N(L)$. This ratio of the second minimum eigenvalue to the largest eigenvalue of the graph Laplacian L is defined as the *eigenvalue ratio*. The network with a larger eigenvalue ratio R has a smaller spectrum radius ρ^*. Therefore, a consensus is achieved more quickly in such a network.

Some important network properties can be derived from the distribution of the eigenvalues of the graph Laplacian. One property is that the number of the zero eigenvalue represents the number of isolated networks. If the network is connected without disconnected agents, except for the minimum eigenvalue of zero, all other $N - 1$ eigenvalues are greater than zero. We also have the following relationship between the eigenvalues of the graph Laplacian (Chung, 1997):

$$\lambda_2(L) \leq \frac{N}{N-1} k_{min} \leq \frac{N}{N-1} k_{max} \leq \lambda_N(L) \leq 2k_{max}, \tag{7.55}$$

where $\lambda_i(L)$ represents i-th minimum eigenvalue of the graph Laplacian, and k_{min} and k_{max} are the minimum and maximum degree of agents, respectively.

The eigenvalue ratio, which determines the convergence speed of consensus, is bounded from the above:

$$R = \frac{\lambda_2(L)}{\lambda_N(L)} \leq \frac{Nk_{min}}{(N-1)2k_{max}} \leq \frac{k_{min}}{k_{max}} \leq 1. \tag{7.56}$$

From eqn. (7.56), the eigenvalue ratio is maximized on the *k-regular* network, and therefore, a consensus can be achieved more quickly in an agent network where each agent has the same degree.

Figure 7.3 shows the distributions of the eigenvalues of the graph Laplacian of some networks. The ring lattice network shown in Figure 7.4 is a kind of k-regular network and each agent has the same degree ($k = 4$). The algebraic connectivity $\lambda_2(N)$ of the ring lattice network is the smallest and the eigenvalue ratio is also the smallest, and $R = 0.0001$. Since the largest eigenvalue $\lambda_N(L)$, the scale-free network, $SF(BA)$, is large, its eigenvalue ratio is small and $R = 0.0091$. The eigenvalue ratio of the random regular network is $R = 0.075$, which is the largest among these networks.

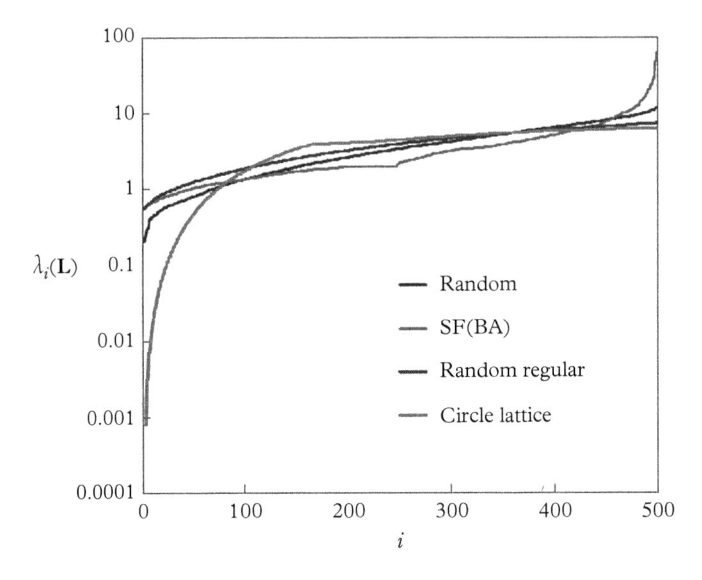

Figure 7.3 *The spectrum of the graph Laplacian. The y-axis represents the i-th minimum eigenvalues in order. Each network has* $N = 500$ *agents and the average degree is* $k = 4$

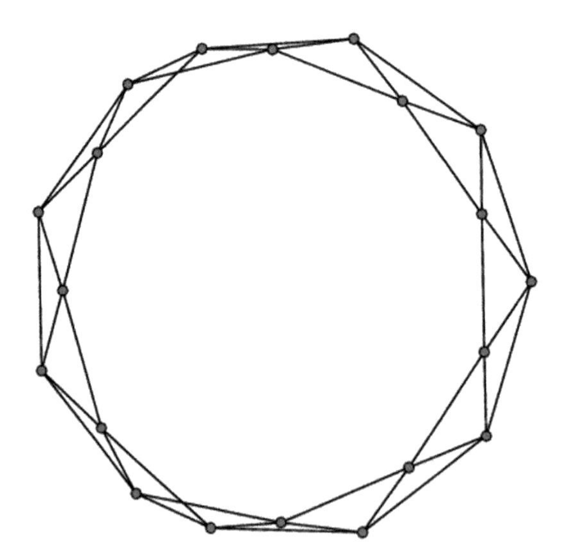

Figure 7.4 *The ring lattice network with the same degree* $k = 4$

A *Ramanujan* graph is a subset of the class of *k-regular graphs* such that the algebraic connectivity, the second smallest Laplacian eigenvalue $\lambda_2(L)$, satisfies the following condition:

$$\lambda_2(L) \geq k - 2\sqrt{k-1}. \tag{7.57}$$

It is known that the fastest consensus is achieved on a Ramanujan graph.

In Figure 7.5, the algebraic connectivity $\lambda_2(L)$ of several networks is compared as the function of the average degree z. In general, the algebraic connectivity $\lambda_2(L)$ increases if the average degree of the network increases. A Ramanujan graph has the largest algebraic connectivity $\lambda_2(L)$ if the average degree is greater than or equal to four. In Figure 7.6, the eigenvalue ratios R of several networks are compared as the function of the average degree z; the Ramanujan graph has the largest eigenvalue ratio R, implying that the fastest consensus is achieved on the Ramanujan graph.

In general, the upper bound of the second smallest algebraic connectivity $\lambda_2(L)$ of any graph Laplacian is given by (Chung, 1997):

$$\lambda_2(L) \leq 2\frac{|E(S, S^c)|}{|S|}, \tag{7.58}$$

where $|S|$ is the total number of nodes in any subset of nodes S satisfying $0 < |S| \leq N/2$, and $|E(S, S^c)|$ is the number of edges between S in the complement set of S.

Let us consider two Ramanujan graphs with 500 nodes and 100 nodes respectively and they are connected by one link, as shown in Figure 7.7. The algebraic connectivity of

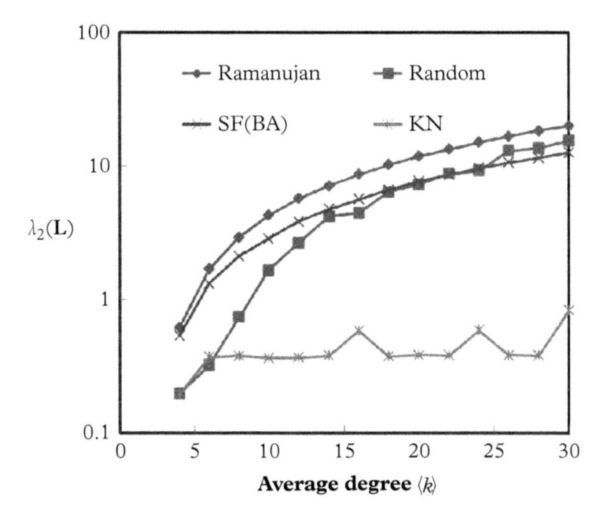

Figure 7.5 *The comparison of the second minimum eigenvalue $\lambda_2(L)$ of several graph Laplacians. Each network consists of N = 500 nodes*

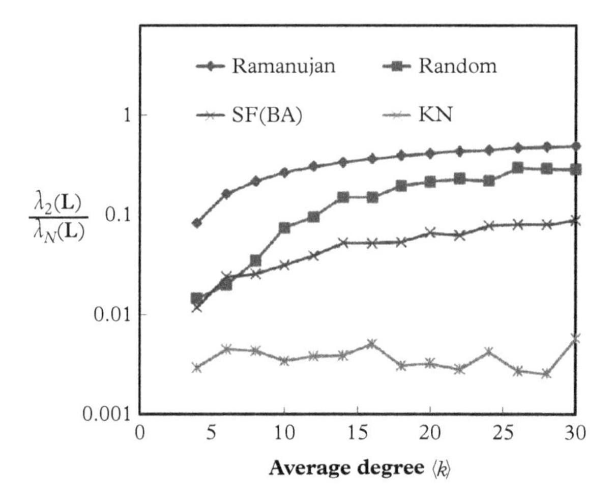

Figure 7.6 *The comparison of the eigenvalue ratio $\lambda_2(L)/\lambda_N(L)$ of several graph Laplacians. Each network consists of $N = 500$ nodes*

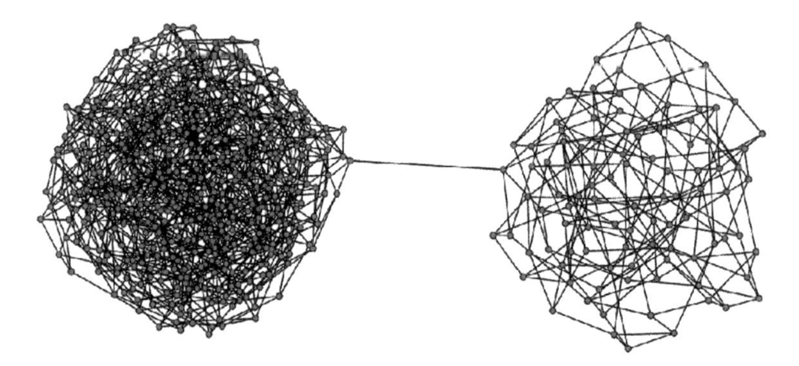

Figure 7.7 *The two Ramanujan graphs are connected by a single link. The Ramanujan graph on the left has $N = 500$ nodes, and the right-hand Ramanujan graph has $N = 100$ nodes. The average degrees of both graphs are the same, and $k = 4$. The algebraic connectivity of the left graph is $\lambda_2(L_{left}) = 0.617$ and that of the right graph is $\lambda_2(L_{right}) = 0.938$. The algebraic connectivity of the interconnected network is $\lambda_2(L) = 0.00698$*

each Ramanujan graph is $\lambda_2(L_{left}) = 0.617$ and $\lambda_2(L_{right}) = 0.938$, respectively. However the algebraic connectivity of the combined network is $\lambda_2(L) = 0.00698$, since $|E(S, S^c)|$ is very small. Although the fastest consensus is achieved in both Ramanujan graphs if they are separated, the consensus is reached very slowly if they are connected by only a few links. Therefore, consensus dynamics is an example that illustrates that the network dynamics are affected by the local topological properties as well as by the global properties of the network.

The network design problems related to achieving faster consensus algorithms have attracted considerable attention. As we have seen, the speed of achieving consensus is determined by the eigenvalue ratio $R = \lambda_2(L)/\lambda_N(L)$ of the graph Laplacian. The fastest consensus can be achieved in a Ramanujan graph, a class of *regular* graph, which satisfies the property of eqn. (7.58). However, for a sparse network with a small average degree, the lower bound of the algebraic connectivity in eqn. (7.58) does not carry any information. For instance, a Ramanujan graph with $z = 2$ is a ring network in which the eigenvalue ratio is very small.

Let us consider a network of a ring with trees, as shown in Figure 7.8. This network has a ring network at the center, and many modularized tree networks are interconnected via the ring network. In Figure 7.9, the eigenvalue ratio R of a Ramanujan graph and that of a heuristically designed network of a ring with trees are compared. The eigenvalue ratio of the Ramanujan graph is smaller than that of a network of a ring with trees. This implies that the consensus can be achieved most quickly in a ring network with trees when the average degree z is restricted to two.

We compared the speed of consensus between a network with a ring–tree topology and a Ramanujan graph. We set the initial state value of each agent as

$$x_i(0) = i, \ (i = 1, 2, \ldots, N). \tag{7.59}$$

When each agent updates her state value according to the rule in eqn. (7.49), the state of each agent converges to a constant value, the average of the initial values of all agents. We compared the time required for achieving consensus. In Figure 7.10, the convergence in two networks is compared. On these sparse networks, the consensus is reached in a ring–tree network much faster than in a Ramanujan graph. On the ring–tree agent

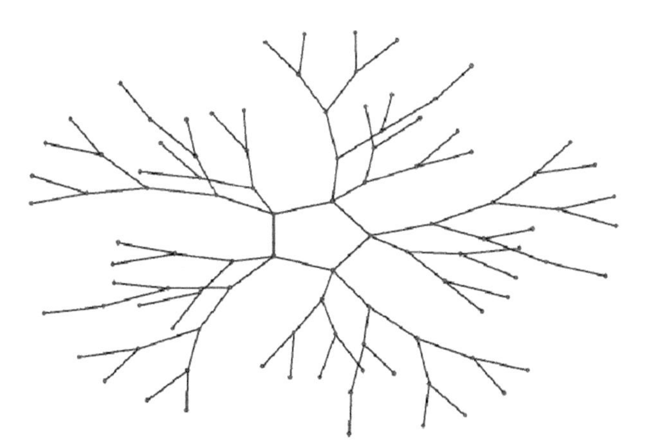

Figure 7.8 *A network of ring with trees: This network has a ring network at the center and many modularized tree networks are interconnected via the ring network. The total number of nodes is $N = 100$ and the average degree is $k = 2$*

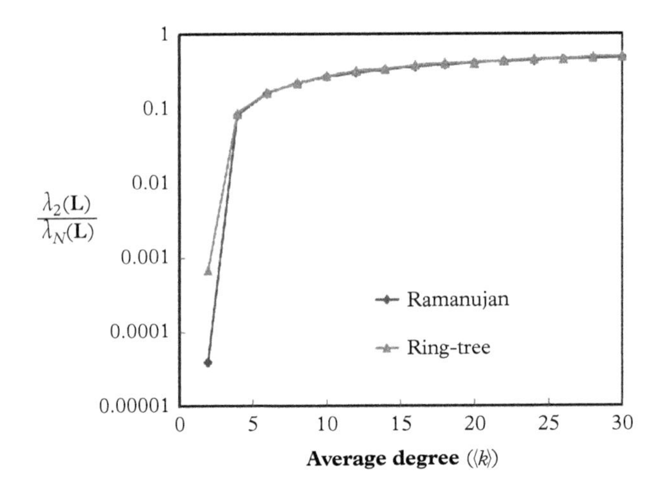

Figure 7.9 *The eigenvalue ratio $R = \lambda_2(L)/\lambda_N(L)$ of the Ramanujan graph and that of a network having a ring with trees, as shown in Figure 7.8*

network, some agents form a tree network as a module, and the global network consists of these modules in tree structures. The consensus is achieved in each module locally, and then the global consensus among modules is achieved via the ring network. Consensus formation is promoted with these two steps.

We now seek the networks on which the consensus formation occurs very slowly. We heuristically designed two combined networks: a clique and a line network, shown in Figure 7.11. The clique has N_M nodes and L_M links. The line network has N_{Line} nodes and M_{Line} links. The total numbers of nodes N and links L of the two combined networks are:

$$\begin{aligned} N &= N_M + N_{Line} \\ L &= L_M + L_{Line} \quad . \\ L_{Line} &= N_{Line} \end{aligned} \tag{7.60}$$

The length of the line network with N_{Line} nodes should meet the following condition:

$$N - N_{Line} - 1 \leq L - N_{Line} \leq {}_{N-N_{Line}}C_2. \tag{7.61}$$

The interconnected network with a clique and a line shown in Figure 7.11 has the smallest algebraic connectivity, implying that the slowest consensus formation is taking place. The consensus formation in this network is shown in Figure 7.12(a). The state value of each agent is plotted over time. It takes more than 100 times more steps than the ring-tree network in Figure 7.8. The changes in the agent states at the beginning are shown in Figure 7.12(b). This figure shows that consensus among agents in a clique

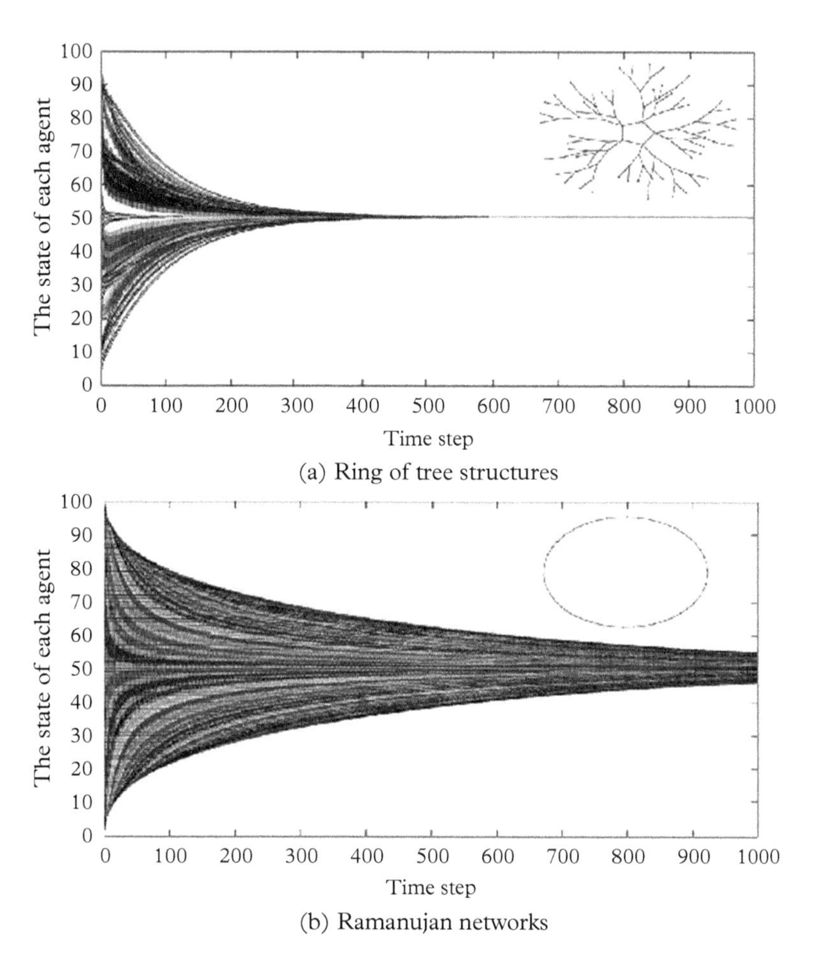

Figure 7.10 *Comparison of the speed of consensus of a network of a ring with trees (a) and Ramanujan network (b). Both networks have $N = 100$ nodes and the average degree is $k = 2$. The exact convergence time is 1200 steps in a network of a ring with trees and over 5000 steps on a Ramanujan graph*

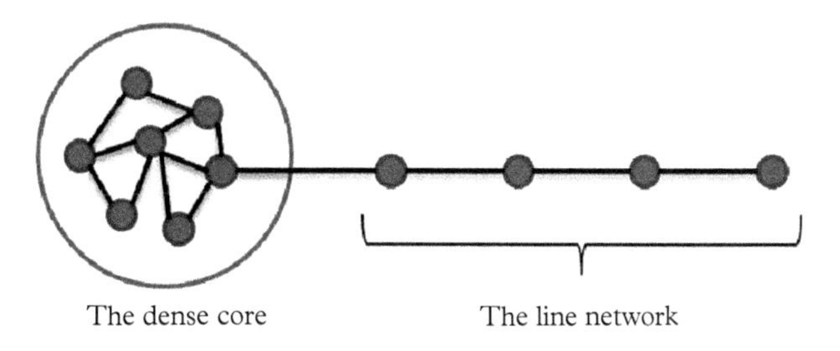

Figure 7.11 *The interconnected network with a clique and a line network*

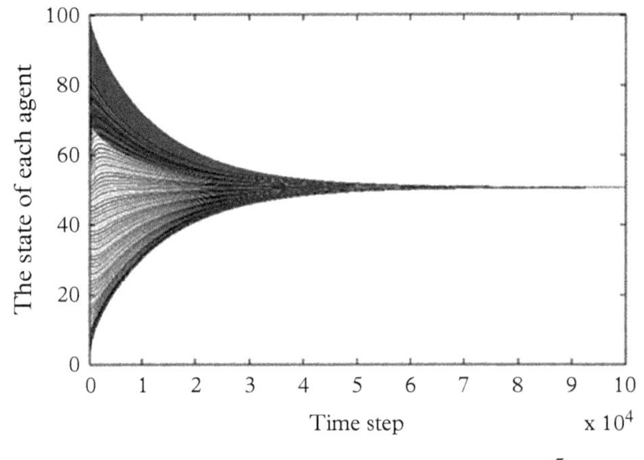

(a) The plot range of time step is [0, 10^5]

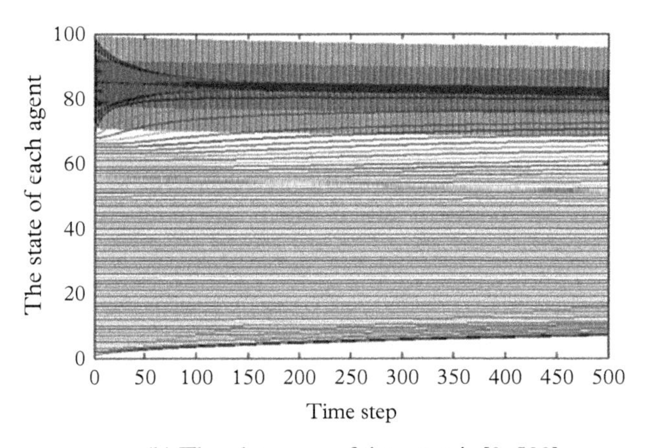

(b) The plot range of time step is [0, 500]

Figure 7.12 *The diagram of consensus dynamics between agents on a clique with a line network with $N = 100$ nodes*

is achieved first, then consensus is achieved including agents located on the line, and finally, global consensus is achieved very slowly.

7.5 Network Intervention with Stubborn Agents

The local interaction rules may involve the simple imitation of a neighbor, an alignment to a local majority, or more involved negotiation processes. The main question in all cases concerns the possibility of the appearance of a global consensus, defined by the fact that all agents' internal states coincide, without the need for any central authority to supervise such behavior. Alternatively, polarized states in which several coexisting states or opinions survive can be obtained, and the parameters of the model drive the transition between consensus and polarization.

In this section, we consider a model for investigating the tension between information aggregation and the spread of biased consensus formation in networks of agents. In our society, people form beliefs on various economic, political, and social issues based on information they receive from others, including friends, neighbors, and coworkers, as well as local leaders, news sources, and political actors. A key trade-off faced by any society is whether this process of information exchange will lead to the formation of more accurate beliefs or to certain systematic biases and spread of misinformation.

Each agent holds an opinion or a belief represented by a scalar. Agents meet pairwise and exchange their beliefs, a situation which is modeled as both agents adopting the average of their pre-meeting beliefs. When all agents engage in this type of information exchange, the society will be able to effectively aggregate the initial beliefs held by all agents. However, there is also the possibility of biased belief, because some of the agents do not change their own belief, or they are forceful and they enforce their beliefs on other agents they meet.

Acemoglu et al. (2013) studied a tractable opinion dynamics model that generates long-run agreements or disagreements with persistent opinion fluctuations. They investigated the conditions under which exchange of information will lead to the spread of biased opinion instead of the aggregation of dispersed opinions. Their model involved continuous opinion dynamics consisting of a society of two types of agents, regular agents and stubborn agents. *Regular agents* update their beliefs according to information that they receive from their neighbors. On the other hand, *stubborn agents* never update their opinions. They may represent leaders, political parties, or media sources attempting to influence the beliefs in the rest of the society. They showed that when a society contains stubborn agents with different opinions, the belief dynamics never lead to consensus, and beliefs in the society almost certainly fail to converge; instead, the belief profile continues to fluctuate.

In recent years, numerous distributed algorithms have been proposed which, when executed by a team of dynamic agents, result in the completion of a joint task. However, for any such algorithm to be practical, one should be able to guarantee that the task is still satisfactorily executed even when agents fail to communicate with others or to perform their designated actions correctly. During the execution of a distributed algorithm, some agents may stop functioning in many ways.

Failure mode 1: An agent may fail by simply ceasing to communicate with other agents.

Failure mode 2: An agent fails by setting his/her internal state value to a constant.

Failure mode 3: An agent alters the control input to set his/her internal state at every time step to an arbitrary value. The sequence of the values can be chosen maliciously so that the other agents are hindered in the pursuit of the cooperative task.

When an agent fails, the control law and the communication law are altered. Therefore, we need to investigate the robustness of any distributed algorithm that is well-suited for general purposes. To evaluate the robustness of a distributed algorithm, we can investigate the manner in which the presence of stubborn agents who never change their opinions interferes with information aggregation.

We quantify the role of the presence of stubborn agents by providing the results for the maximum distance of the consensus value from the benchmark without stubborn agents. Under the assumption that even stubborn agents obtain information from other agents, we investigate the following questions: What type of agent network will foster convergence to a consensus among all other agents except a stubborn agent? In what agent network will the worst outcome be obtained so that the opinion of almost all agents converges to a value close to the opinion of a stubborn agent? How is opinion dynamics affected by the existence of multiple stubborn agents holding completely opposite opinions?

The structure of the graph describing the manner of interactions among agents and the location of the stubborn agents within the agent network shapes the opinion dynamics. Fagnani (2014) showed the analytical framework of consensus dynamics over networks including regular agents and stubborn agents. Consider a symmetric connected graph $G = (V, E)$ with the total number of agents $|V| = N$ and the total number of links $|E| = L$. We divide N agents into $V = S \cup R$, the sets of stubborn agents and regular agents. The agents in S are stubborn agents who do not change their opinions, while those in R are regular agents who modify their opinions over time according to the consensus dynamics. The whole consensus dynamics is described by.

$$x(t+1) = Wx(t). \tag{7.62}$$

If we order the agents in V such that the regular agents in R come first, the interaction matrix W exhibits the block structure:

$$W = \begin{bmatrix} W_1 & W_2 \\ 0 & I \end{bmatrix}. \tag{7.63}$$

Dividing accordingly the opinion vector $x(t)^T = (x_R(t), x_S(t))^T$, we have the dynamics:

$$\begin{aligned} x_R(t+1) &= W_1 x_R(t) + W_2 x_S(t) \\ x_S(t+1) &= x_S(t) \end{aligned}. \tag{7.64}$$

It is worth noting that W_1 is a substochastic matrix, i.e., all row sums are less than or equal to one. Moreover, at least one row exists. the sum of which is strictly less than one, which is the row corresponding to a regular agent connected to a stubborn one. Using the connectivity of the graph G, this easily implies that there exists t such that $(W_1)^t$ has the property that all its rows have a sum strictly less than one, and therefore, the matrix is asymptotically stable. Then, the long-run opinions of regular agents satisfy the relation:

$$x_R(\infty) = W_1 x_R(\infty) + W_2 x_S(0), \tag{7.65}$$

where we denote the convergence of the opinion vector of regular agents as:

$$x_R(t) \to x_R(\infty) \ \ (\textit{for } t \to +\infty). \tag{7.66}$$

From eqn. (7.65), the long-run opinions of regular agents are obtained as:

$$x_R(\infty) = (I - W_1)^{-1} W_2 x_S(0). \tag{7.67}$$

Therefore, the asymptotic opinions of regular agents are obtained as the convex combinations of the opinions of stubborn agents. In particular, if all stubborn agents are of the same opinion or there is only one stubborn agent, then the consensus reached by all agents is the opinion of the stubborn agents. We studied the manner in which consensus is reached if multiple stubborn agents with different initial opinions exist through a simulation.

In Figure 7.13, we observe two cliques given by the sets of agents $\{1,2,3,4\}$ and $\{5,6,7,8\}$. All the agents in the same clique influence each other, but there is no between-clique influence. Further, a single bridge exists between agents 4 and 5. A single communication class exists because of the two bridge agents 4 and 5. Setting the

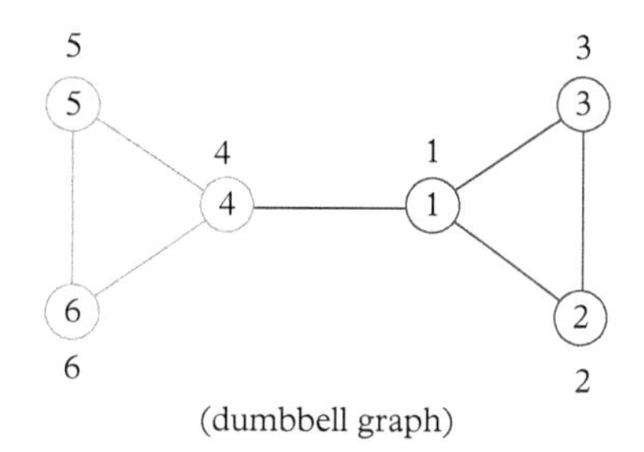

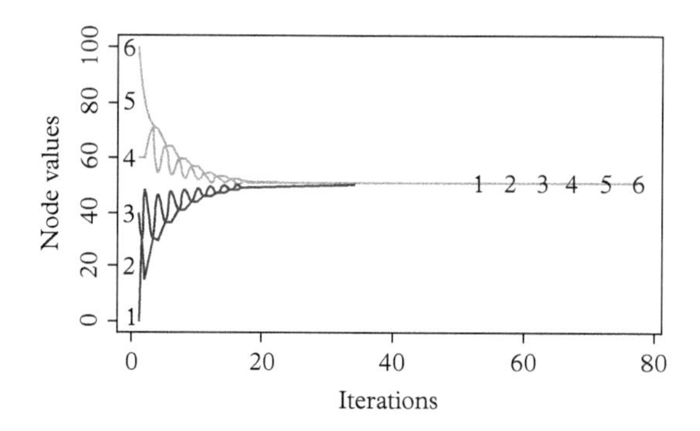

Figure 7.13 *Shows how agents initial beliefs converge when a hub agent (with the highest degree) fails to communicate and retains his/her initial belief*

vector of initial opinions and the social influence parameter, we can obtain the long-run opinions through iteration. We set the initial state value of each agent as:

$$x_i(0) = i - 1, \quad (i = 1, 2, \ldots, N). \tag{7.68}$$

In this example, differences of opinion may persist even when all the agents are influenced by each other. However, we see that the opinions of individuals who hold very similar network positions within the agent network do become more similar over time. It is interesting to note that, while agents 4 and 5 initially hold the same opinion, their opinions diverge over time. Intuitively, agents 4 and 5 are drawn toward the other members of their respective cliques. At the same time, we also observe the convergence to similar opinions within cliques.

We now set agent 1 as a stubborn agent, that is, she never updates his/her state value; all other agents are set as regular agents and update their state value according to the rule in eqn. (7.49). The state of each agent converges to 0, the initial value of the stubborn agent 1, as shown in Figure 7.14. We set two stubborn agents in the sets of agents $\{1, 2, 3\}$ and $\{4, 5, 6\}$: agent 1 with the initial state 0 and agent 6 with the initial state 8; they never update their initial state value as stubborn agents. The state of each regular agent in the set of agents 2, 3 converges to close to the initial values of the stubborn agent 1, and the state of each regular agent in the other set of agents 5, 6 converges to close to the initial values of the stubborn agent 8.

We now set two stubborn agents in the same set of agents $\{1, 2, 3\}$: agent 1 with the initial state 0 and agent 2 with the initial state 8. In this case, the state values of agent 3 and agent 4 in the same clique converge to the convex combination of the initial state values of the two stubborn agents. However, the state value of each agent in the other clique converges to the average of the initial state values of all regular agents in the same clique. This is evidence of a sparse influence between two cliques caused by the two bridge agents 1 and 4 preventing the influence of stubborn agents.

The term *network intervention* describes the process of using agent network data to accelerate the behavioral changes of agents or improve network performance (Valente, 2012). Network interventions are based on accelerating influence, which requires a deeper understanding of the social mechanisms driving behavior change. The science of using networks to accelerate behavior change and improve network performance is still in its infancy. We may consider using some algorithms or methods as network interventions. Consequently, we have many intervention choices at our disposal. The selection of the appropriate network intervention depends on the availability and character of the network data, the perceived characteristics of the behavior. Research is clearly needed to compare different network interventions to determine which are optimal under what circumstances. In the most basic network intervention, network data are used to identify individuals to act as champions. The most frequent intervention of this type is the use of opinion leaders, who usually do not change their opinions.

We can generalize the model so that each agent's initial opinion continues to exert some influence on the agents' later opinions, even in the long run. In order to generalize the observation from the toy network, we consider an agent network of a dumbbell graph

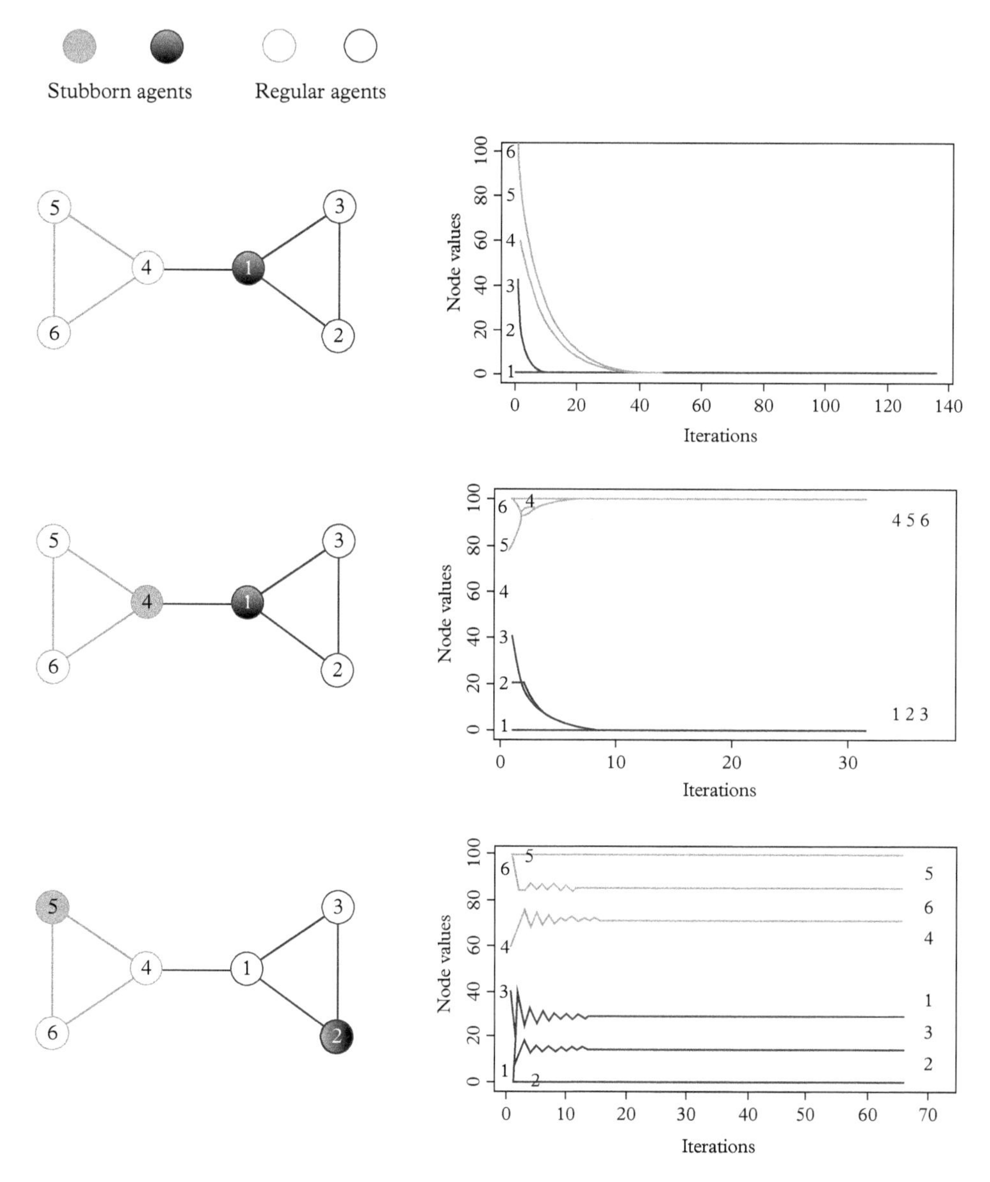

Figure 7.14 *The diagram of a dumbbell graph with two cliques (complete graph) and one link between cliques.*

class, which consists of two disjoint cliques, each consisting of $N/2$ agents, and a single edge in the middle connecting the cliques, (Kitamura and Namatame, 2015).

Consensus problems rely on a small set of micro-prerequisites of mutual influences leading to the emergence of macro-outcomes. These macro-outcomes may refer to polarization or fragmentation in opinion formations, according to the structure of the agent network. We investigated the impact of both the regulation of intrapersonal discrepancy and the interpersonal variability of opinions on network dynamics. The findings can facilitate the identification of specific topological characteristics, such as the average degree and *critical agents* that affect the consensus dynamics. Nonetheless, some topological characteristics that affect the system dynamics and some essential factors, such as the network topology and critical agents, *change agents*, can be identified. This knowledge can be used to enhance the predictability of outcomes and manage possible conflict opinions in communities.

Moore et al. (2011) extended the basic opinion dynamics model to include two processes that are important for the analysis of diseases caused by unhealthy behaviors. The first is an antagonistic reaction that drives individuals further apart in opinion space; the second is the addition of hysteresis representing the constraint addiction places on an individual's behaviors. We can interpret the opinion value of an individual to represent the individual's opinion about smoking in this opinion dynamics investigation. Using a continuous range of opinion over $[0, 1]$, they interpreted an opinion value of 0 to be extremely anti-smoking, an opinion of 1 to be very favorably disposed toward smoking, and other values in the range to represent neutrality on the topic.

Research on voter participation in elections has revealed the importance of opinion dynamics (Krebs, 2005). What influences people to participate in the voting process? How do voters choose which candidates and issues to support? These questions are constantly in the minds of both political scientists and activists. Various models of human behavior have influenced the answers to these two important questions. These models of human behavior span the extremes from very simple and easily influenced behavior to very complex, interdependent behavior where nothing is linear or easily controllable. Voter turnout is also highly correlated among family, friends, and coworkers. Political scientists studying the voting behavior proposed the *social voter model*. In this model, citizens do not make decisions in a social vacuum: *who we know* influences *what we know* and *how we feel* about it. Most voters have interactions with a diverse set of others, including family, friends, neighbors, coworkers, members of their place of worship, and acquaintances. After controlling for personal attitudes and demographic membership, researchers found that the social networks in which voters are embedded exert powerful influence on their behavior.

As agents interact with others in their agent networks, they exchange beliefs, ideas, and opinions both directly and indirectly. A simple discussion of ideas and opinions frequently leads to some agents convincing other agents to modify their opinions about the concepts or beliefs being discussed, while extended interaction can result in each agent gradually modifying her opinion toward a more consensual view in search of common ground. Direct and indirect exchange of opinions and ideas within agent networks may result in changes in agents' behaviors. To the extent that an agent's behavior is influenced by the opinions of his/her colleague, it can be seen that changes in opinions may result in changes in behaviors. If opinions can be seen as propagating through agent networks and opinions influence behaviors, then one of the most direct observable results would be a tendency of the resulting behaviors to cluster in agent networks, forming smaller subnetworks of agents with similar opinions and behaviors.

The society of agents is self-organized into some clusters, different opinion groups that *emerge* at the macro level through properties and interactions at the micro-level. Namely, both the agents' properties and agent network structure influence the consensus dynamics, which is insensitive to the initial states and predictable, resulting in a steady state. The impact of agent network structures can contribute to guiding the behavior and speed of convergence toward a desired state. Agent networks are considered more difficult to manipulate or control as compared to physical and technological systems, and control attempts may lead to outcomes very different from the intended or predicted ones.

The concept of *homophily* states that an individual is more likely to develop social relationships with people who are like him or her in some respect. Homophily is one of the most fundamental characteristics of social networks. This suggests that an agent in the social network tends to be similar to his/her friends or connected neighbors. This is a natural result, because the friends or neighbors of a given agent in the social network are not a random sample of the underlying population. The neighbors of a given agent in the social network are often similar to that agent along many different dimensions, including their interests and beliefs. McPherson et al. (2001) provided an extensive review of research in the long and rich history of homophily. Many experiments confirmed the existence of homophily at a global scale in social networks. Homophily can predict that ties between people with similar cultural profiles are much more likely to exist than ties between people with different cultural profiles.

8

Economic and Social Networks in Reality

Increasingly over recent years, researchers have used network concepts to model and understand a broad array of important environments. Examples include labor market participation, industry structure, the internal organization of firms, and sociological interactions covering such broad phenomena as social norms, peer pressure, and the attainment of status. In this chapter, we shall review the three most important developments in this direction; they are the buyer–seller network (Section 8.1), the labor market network (Section 8.2), and the world trade network (Section 8.3).

8.1 Buyer–Seller Networks

The buyer–seller network stands in a critical position in the recent development in economics, particular in its relation to sociology or the economic sociology. The significance of trust and trustworthiness to the smoothness of the operation of markets becomes part of the literature on the buyer–seller network. In a nutshell, it poses a fundamental question for economists from the sociological viewpoint, i.e., what is the role of personal relations in the market? More specifically, what is the role of personal relations under different trading mechanisms, such as auctions and bilateral trading? In the 1970s and 1980s when institutional economists suggested the appearance of the hierarchy as a solution to market failures, economic sociologists responded to it by pointing out the ignored role of social networks in economics. The buyer–seller network then stands in a central place in the development of this literature. Its appearance blurs the boundary between markets and hierarchies, but further bridges the gap between economics and sociology.

In the standard economic theory of markets, buyers and sellers are basically anonymous to each other. The focus is on the price and the quality of goods or service. Subject to the trading mechanism and buyers' shopping behavior, buyers and sellers are randomly matched. This picture may be realistic in some scenarios; for example, if the trade is carried out in a (double) auction market, buyers normally do not care who the sellers are, and vice versa, even though the trade is not one-shot and will be repeated indefinitely. In this context, there is little room for developing a meaningful buyer–seller

Agent-Based Modeling and Network Dynamics. First Edition. Akira Namatame and Shu-Heng Chen.

network. However, many markets are not centralized, and the trade is carried out in a distributed bilateral manner; the price is either determined by the seller's posted price or is open to bargaining. Under these situations, it is possible to see the formation of the buyer–seller network.

A number of earlier filed studies illustrated the function of the buyer–seller network in different industries: garments (Lazerson, 1993), electronics (Nishiguchi, 1994), apparels (Uzzi, 1996), entertainments (Sorenson and Waguespack, 2006), and fishery (Weibuch, Kirman, and Herreiner, 2000). From these studies, one can clearly see that the buyer–seller networks are not necessarily restricted to the relationship between suppliers and consumers, but probably are more prevalent in the relationship between suppliers and suppliers. Hence, the buyer–seller network can exist in different forms. One of the prominent buyer-seller networks is the *loyalty network*. In the loyalty network, while the buyer can get access to many sellers, he may disregard the potential gains (cheaper prices) from shopping around and only trade with a few or even just one single seller.

One of the famous examples of the loyalty network is the Marseille fish market (Weibuch, Kirman, and Herreiner, 2000; Kirman, 2001). One intuitive explanation for this stable relationship is that it leads to a more predictable service to the loyal buyers, specifically in their demand for fish being satisfied when fish catches are in short supply. More generally, a long-term buyer–seller relationship can help both sides to cope with market uncertainty more efficiently. To some extent, a long-term relationship cultivates a sense of *trust* between the two, which can smooth the business operation when a complete contract is not available. Dulsrud and Gronhaug (2007) studied the network of the Norwegian exporters (sellers) and Danish importers (buyers) in the market for whitefish. Not only did they find a stable relationship between exporters and importers, but also used the term "friendship" to indicate that this connection has a deeper social meaning than just a mechanical link as part of a graph.

As mentioned above, the significance of social networks in markets can also depend on the trading mechanism. When a good can be simultaneously traded in markets with different trading mechanisms, for example, one through auction and one through bilateral trade, then it may be interesting to see the effect of the social network on the reliance on these two mechanisms. In Van Putten, Hamon, and Gardner (2011), they study the effect of fishery policy on the buyer–seller networks in Australia. In 1998, individually transferable quota management was introduced in the rock lobster industry in Tasmania, Australia. To go fishing, fishers need to have allocated quota units. Fishers without quota units can get it from the quota lease market. Fishers who have quotas but do not want to go fishing can also lease out their allocated quotas in the lease market. In addition to the lease market, called the "true market," quotas can also be acquired outside the lease market through individual trades, called the "social market."

One may expect that the underlying social network could have implications for the "competition" of the two markets. Strong personal relations in quota trades may make the "social market" more active than they would do for the "true market." Based on their sample period from 1998 to 2007, Van Putten, Hamon, and Gardner (2011) actually found that the quota lease market has expanded gradually since the inception of the new mechanism in the sense that more quota owners trade in the "true market." Their analysis shows that with the emergence of larger "market makers (brokers)" the quota lease market has become more complex and that the direct channels and connections that dominated the quota lease market in the past are now in decline, indicating that the market has become less "personal." Among many studies of the buyer–seller network, Van Putten, Hamon, and Gardner (2011) is one of the few actually showing how the public policy may impact the existing networks and bring in a transition toward another.[1]

8.1.1 Economic Theory and Beyond

The review of the buyer-seller network above has a strong favor of economic sociology. Indeed, the founder of socioeconomics, Mark Granovetter, put more emphasis on the substantial meanings of links. As Granovetter (1973) states,

> Most intuitive notions of the strength of an interpersonal tie should be satisfied by the following definition: the strength of a tie is a (probably linear) combination of the amount of time, the emotional intensity, the intimacy (mutual confiding) and the reciprocal services which characterize the tie. (Ibid, p. 29)

Similarly, "people who deal with others frequently may not be able to help coming to like one another. (Granovetter (1993), p.29)"

For economists, the theoretically underpinnings of the buyer–seller network, like many others, is a result of rational choices. A pioneering work on the theory of the buyer–seller networks is given by Kranton and Minehart (2001). In their model, the economic reason that buyers and sellers would like to form the networks is, basically, *economies of sharing*.[2] This is because in the supply side there is a fixed cost and in the demand side there is a variation in demand. In this situation, the networks can enable sellers to pool uncertainty in demand. Hence, the number of connections that buyers or sellers choose and the resultant topologies are the optimal trade-off between possible gains and costs. The formed buyer–seller network can be efficient with great social surplus being created. However, the efficient buyer–seller network may not be formed since establishing a link is assumed to have a cost.

We believe that risk sharing or cost sharing can be a rather important rationale for the formation of the buyer–seller network. Roughly speaking, the decision to form a personal

[1] For a more grand review of the impacts of the economic transformations on social networks, which is far beyond the buyer–seller network and comes from a large scale of personal networks, the interested reader is referred to Lonkila (2010).

> This book has similarly proposed that the formation and significance of personal networks in post-Soviet Russia is not only a legacy of connections dating from the Soviet era but also an unintended result of the very process of transformation of the Russian society and economy: the turmoil of the post-Soviet transition forced the newly emerging Russian entrepreneurs to turn to their trusted social ties such as family, kin, friends, and acquaintances. (Lonkila (2010), p. 142).

[2] In fact, this research methodology is closely related to the literature of the formation of a risk sharing group.

trader network is similar to buying insurance. However, there is a substantial difference between buying insurance and forming a network. In the real world, for the former we know exactly what we are up against, whereas for the latter we are not quite sure exactly what we are up against. There are simply too many events that we do not know that we do not know, the so-called true uncertainty. Uzzi (1996) shows that the number of close relationships a firm has is positively correlated with their survival probability. This is certainly an important result, while not surprising. Nonetheless, this statistical pattern is built upon the fact that the number of connections is heterogeneous among firms. Since some ill-connected firms may exit, the buyer–seller network may constantly evolve, and the possible evolving heterogeneity of the buyer–seller network is beyond the scope of the Kranton-Minehart model.

From the perspective of agent-based modeling, we may have different approaches toward the network formation. We can use Uzzi's result (Uzzi, 1996) to motivate a very simple heuristic, i.e., to survive, one needs "friends." However, making friends is costly, so the next question is how many friends are needed. This decision can be affected by many factors, such as risk attitude, personal traits, cognitive capacity, and past experiences, if we are still talking about individual decision-makers. Different agents may, therefore, come up with different decisions and have a different number of connections. This implies a heterogeneity of links: some have many links, and some may have only one. However, this is just a snapshot, as times goes on agents may learn from their own or others' experience, and they may revise their choices: some decide to have more links, but some decide to have fewer links. What the theory needs to explain or replicate is this dynamic process of network formation, such as the empirical patterns, as shown in Innes et al. (2014)[3]. We have not seen any agent-based model being devoted to this area.[4]

8.1.2 The H Buyer–Seller Network

The Kranton–Minehart model can be extended to allow us to consider more variants of the buyer–seller network. First, in the Kranton–Minehart model, links between buyers and sellers are formed on the individual base, i.e., links in the standard bipartite graph. However, in reality, we may have sellers who form a group, alliance, coalition, or association, and collectively promulgate a treatment that applies to all buyers (customers) who are linked to any single seller of the group.

Second, in the Kranton-Minehart model, the good traded in the buyer–seller network is assumed to be homogeneous, which makes sense if we only consider the buyer–seller network on the individual base. However, if we allow sellers to form an association, then it does not have to be the case that the goods or services provided by sellers of the same association have to be homogeneous. Not only can they be heterogeneous, but they can also come from different industries. A familiar example is that a group of airlines, hotels, and car rental companies form an association, then a buyer can have a link to one specific

[3] See their Figure 4, p. 13.

[4] Maybe we are too ignorant and will not be surprised if someone points out that we are wrong on this.

airline (hotel, car rental company) of the group or they can simultaneously have one link or multiple links to the sellers of each industry.

Third, buyers may also form their own group, and have one or multiple links to sellers or associations of sellers. Hence, in these directions, the buyer–seller network has added a hierarchical structure, which may be called the *hierarchical buyer–seller network* or the *H buyer–seller network*. While currently we can already find many real-world examples of the H buyer–seller network, and in the future we may expect more, the economic theory that can account for this complex form of networks is limited.[5] This is another area to which agent-based models may contribute.

8.1.3 Price Dispersion

The buyer–seller network can be exogenously given or endogenously determined. For example, due to spatial reason or limited information or search costs, a buyer can only get access to a number of sellers (restaurants, barber shops, gasoline station, etc.). Therefore, he has a limited number of connections, and chooses the one that is the most competitive. In this network, the main concern is still on the competitiveness of the price; sellers and buyers do not necessarily know each other and their relation, to a large degree, is still anonymous. Kakade et al. (2004) is the first study addressing a buyer–seller network of this kind. Extending the preferential attachment mechanism to a bipartite graph and certain assumptions, they have shown that the price dispersion (price variation among different sellers) can depend on the network topologies.

8.2 Labor Market Network

The research on the labor market network is mainly motivated by some earlier work by Albert Rees (Rees, 1966) and Mark Granovetter (Granovetter, 1973), who provide empirical evidence supporting the significance of the *informal labor market*, which nowadays is characterized by social networks.

> We may divide information networks in the labor market into two groups: formal and informal. The formal networks include the state employment services, private fee-charging employment agencies, newspaper advertisements, union hiring halls, and school or college placement bureaus. The informal sources include referrals from employees, other employers, and miscellaneous sources, and walk-ins or hiring at the gate. (Rees (1966), p. 559)

As a part of the coherent body of the labor market, the issues studied in the informal labor market are basically the same as the formal labor market, which includes job search, the unemployment rate (specifically, the friction unemployment rate), the wage, the wage differentials, the matching efficiency, and the unemployment insurance and other policy design issues. In addition, just like the buyer–seller network (Section 8.1),

[5] The only study known to us that makes a step toward this direction is Wang and Watts (2006).

the informal (labor) market can compete or cooperate with the formal (labor) market; therefore, an additional issue is the behavioral underpinning of the coevolution of the two markets.

The essential element treading through these issues is the social network topology, but social network coevolves with the labor market. Hence, a partial analysis will assume a fixed parametric network and see how the change of the parameters may affect the labor market operation. A general analysis will take the social network as a part of the market and examine its coevolution with the market. Given the general picture above, we shall briefly review the work that has been done with a direction pointing to the future. The focus of this literature is on the flow of information about job opportunities through social networks (Montgomery, 1991).

8.2.1 Empirical Evidence

Before we formally look at the labor market network or, more specifically, the job contact network, it will be useful to give a glossary of the empirical evidence of social networks in labor markets. The list of studies below gives a good summary on how social networks, job contact networks, or weak ties may help the prospective employees to beef up their job-finding opportunity and earnings, (1)–(3). Hence, the personal network (job contact network) that an individual shall own is a pertinent economic decision and a part of "rational" choices, (4)–(8). The references also show that a labor contact network is important for employers to fill their vacancies, through, for example, the referral systems, (9)–(12). In addition to macroscopic implications, (22)–(23), these findings have mesoscopic implications on the employment heterogeneities among different locations, ethnicities, and social communities (13)–(21).

1. Granovetter (1973) found in a survey of residents of a Massachusetts town that over 50% of jobs were obtained through social contacts.
2. For the individuals who find jobs through social connections, many of these connections are weak ties or random connections (Granovetter, 1995).
3. In a survey of 24 studies, Bewley (1999) concluded that 30–60% of jobs are found through friends or relatives.
4. Individuals with more weak ties or random connections in their social networks should earn higher wages or have higher rates of employment (Calvo-Armengol, 2004; Calvo-Armengol and Jackson, 2004).
5. Increasing breadth of social networks due to less overlap of friendships increases individual earnings (Tassier, 2006).
6. The probability that a worker finds a new job through his network of contacts is positively correlated with the separation rate, when it is low; this correlation becomes negative when the separation rate is high (Galeotti and Merlino, 2010).
7. The investment in the network is high when there is a moderate separation rate in the labor market, whereas it is low when the job separation rate is either high or low (Galeotti and Merlino, 2010).

8. Gemkow and Neugart (2011) found that a more volatile labor demand reduces the number of friends a worker wants to have in order to "insure" against relatively long-lasting unemployment spells.
9. Approximately half of all American workers report learning about their job through their social network (friends, acquaintances, relatives, etc.) and a similar proportion of employers report using the social networks of their current employees when hiring (Topa, 2001; Ioannides and Loury, 2004).
10. Both employees and employers use referrals extensively when searching for a job or trying to fill a vacancy, respectively (Galenianos, 2014).
11. Increasing access to referrals increases a worker's job finding rate (Galenianos, 2014).
12. Referred applicants are statistically different from non-referred ones in terms of productivity, wages, and separation rates (Galenianos, 2014).
13. Topa (2001) argued that the observed spatial distribution of unemployment in Chicago is consistent with a model of local interactions and information spillovers, and may thus be generated by agents' reliance in informal methods of job search such as networks of personal contacts.
14. Topa, Bayer, and Ross (2009) used micro-level census data for Boston, and found that those who live on the same block are more than 50% more likely to work together than those living in nearby blocks.
15. Calvo-Armengol and Jackson (2007) showed, by comparing two groups with identical information networks, that initial differences in wages and employment status breed long-run differences.
16. Immigrant groups with a higher level of country assimilation (a proxy for higher hosting country language proficiency) experience a higher network matching rate (Battu, Seaman, and Zenou, 2011).
17. Many minority groups tend to have less random social networks than majority groups (Patterson, 1998).
18. Farley (1998) found that individuals of Belgian, Hungarian, Russian, and Yugoslavian ethnicity outperform other individuals in terms of income even when accounting for education levels and other relevant characteristics.
19. Calvo-Armengol and Jackson (2004) showed how particular network structures may preclude equality of opportunity in the labor market across groups.
20. Krauth (2004) showed that an individual's likelihood of employment is increasing in the number of his social ties and the employment rate that these group members have.
21. Laschever (2005) relied on the random assignment of American WWI (World War I) veterans to military units. Using a data set of 1295 observations, he was able to show that an increase in peers' unemployment decreases a veteran's likelihood of employment.

22. Calvo-Armengol and Jackson (2004) also showed that, if information about job opportunities spreads through social networks, transition rates into employment for those who are unemployed decrease with the duration of unemployment.
23. Networks induce persistence in unemployment (Calvo-Armengol and Zenou, 2005; Fontaine, 2008; Galenianos, 2014).

8.2.2 Job Information Transmission

To discuss the effect of social networks on the labor market, let us consider the following job information transmission channel, which has been shown in Figure 8.1. Suppose an employer, say *A*, has a job vacancy and wants to fill it. He can make this information public, for example, through the internet and social media. In this case, a formal job matching will be carried out by the (internet) job market. This is the conventional way that economists think about the operation of labor markets and job matching, which has no role for the social network. Like the standard way that economists think of impersonal markets, buyers and sellers do not know each other, and they do not bother to know that. In the standard labor market, employers and employees do not have to know each other ex ante personally (they may know each other by reputation), but they care less about who they are than what they can do or what they can offer. The only crucial variables there are productivity and the wage contract.

Alternatively, the employer can go through the *referral* channel and release this job information to his most capable incumbent employee or the most trustworthy employee, say, B, who is assumed to know well of the ability required by the job, and ask him to recommend anyone whom he knows. Upon the request, one natural way for employee A to recommend a candidate is to find a person who has a link with him and can do the job and needs the job. By a priority rule, he may reach his "friend" B1, and if B1

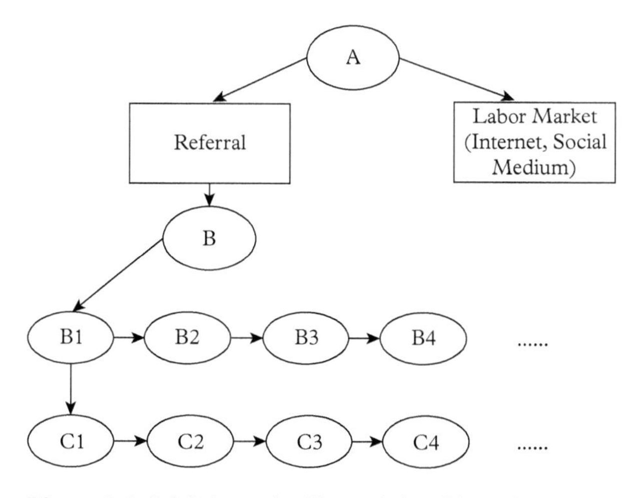

Figure 8.1 *Job Information Transmission Channel*

declines, he reaches B2, and if B2 declines, he reaches B3, and so on. He may continue this iteration until he finds one or has run out of all his personally known candidates. As shown in Figure 8.1, this is a horizontal sequential search or, in terms of social network, the search is subject to one degree of separation.

Apart from the horizontal search, there is a vertical search. For simplicity, let us assume that, by the same priority rule, the incumbent employee B will reach only one of his friends, say $B1$, and ask $B1$ to recommend one of his friends to him if $B1$ has no interest in applying for the job. $B1$ may help B as the way he helps the employer A, taking either the horizontal sequential search or vertical sequential search. In this manner, the job information is transmitted through the social network with a possible chain effect. The search is now extended to multiple degrees of separation.

The job information transmission channel described above is, of course, highly schematic; it can be altered in many ways. Nevertheless, we can use it as a point of reference to see how economists have incorporated social networks into parts of labor market modeling since Mark Granovetter published his pioneering and influential work on the strength of weak ties (Granovetter, 1973). Even though the central role of personal contacts in labor markets has been recognized for three decades, there are few formal models that analyze them (Boorman, 1975; Montgomery, 1991; Calvo-Armengol, 2004; Calvo-Armengol and Jackson, 2004; Calvo-Armengol and Zenou, 2005; Fontaine, 2008; Galenianos, 2014).

8.2.3 The Boorman Model

Similar to what we have seen from the buyer–seller networks, the initial attempt is to see whether we can endogenously determine an employer–employee network or leave the employer aside, to determine an employee network, more generally known as the *contact network*. In other words, the network underlying the labor market operation is a result of optimal choices. By the job information transmission mechanism specified above, the payoff of having a number of friends in the market is that if there is information about a job vacancy, there is a higher probability of being informed and even getting the job. Since everyone in the market may have the chance of being unemployed, it is important to invest in the friendship network to enhance his rehired opportunity. Hence, in this sense, the motivation to have some connections is very similar to the buyer–seller network, i.e., reducing risk and enhancing job security. On the other hand, there is a cost to maintain the link; for example, one needs to spend more resources to manage his social life.

Suppose the available budget for agent i at time t is $y_i(t)$, and he adopts a uniform portfolio for all the agents in the network, then his "investment" to each agent j ($j \in \mathbf{N}/i$) is only $w_{i,j}(t) = \frac{y_i(t)}{N-1}$. In this way, he can maintain a relationship with all the agents in the network, but the intensity of the link characterized by $w_{i,j}(t)$ is thinner (weaker). If other agents, say k, adopts a concentration policy, say, only investing in a few chosen agents, then it is quite likely that, for each j, there exists a k, such that

$$w_{k,j}(t) > w_{i,j}(t).$$

In this case, agent i may have a lower priority of being informed of the job vacancy. This reduces his employment opportunity. Therefore, the uniform strategy may not be the best one; sometimes, like we saw from the buyer–seller model, agent i may keep a few links, but not many. Boorman (1975) wanted to find the best investment policy in a simplified model. He treated the intensity of the link as a discrete choice, a strong contact (strong tie), or a weak contact (weak ties). Under the limited resources, agents then have to decide whether they would like to have strong ties with a few other agents, or have weak ties with many agents.

Boorman (1975) showed that the network formation will depend on the probability of losing a job (the separation rate). This probability is exogenously given in the model. In other words, labor demand is held exogenously. If the probability is low, then an "all-weak network" will be formed, i.e., all agents take a weak tie strategy. Boorman (1975) further showed that this network is both (Nash) stable and efficient (Pareto optima). However, if the probability of losing a job is high, then the formed network topology will change to "all-strong network," which is again stable but not efficient.

By deriving the results above, Boorman (1975) gives the first network formation model in the labor market, while only on the employee side. The derived network topologies are the consequences of the assumed priority rule. The priority rule used by Boorman (1975) is a vertical sequential search, as we have outlined in Section 8.2.2. Basically, the agent who receives the job information will consider his strong ties first and then weak ties. Notice that in Boorman (1975) the job information does not come from employers as our general description shows (Section 8.2.2), but has been modeled as an external signal randomly assigned to a "lucky" agent.

Given the absence of employers, the Boorman model can be roughly understood as a kind of birth-and-death process. At each point in time, say t, there are N agents; each of them can be one of two states: either employed (1) or unemployed (0). Hence, together, the state of the economy can be represented by a N-tuple binary vector.

$$\mathbf{X}(t) = \{x_i(t)\}_{i=1}^{N}, \quad x_i(t) = 0 \text{ or } 1.$$

With a birth arrival rate, μ_1, a state of "1" will occur in a random increment of Δt, and is randomly assigned to one entry of the $\mathbf{X}$, say i. Let $j^*(t)$ be defined as

$$j^*(t) = \sup\{j : j \in \mathbf{N}_i(t) \text{ and } x_j(t) = 0\}. \tag{8.1}$$

The *sup* operator is to output the agent who has a connection with agent i and, among all of i's unemployed friends, has the highest priority of being notified of the job information. Then

$$(x_i(t + \Delta t), x_j(t + \Delta t)) = \begin{cases} (1, 0), \text{ if } x_i(t) = 0 \\ (1, 1), \ \text{ if } x_i(t) = 1 \end{cases}. \tag{8.2}$$

The Boorman algorithm does not allow the chain effect (the vertical dissemination of information); hence if all agent i's contacts already have a job, the job vacancy will be unfilled even though there are still unemployed workers in the market. "Note also that the present rule does not exclude the possibility that a vacancy may exist for more than one time period. (Boorman, 1975, p. 19)" The non-disseminated job information may stay in agent *i*'s circle for a while, but could disappear if the employer finally decides to close this job after a long period of silence.[6] If that happens, a mismatch in a very simple form appears. This just confirms our expectations that the network topology may impact the friction unemployment rate, an issue which has not been addressed in Boorman (1975), but by the labor market network models coming later.

8.2.4 The Montgomery Model

There are some recent studies reporting that a substantial proportion of employers recruit their employees using social networks, specifically, from the recommendation of their current employees. In Figure 8.1, we have shown that the referral system provides an alternative for employers to fill their job vacancy. Since the Boorman model does not have employers, this referral system is not part of the model. Montgomery (1991) is the first network model taking the referral system into account.

Generally speaking, in the Montgomery model, the employer is also part of the network and has connections with each of his employees, assuming that there is no hierarchy in the middle. The employer can distinguish between those who are productive and those who are not, and expect that the productive ones can recommend equally productive agents from their personal network. Hence, when there is a job opening, word-of-mouth search will start only from these productive employees. Once, after the referrer channel is used, the random arrival process of job information in the Boorman model is not applicable, since the job information may not be available in the market.

Montgomery (1991) did not consider networking strategies, which is a core of Boorman (1975). Whether or not the networking strategy should be considered in the labor market is a disputable issue that we shall come back to later. Nonetheless, without a networking strategy, the social network is inevitably exogenously given. Most of the literature on the labor market and social networks assumes that the use of job contact networks is exogenous to labor market conditions (Montgomery, 1991; Calvo-Armengol and Zenou, 2005; Fontaine, 2008; Cahuc and Fontaine, 2009; Galenianos, 2014).

An interesting feature of the Montgomery model is that all employees, regardless of their productivity, have the same size of social networks.[7] There is, however, no direct evidence on the relationship between productivity and the size of the personal network. Since networking can be costly, one could have a trade-off between productivity and sociability, at least in term of the time to spend. It is possible that productivity may be related to the personal networks, not just in their topologies and size, but also the heterogeneities of the connections. One plausible hypothesis is that the social networks of highly professional employees may be less heterogeneous than those of "ordinary" employees. Therefore, it is not immediately clear how word-of-mouth search should be

[6] This possibility of involuntary "no-vacancy" is not considered by Boorman (1975).

[7] In the Montgomery model, the link is randomly determined, but the probability is the same for all agents.

designed. Of course, any abstract analytic model has to leave out these details, but the end of the analytic model is the beginning of the agent-based computation model.[8]

8.2.5 Economic Theory of Ties

In this section, we shall come back to the question that we encountered earlier. To discuss the labor market network formation, we need to address whether we can apply the game-theoretic formation model to the labor market in a sensible way. In Section 8.1, we saw the endogenous formation of the buyer–seller networks. If buyers and sellers have rationales to form complex networks, one might expect that employers and employees shall do that with similar rationale; either risk sharing or uncertainty management. From what we can see from the list of empirical evidence in Section 8.2.1, a central issue here is that employers and employees may use social networks (the informal channel or the referral systems) to fill their vacancy or find their jobs. This has been known since Mark Granovetter's influential work on the strength of weak ties (Granovetter, 1973).

The weak ties are distinguished from the strong ties. In Granovetter (1973), these weak ties refer to "acquaintances." There are some concerns for applying any networking strategy to form this group of agents, in terms of its costs and payoffs. While most studies on labor market networks take network topologies as exogenously given, their reason is for technical tractability. The notable exception that expresses a legitimate concern is Fontaine (2008).

> However, remark that, in real life, the possibilities for the individuals to choose their acquaintances are only limited since the *social contacts relevant for the job search are the output of a process where the level of education, the gender and the social origins play a large role.* From a more practical point of view, the theoretical literature on endogenous networks does not offer any simple way to deal with strategic network formation process... (Fontaine (2008). Italics added.)

However, for other studies that leave social networks endogenously formed, specifically those using agent-based modeling, weak ties are replaced by strong ties, and "acquaintance" is replaced by "friends."

> Additionally, agents may choose to expend effort by making friends. Friends are valuable because agents tell their friends about the jobs they find through direct search. We use the term "friend" to refer to all activities by which a person may learn about a job from another person. Examples include: social events, telephone calls, electronic mail, or personal conversations. (Tassier and Menczer (2001), p. 483)

[8] Unlike the buyer–seller network, agent-based modeling has been applied to address the formation and operation of the job contact network or labor market network. Tassier and Menczer (2001) is probably the first study in this line of research. They studied endogenous network formation in a labor market with referral hiring, in which agents choose their effort level so as to maintain links based on an evolutionary algorithm. They find that there is too excessive (costly) competition for job information causing inefficiencies. Other agent-based models of labor market networks can be found in Tassier and Menczer (2008), Thiriot et al. (2011), and Gemkow and Neugart (2011).

How do we reconcile these two different views in modeling? First, we have to notice that strong ties and weak ties are not the standard terms used to characterize network topologies. The terms are somewhat causal. Implicitly, weak ties (acquaintances) can be agents' college friends. We can hardly expect that an agent in his adolescence had already begun to "choose" his connections for his future career. In other words, those acquaintances were not chosen for that purpose. In this regard, Fontaine's concern seems to be reasonable.

Second, while weak ties may have their strength, both weak ties and strong ties are part of agents' social networks. The final determination, according to Granovetter (1973), is a "combination of the amount of time, the emotional intensity, the intimacy (mutual confiding) and the reciprocal services which characterize the tie. (Granovetter (1973), p.29)." Hence, indeed, there is a cost for a link, either physically or pecuniary. In this regard, links are chosen and are economically motivated. As Galeotti and Merlino (2010) state,

> The incentives to form connections therefore depend on the probability of becoming unemployed and the effectiveness of connections in transmitting information, which are both related to the conditions of the labor market. (Galeotti and Merlino (2010), p. 2)

The statements made by both Tassier and Menczer (2001) and Galeotti and Merlino (2010) are also plausible.

Third, it is, therefore, reasonable to have both strong ties (more expensive ties) and weak ties (less expensive ties) in the model, as Boorman (1975) has done. Nonetheless, as we shall argue below, if we want to do so, a proper network should be a weighted network instead of a binary network, very similar to the one we studied in the Skyrms-Pemantle model (Sections 4.5.1 and 4.5.2) or the network-based trust game model (Section 4.9).

Generally, the mode is very similar to the network-based trust game we reviewed in Section 4.9. There is a reciprocal relation between an agent (trustor) i and his friends or acquaintances (trustees), collectively characterized by the $\mathbf{N}_{i,S}(t)$. As in the trust game, $\mathbf{N}_{i,S}(t)$ is selected by agent i based on his strategy. At each point in time, i will make an investment in agent j ($j \in \mathbf{N}_{i,S}(t)$) to maintain a connection with agent j. The purpose of this investment is to strengthen the connection so that when agent i becomes unemployed at some uncertain time, he may receive job information from agent j. The reciprocity here means the provision of job information from agent j. The trustworthiness of agent j to agent i is the priority or probability that every time j's employer, say E_j, has a vacancy, agent j will forward this information to agent i.

Different from the network-based trust game, agent i does not need to invest in his trustees every period in time; likewise, his trustees need not reciprocate agent i regularly. The strength of the trust relationship between i and j, (i,j), is characterized by a weight $w_{i,j}(t)$ which depends on the magnitude of investment of agent i in agent j *over time*. It is accumulative but also decayable. The acquaintance j ($j \in \mathbf{N}_{i,S}(t)$) that agent i has long not invested in to maintain the link $w_{i,j}(t)$ may have decreased over time and, when it is below some threshold, the channel is no longer able to transmit information.[9] On the other hand, similar to the trust game, agent *i*'s investment in agent *j* may inevitably have some strategic considerations, which include his estimation of agent *j*'s trustworthiness. That estimation can be based upon how frequently agent *i* received information from agent *j* in the past as well as agent *j*'s current network position, for example, his network centrality.[10] This estimated trustworthiness is similar to $\kappa_{i,j}(t)$ in our earlier network-based trust game (Section 4.9).

Like in the trust game, at each point in time, to ensure his job security, agent *i* will allocate some of his income to networking (saving in a form of buying insurance); the portfolio of his investment depends on the expected return (expected amount of information received). Based on those expectations, some connections are heavily invested in, but some are not. By ranking these investment portfolios accumulated over time from high to low, we can clearly visualize who are agent *i*'s strong ties and who are agent *i*'s weak ties.

Despite this, the strong ties and weak ties are *directed*. Even though agent *i* considers agent *j* as his strong tie in terms of the portfolio $w_{i,j}$, that does not mean that agent *j* will also treat agent *i* as his strong tie. This can happen when agent *j* stands in a more prestigious position in the labor market than agent *i*. Furthermore, even though agent *j* has received agent *i*'s investment, it does not mean agent *j* will fulfill agent *i*'s expectations of him. The notion of strong ties and weak ties introduced through the network-based trust game may, therefore, have be subtly different to the one generally known in the literature encouraged by Granovetter (1973).

Having said that, we notice that the notion of strong and weak ties used in Boorman's model is also not the same as the notion used in Granovetter (1973), but is consistent with the subsequent literature on endogenous contact networks, specifically those contact networks that are formed by the trust game mechanism.

> Assume that each individual in the system has the same fixed amount of time $T > 0$ which he individually allocates between strong and weak ties. Let 1 be the (numeraire) time unit required to maintain a single weak tie, and let $\lambda > 1$ be the corresponding amount of time required to maintain a single strong tie. (Granovetter (1973), p. 220)

Granovetter's notion of ties is not purely based on the cost-payoff calculations that interests game theorists. This has been already pointed by Fontaine (2008), as we have seen. It is also partially based on a *time-space structure*. As time went on and an agent travelled through different space and time, he actually experienced different clusters of contacts, from classmates to colleagues. Therefore, these different clusters that were formed in a different time may not overlap much, and when we plot these connections in a map of the network, we can see some bridges emanating from agent *i* connecting different components or cliques around him. It is this geometric structure (overarching bridge) which motivates what we know as a weak tie.

[9] Under that circumstance agent *j* probably has already forgotten *i*, and, even if he has not, he may not have agent *i* in his list of top priorities.

[10] For example, agent *j* may have just been promoted to a new position that enhances his accessibility to more job information.

However, when discussing strong ties, Granovetter gave a resources-based description in terms of involved time, emotion, intimacy, etc. Therefore, an inconsistency emerges here: the notion of strong ties is resources-based, which is consistent with the game-theoretic approach, but the definition of weak ties is topology-based, and that topology can be facilitated through historical time. Therefore, to further reconcile the difference between the sociological approach and the game-theoretic approach, one needs to integrate these three elements, namely, resources, space (topology), and time (history).

The notion of (weak) ties in Granovetter (1973) requires a time structure. The acquaintance refers to an agent j whom agent i has known for some while (by definition) and has either constantly invested in or had intensively invested in for a while before resting. The latter case is more applicable.

> In many cases, the contact was someone only marginally included in the *current network of contacts*, such as an old college friend or a former workmate or employer, with whom sporadic contact had been maintained... (Granovetter (1973), p. 1372; Italics added.)

Even without enduring investment, $w_{i,j}(t)$ may decay slowly over time; hence, it remains a strong tie, *not a weak tie*. It is a weak tie only if we restrict the attention to agent i's recent investment ("*recent network of contacts*"), say, over the interval, $[t-\tau, t]$. Since agent i may no longer invest in agent j after $t-\tau$, presumably, the weight of the investment calculated from $t-\tau$ to the present, $w_{i,j,\tau}(t)$, is zero. Hence, what we have is a stark difference between $w_{i,j}(t)$ and $w_{i,j,\tau}(t)$, namely,

$$w_{i,j}(t) \gg w_{i,j,\tau}(t) \approx 0. \tag{8.3}$$

Therefore, this becomes a way to incorporate the memory (history)-based notion of weak ties into the resource-based framework; those acquaintances are no longer invested in, but memory remains. Although their temporal weights are low, the accumulated weights remain high (eqn. 8.3).

Our proposed synthesis may look complicated, but it simply points out the role of time, history, or memory in the social network analysis. Many networks have been indexed by time, such as the buyer–seller network of a fish market in a specific time, say $[t - \tau, t]$. If τ is small, such as a day or even a month, the constructed network can be considered a "snapshot" reflecting market recency. However, in this short interval, links that do not appear frequent enough to be captured by the network may be absent; hence, the constructed network is not complete. On the other hand, although the network constructed over a long period can well capture active links, a binary network that treats each link equally in time and is space is misleading. Therefore, characterizing each link with intensity, weight, or probability can better capture the network in time and in space.

In the context of the weighted network, the idea of bridges will become more problematic. It depends on how we deal with those connections with low frequency. If we use the popular truncation method, and delete those low-frequency or low-weight links, then some bridges may be deleted. Of course, those links satisfying property (8.3) will remain. This can particularly happen when the personal network of an agent has been evolving for a long time; hence, for those old acquaintances, even though the recent connecting frequency to them is low, the ties with them still sustain. These are the typical weak ties known to us, but not all weak ties are this type. In particular, when weak ties an used as a networking strategy. For example, an agent can reallocate his connections so as to reduce his own cluster coefficient. This behavior will lead him to find some new connections (bridges) and invest heavily. In this case, these bridges are neither old nor weak, but by no means are they acquaintances. Therefore, the topological notion of weak tie is hard to be reconciled even in our synthesizing framework. After all, the bridge as a topology property is not something one can directly control; it is an emergent property based on the interactions of a large group of agents whose networking strategies are heterogeneous.

In sum, to study the significance of the social network in the labor market, we do need to endogenize the social network formation process; however, to do this properly, a weighted network is preferable to the binary network. The binary network is memoryless, and cannot be used to give a proper test for the weak tie hypothesis as formulated by Granovetter (1973). However, most models dealing with endogenous network formation consider only binary networks. Using weighted networks in this study is a direction for the future.

8.2.6 Multiplex Networks

The social network related to the labor market is much more complex than we have reviewed in this section. A key assumption that we make to simplify the social network pertinent to the labor market is to assume that the sole purpose for networking is for job information. In reality, nothing can be further from the truth. In this regard, we totally concur with what Tassier and Menczer (2008) said.

> ... note that real world social networks are chosen for reasons that are not exclusively job search related. We choose to form social connections for many reasons: similar interests, hobbies, social activities, geographic proximity, social affinity, and so on. (Tassier and Menczer (2008), p. 526.)

Hence, before we close this section, it is important to make a remark on this assumption.

First of all, our purpose is not to show how the contact network can be embedded with a "naturally" formed, and possibly larger, social network; instead, we want to show how we can use agent-based model to simulate agents' proactivity to create or choose their own networks (the contact networks) to facilitate their accessibility to job information.

Second, if we model the formation of the "naturally" formed social networks, as we quote from Tassier (2006), we need to know more about the attributes of each agent, from their culture inheritance, ethnic group, personal preference, to their personality. For example, we can include the Schelling model to generate an ethnic (segregation) network (Schelling, 1971) or a Schelling–Axelrod model (Axelrod, 1997) to generate a homophily network. Therefore, individuals can be connected with different motives in this context. Roughly speaking, these "naturally" formed connections can be treated as outcomes of the pleasure-creation purpose. In parlance of economics, these networks are the side benefits from consumption activities. The costs of these connections are very low or even zero. On the other hand, the network pursued by the agent for work-related purposes can be considered investment, since they are costly to build and maintain. If the latter (work-related) network is already encapsulated into the former (non-work-related) network, then the modeling strategy can focus on the creation of a general social network first, and use it as the channel for job information transmission. In fact, this is another justification for why the contact network can be treated exogenously (Fontaine, 2008).

However, assuming the work-related social network will be included as a part of the "naturally" formed network is equally implausible. If the two are different, then some resources have to be appropriate from consumption to investment. In other words, we can no longer leave our personal preference, ethnic characteristics, or any affective reason to determine the work-related network; in some cases, we have to reduce the time spent with family, playing basketball, and visiting old friends so as to increase the time to get engaged in many seminars, trade fairs, and party with colleagues. When we come to this point, networking is basically an economic-oriented problem.

Some studies that try to separate the two networks above suggest the connections belonging to the work-related network as strong ties, and those from the naturally formed network as weak ties. This taxonomy may be misleading and may make the problem even more blurred. Based on what we have discussed at length in Section 8.2.5, all ties are resources-based, regardless the resources being allocated to consumption or investment. If a new generation of young people spends the majority of their time with their peers, and less with their parents, thanks to social media, then their parents are no longer their strong ties, and peers are no longer their weak ties.

Of course, the work to combine the two differently purported networks together is hard.

> Nevertheless, because of the difficulty to generate plausible networks including individuals with several interdependent attributes, these "more detailed" networks remain very simplified compared to real ones. (Thiriot et al. (2011), p. 16.)

8.3 World Trade Network

8.3.1 From Trade to Networks

What is a network of world trade? How can we imagine such a network? Consider the following data from International Monetary Fund (IMF) in 2008. According to the IMF there was \$16 trillion in exports during 2008. There are 181 countries, and hence 181 vertices in a graph. Needless to say, the superset of possible connections $\{g : (i,j), i, g \in N\}$ is incredibly huge, and we are interested in just one of them.

If we consider imports as in degrees and exports as out degrees, it is possible to construct a binary *directed network* where vertices represent countries and directed links represent the import and export relations between them. A directed link between country i and country j is in place, $\delta_{ij} = 1$, if and only if i exports to j. Alternatively, even simpler, we can consider an *undirected network* in which $\delta_{ij} = \delta_{ji} = 1$ if i and j trade in at least one direction.

A binary approach may not be able to fully extract the wealth of information about the intensity of the trade relationship carried by each edge and therefore might dramatically underestimate the role of heterogeneity in trade linkages. Hence, it is desirable to consider a *weighted directed network*, where directed links represent the value trade (export) flows between countries.

Let $X_{i,j}$ be the exports from country i to country j, then the weighted world trade network can be constructed based on export matrix

$$\mathbf{X} = [X_{i,j}]_{i,j=1}^{N}.$$

One direct way to do so is to set

$$w_{i,j} = X_{i,j}, \tag{8.4}$$

or

$$\mathbf{w} = \mathbf{X}. \tag{8.5}$$

To make $w_{i,j}$ lie between 0 and 1, the entire $\mathbf{X}$ can be divided by the maximal value of $\mathbf{X}$, i.e.,

$$w_{i,j} = \frac{X_{i,j}}{X^*}, \quad \text{where} \quad X^* = \max \mathbf{X}. \tag{8.6}$$

One can also make the above weighted directed network into a weighted undirected network by taking the average of both exports and imports between any pair of countries, say, i and j:

$$w_{i,j} = w_{j,i} = \frac{X_{i,j} + X_{j,i}}{2}. \tag{8.7}$$

8.3.2 Why the World Trade Network?

The interest in the world trade network is not just for the understanding of the structure of international flow of services and goods per se, a subject intensively studied in international trade theory, but, from an international macroeconomic viewpoint, it is also for the purpose of macroeconomic stability enhancement. For the latter purpose, the world trade network can help us see how a sources of macroeconomic instability, such as energy shocks, international conflicts, regional strikes, local political turmoil, and financial institutions' bankruptcies, can propagate to other countries through the channels of trading partners (Kali and Reyes, 2007, 2010).

In addition, the constructed world trade network allows us to explore not only first-order phenomena associated with the degree of openness to trade of any given country, but also the second-order and higher-order empirical facts. The former concerns the *disassortativity* of the network, i.e., the extent to which highly connected countries tend to trade with poorly connected ones (Newman, 2002), whereas the latter concerns the *degree-clustering coefficient*, i.e., the likelihood that trade partners of highly connected countries are themselves partners. Network features like assortativity and clustering patterns do instead depend on indirect trade relationships, i.e., second or higher-order links between any two countries not necessarily connected by a direct-trade relationship.

8.3.3 Analysis of the World Trade Network

In recent years, an extensive research effort has been devoted to analyzing the structure, function, and dynamics of the world trade network from a social network perspective. The research on the construction of the world trade network started in the early 2000s. It began with the construction of binary directed networks (Kim and Shin, 2002; Serrano and Boguñá, 2003; Garlaschelli and Loffredo, 2004, 2005). Serrano and Boguñá (2003) and Garlaschelli and Loffredo (2004) focused on the undirected description of a single snapshot of the world trade network, whereas Garlaschelli and Loffredo (2005) placed this study in a dynamic context and studied the topological properties of the world trade network as a directed and evolving network.

The binary directed or undirected network analysis treats all links in the world trade network as if they were completely homogeneous. This is rather odd since actual import and export flows greatly differ either when they are evaluated in their levels or when they are computed as shares of gross domestic product (GDP). Thanks to progress in the weighted network analysis (Barrat et al., 2004), various characterizations of nodes in the binary network are extended into the weighted network versions; for example, node degree was generalized to *node strength* (De Montis et al., 2005), the nearest-neighbor degree was generalized to the *nearest-neighbor strength* (De Montis et al., 2005), and the binary clustering coefficient was generalized to the *weighted clustering coefficient* (Saramäki et al., 2007). With these generalizations, in the late 2000s, a number of studies were devoted to the construction of weighted world trade networks (Bhattacharya et al., 2008; Fagiolo, Reyes, and Schiavo, 2008, 2009, 2010).

The interesting thing is that when the world trade networks are extended from the binary version to the weighted version, one actually obtains quite different properties. The basic characteristics of the binary directed world trade network are two-fold. First, it is characterized by a *disassortative pattern* (Newman, 2002), i.e., countries with many trade partners are on average connected with countries with few partners. Second, it is characterized by a *hierarchical pattern*, i.e., partners of well-connected countries are less interconnected than those of poorly connected ones. In terms of network statistics, the above two stylized facts can be summarized as two negative correlations, namely, the negative correlation between node degree and average nearest-neighbor degree and the negative correlation between node degree and clustering coefficient. Garlaschelli and Loffredo (2005) further showed that the above two properties are quite stable over time.

Statistical properties of the world trade network perceived as a weighted undirected network crucially differ from those binary constructions. Many findings obtained by only looking at the number of trade relationships that any country maintains are completely reversed if one takes into account the relative intensity (strength) of trade links. First, the distribution of trade intensity (the strength of link) is right-skewed (Bhattacharya et al., 2008; Fagiolo, Reyes, and Schiavo, 2008), indicating that a few intense trade connections coexist with a majority of low-intensity ones. Second, the countries of the world trade network holding many trade partners and/or having high trade intensity are also the richest and most central countries. They typically trade with many partners, but very intensively with only a few of them, who turn out to be themselves very connected; hence, they form few but intensive-trade clusters (triads) (Fagiolo, Reyes, and Schiavo, 2009, 2010).

8.3.4 "Theory" of the World Trade Network

Given the topological properties of the constructed world trade network, the next question is the formation mechanism of this construction. So far, there is no unique answer for it, and it is still an ongoing research topic.

> Despite we know a great deal about how economic networks are shaped in reality and what that means for dynamic processes going on over networked structures. . . , we still lack a clear understanding of why real-world network architectures looks like they do, and how all that has to do with individual incentives and social welfare. ((Fagiolo, Squartini, and Garlaschelli, 2013), p. 77)

Gravity Model

One attempt is to seek an answer from international trade theory. At the point that the network analysis of world trade became available, scholars in international trade theory were using the *gravity model* to describe international trade and had done so already for over four decades (Tinbergen, 1962).

The gravity model aims to analyze spatial interactions among different kinds of variables using the general idea of the theory of gravity in physics. According to the gravity model, the flow between any two locales increases in direct proportion to their combined economic mass between these locales and in inverse proportion to the distance between the locales. In its simplest form, the gravity model predicts that the volume of trade between countries i and j is:

$$X_{ij} = \alpha \frac{GDP_i^{\beta} GDP_j^{\beta}}{d_{ij}^{\gamma}}, \tag{8.8}$$

where GDP refers to the gross domestic product and "d" refers to the mutual geographical distance. Different versions of the gravity model have been applied to replicate the world trade web (Bhattacharya et al., 2008; Dueñas and Fagiolo, 2013). Existing works using the gravity model to account for the empirically constructed world trade network have suggested that a gravity specification is not always able to replicate the topological properties of the trade network, especially at the binary level (Dueñas and Fagiolo, 2013). Although several improvements and generalizations of the standard gravity model have been proposed to overcome this problem (Dueñas and Fagiolo, 2013), so far none of them succeeded in reproducing the observed complex topology and the observed volumes simultaneously.

The Fitness Model

The gravity model is widely used by economists, although in econophysics an alternative approach known as the *fitness model* is more popularly known. The idea of the fitness model is that the connection probability of the pair of two vertices i and j is assumed to be a function of the values of some "fitness" characterizing each vertex. The idea was first proposed by Caldarelli et al. (2002) as an alternative mechanism, which they called the "good get richer" mechanism, to replace the original preferential attachment, the so-called "rich get richer" mechanism.

The fitness model has a stronger feavor of the agent-based model, since in general each individual can be given a set of attributes, $\mathbf{v}_i(t)$, which may change over time. The attributes together will determine the probability that agent i and agent j will have a connection between them, which can be written as

$$w_{ij}(t) = Prob(\delta_{ij} = 1 \mid \mathbf{v}_i(t), \mathbf{v}_j(t)). \tag{8.9}$$

The network-based investment game studied in Section 4.9 can be reformulated in this way; also, the celebrated Schelling-Axelrod model (Axelrod, 1997) of culture transmission can also be formulated in this manner. In fact, this formulation can be so general that it can go without the notion of "importance" of a node and does not have to be attached with the "good get richer" interpretation. In fact, in the more general setting, the importance, value, or fitness of an agent can be subjective and vary among different agents (Axelrod, 1997).

The fitness model was then applied to modeling the formation of the world trade web, first by Garlaschelli and Loffredo (2004). They used the real-world data to analyze an undirected binary version of the world trade network. They have shown that the total GDP of a country, GDP_i, can be identified with the fitness variable that, once a form of the probability of the trade connection between two countries is introduced, completely determines the expected structural properties of this network. This leads to a more economic interpretation where the fitness parameters can be replaced with the GDP of countries, and used to reproduce the properties of the network. From the viewpoint of statistical mechanism (Park and Newman, 2004), this also implies that the world trade network viewed as a binary network is a typical representation of *exponential random graphs*, i.e., the distribution over a specified set of graphs that maximizes the entropy subject to the known constraints. The fitness model, therefore, is capable of giving predictions for the network based only on macroeconomic properties of countries, and reveals the importance of the GDP to the binary structure of the world trade network.

In addition to the purely binary network, the fitness model is also applied to the weighted network (Fronczak and Fronczak, 2012). What clearly emerges is that both topological and weighted properties of the network are deeply connected with purely macroeconomic properties (in particular the GDP) governing bilateral trade volumes. However, it has also been clarified that, while the knowledge of the degree sequence allows us to infer the entire binary structure of the network with great accuracy (Squartini, Fagiolo, and Garlaschelli, 2011a; Fagiolo, Squartini, and Garlaschelli, 2013), the knowledge of the strength sequence (i.e., the total volume of trade for each country) gives a very poor prediction of all network properties (Squartini, Fagiolo, and Garlaschelli, 2011b; Fagiolo, Squartini, and Garlaschelli, 2013).

Maximum Entropy Model

This section starts with a review of the use of the gravity model and the fitness model as a formation mechanism of the world trade network. However, probably partially under the influence of Park and Newman (2004), the subsequent development of the literature has been given a strong flavor of statistical mechanics. The formation mechanism, or the theory accounting for the observed world trade network, has been questioned first on whether the data is not random but can suggest a theory. Do we need a theory to account for various high-order patterns? Are the observed high-order patterns spurious? A historically-long standing approach to address this question is the entropy maximization approach; indeed, this approach has been used to establish the *null model* to filter out the spurious patterns. Intuitively, if what we observe can be replicated by a random mechanism, then there is no need to search for a theory for it. This entropy-maximization argument is exactly the fundamental message of Squartini, Fagiolo, and Garlaschelli (2011a), Squartini, Fagiolo, and Garlaschelli (2011b), and Fagiolo, Squartini, and Garlaschelli (2013).

Instead of building economically-based (such as the gravity model) or stochastically-based (such as the fitness model or the preferential attachment model) micro-foundations for explaining observed patterns, the entropy-maximization approach tries to ask the question: whether observed statistical network properties may be simply reproduced by simple processes of network generation that only match some (empirically-observed) constraints, but are otherwise *fully random*. If they do, then the researcher may conclude that such regularities are not that interesting from an economic point of view, as no alternative, more structural, model would pass any test discriminating against the random counterpart. Using the entropy maximization principle to make the networks randomly grow, subject to some constraints of degree and strength sequences, Fagiolo, Squartini, and Garlaschelli (2013) showed that, in the binary world trade network, node-degree sequences are sufficient to explain higher-order network properties such as disassortativity and clustering-degree correlation. Conversely, in the weighted world trade network, the observed sequence of total country imports and exports are not sufficient to predict higher-order patterns of it.

The entropy maximization formulation requires solving a constrained optimization problem. The constraints represent the available information and the maximization of the entropy ensures that the reconstructed ensemble of networks is maximally random, given the enforced constraints. What are these constraints for generating the world trade network? Mastrandrea et al. (2014) recently showed that the joint specification of the strengths (trading volume) and degrees (number of trading partners) cannot be reduced to that of the strengths alone. The knowledge of the strengths alone does not merely include or improve that of the degrees, since the binary information is completely lost once purely weighted quantities are measured. Hence, the reconstruction of weighted networks can be dramatically enhanced by the use of the irreducible set of joint degrees and strengths as the constraints.

However, when both the degrees and strengths are preserved, their implications for the high-order correlation structure, such as the average nearest neighbor degrees (and strengths) or the average clustering coefficients in degrees (and strengths) become rather sophisticated. Garlaschelli and Loffredo (2009) proposed the Bose-Fermi statistics and showed that these statistics completely describe the structural correlations of weighted networks. Built upon these mathematical results, Almog, Squartini, and Garlaschelli (2015) adopted a maximum-entropy approach where both the degree and the strength of each node are preserved. Their proposed model can be regarded as a weighted generalization of the fitness model, where the GDP still plays the role of a macroeconomic fitness shaping the binary and the weighted structure of the world trade network simultaneously. They called this new formulation of entropy maximization the *enhanced configuration model* (ECM), and the biases in the original configuration model can be successfully handled through the ECM.

8.4 Summary

Network thinking has substantially reshaped the recent developments in economics. It enhances economists' understanding on how markets actually work. It has been long challenged by both economists and other social scientists that individuals, rationality and self-interest motive cannot be the backbone of a well-functioning market. Between individuals and markets, there must be something in the middle that makes markets beyond just a total of individuals or a sum of a myriad of utility-maximizing "machines." Over the last two decades, under the influences of sociologists or the socioeconomists (a neologism), economists become increasingly aware that the middle layer is trust, cooperation, and altruism, instead of just pure cold-blooded competition. The framework that can place all of these elements into a coherent body and enhance market operation is now known as networks. From conventional commodity markets, job markets, to modern internet markets, to customer relation management, to social security net, to financial market stability, their existences and functions are prevalent. We can no longer ignore them. Their incorporation into economics will become a new chapter for almost all of economics' branches, from micro, macro, behavioral, institutional, evolutionary, industrial, and financial economics. Econometricians also need to learn how to deal with bid data coming from networks.

9

Agent-Based Modeling of Networking Risks

Networked systems offer multiple benefits and new opportunities. However, there are also disadvantages to certain networks, especially networks of interdependent systems. Systemic instability is said to exist in a networked system if its integrity is compromised regardless of whether each component functions properly. This chapter investigates how the complexity of a networked system contributes to the emergence of systemic instability. An agent-based model, which seeks to explain how the behavior of individual agents can affect the outcomes of complex systems, can make an important contribution to our understanding of potential vulnerabilities and the way in which risk propagates across networked systems. Furthermore, we discuss various methodological issues related to managing risk in networking.

9.1 Networking Risks

Risks are idiosyncratic in the sense that a risk to one system may present an opportunity in another. Networks increase the level of interdependence for their components, and this affects the integrity of the system. Over the last few decades, we have witnessed the dark side to increased interdependency. A crucial question follows: Is an interconnected world a safer place in which to live, or is it more dangerous? This is especially relevant to financial institutions and firms (or agents), where the actions of a single agent can impact all the other agents in a network. In this section, we present an overview of the concepts, ideas, and examples of systemic risks inherent in a network—that is, of "networking risks."

People are not very good at dealing with risk. Psychologists have demonstrated that events that are not vivid in our minds are assigned much lower importance than the facts warrant (Tversky and Kahneman, 1992). We often use the terms "risk" and "uncertainty" interchangeably. Risk exists where the outcome is uncertain, but where it is possible to determine the probability distribution of outcomes. Therefore, we can manage risk using a probability calculus. By contrast, uncertainty exists in a situation where we do not know the outcome, or the probability cannot be determined. Most systems

Agent-Based Modeling and Network Dynamics. First Edition. Akira Namatame and Shu-Heng Chen.

we deal with in the real world contain uncertainties, rather than risks. Risk can be understood as the probability of an adverse outcome with consideration to the severity of the consequences should it occur. This definition is sufficient in many cases, but it is inadequate for situations of the greatest interest—for pivotal events and large-scale disturbances. Because pivotal events are generally rare and unprecedented, statistics are powerless to give any meaningful probability to the occurrence of such events.

Most natural and social systems are continually subjected to external disturbances that can vary widely in degree. Scientific research has focused on natural disasters, such as earthquakes, or on failures to engineered systems, such as electrical blackouts. However, many major disasters affecting human societies relate to the internal structure of networked systems. Risk can be categorized as either exogenous or endogenous. When we explore the sources of risk, it is important to distinguish between these types. Exogenous risk comes from outside a system, and is basically a condition imposed on it. For instance, disaster preparedness involves considering exogenous risk. Other examples include the threat of avian flu, terrorism, and hurricanes. Endogenous risk, on the other hand, refers to the risk from disturbances that are generated and amplified within the system. The amplification of risk proceeds from its systemic nature, and challenges the integrity of the system. As systems increase their interdependency on other systems, we must consider the risk inherent to such interdependency. As a disturbance propagates through a network, it might encounter components known as amplifiers—that is, components that increase the risk to other components in the network. Amplification occurs when such an interaction results in a vicious cycle, reinforcing the effects of amplifiers. Amplifiers are the mechanisms that boost the scale of a disturbance in a particular system, and also the means by which hazards are spread and intensified throughout the system. This phenomenon is especially prevalent in interdependent systems.

Empirical evidence shows that whereas a network's performance improves as its connectivity increases, there is also a corresponding increase to the risk of contagion. Failures to networked components are significantly more egregious than the failure of any single node or component. Network interdependency is a main reason for a series of failures. Examples include disease epidemics, traffic congestion, and electrical blackouts. These phenomena are known as cascading failures, and more commonly as chain reactions or domino effects. The definitive feature to cascading failures is that a local failure results in a global failure on a larger scale. The result is that networked risk, which can lead to the failure of the networked system as a whole, is not related in any simple way to the risk profiles for each component. Such networked risks are common with nonlinear interactions, which are ubiquitous in networked systems (Helbing, 2010).

Upon experiencing a failure, it is common to look for the direct causes and effects in order to fix the problem. However, there is no simple way to understand cause and effect in interdependent systems. Recently, the u.s. has experienced widespread blackouts. In 1996, a single power line failure in Oregon led to a massive cascade of power outages that spread to all the states west of the Rocky Mountains, leaving tens of millions of people without electricity. The basic mechanism behind cascading failures is a vicious circle that often induces undesired side effects. A large-scale blackout may begin as a fairly routine problem in a particular area. This type of a failure may occur with relative frequency in the network, where failures in one node are absorbed by the neighboring nodes. Under certain conditions, however, failures can propagate to other nodes. When that happens, matters worsen along the chain until the network sustains widespread failure. We can reliably assess the risk that any single power generator in the network will fail, under a given set of conditions. Once failures begin to cascade, however, we can no longer recognize what those conditions will be for a particular power generator, because conditions will change dramatically depending on what else is happening in the network. Another property is pertinent to cascading failures: there is a critical load at which risk sharply increases toward the threshold for a cascading failure. We shall discuss the approach of agent-based modeling and simulations for computing the likelihood of a cascading failure. Modeling a cascading failure as a dynamic process supplies the quantitative methods used to compute and monitor criticality, by quantifying how failures propagate.

Over the past year a different kind of cascading failure was encountered, one sustained by the financial system, which produced the equivalent of a global blackout. The fault lines for such a failure are invisible, because they are systemic in nature and result from the unpredictable interplay of the myriad of parts in the system. In the financial system, for instance, there is a tendency for crises to spread from one institution to another. As a result, these phenomena are subject to systemic risks and, at a large enough scale, they can lead to systemic failures. Some studies of systemic failures pursue an explanation of the mechanisms involved in bank failures during financial crises (May et al., 2008). Mitigating risk at the systemic level requires large-scale changes to a system, and often the threats to it are uncontrollable. The financial crisis revealed important weaknesses to traditional risk models for understanding and responding to risks inherent to the financial system. These models focus on the risk encountered by individual financial institutions and are unable to model financial vulnerabilities, because only specific triggers can expose these vulnerabilities; the process by which such triggers are propagated through the financial system is not well understood. Rectifying these weaknesses is the first step in developing the capability for dealing with threats to the financial system.

Networking risks are idiosyncratic, and they are an inherent characteristic of interdependent networks. In a networked society, the risks faced by any one agent depend not only on the actions of that agent, but also on those of other agents. The fact that the risk a single agent faces is often determined in part by the activities of other agents gives a unique and complex structure to the incentives that agents face as they attempt to reduce their exposure to networked risks. The term *interdependent risk* refers to situations in which multiple agents decide separately whether to adopt protective management strategies, such as protection against risk, and where each agent has less of an incentive to adopt protection if others fail to do so. Protective management strategies can reduce the risk of a direct loss to any agent, but there is still a chance of suffering damage from other agents who do not adopt similar strategies. The fact that the risk is often determined in part by the actions of other agents imposes independent risk structures for the incentives that agents require to reduce risk by investing in risk mitigation measures.

Kunreuther and Heal (2003) analyzed such situations by focusing on the interdependence of security problems. An interdependent security setting is one in which each agent whom is part of a network must decide independently whether to adopt protective strategies to mitigate future losses. Their analysis focused on protection against discrete, low-probability events in a variety of protective settings with somewhat different cost and benefit structures, such as airline security, computer security, fire protection, and vaccination programs. Under some circumstances, the interdependent security problem resembles the familiar dilemma in which the only equilibrium is the decision by all agents not to invest in protection even though each would be better off were they all to decide to incur this cost. In other words, a protective strategy that would benefit all agents if adopted widely may not be worth the cost to any single agent, even if all other agents adopt it. In this case, any single agent is better off simply taking a free ride on the others' investments. In other circumstances, however, a protective measure is indeed cost-effective for each agent, provided enough other agents also invest in risk reduction. These independent security problems have the potential for reaching a tipping point, given that one or several agents' actions can determine the outcome, and for cascading, in which there is a set of agents for whom the initial decision regarding whether to protect will incite other agents to follow suit. Therefore, a key issue is how to provide incentives for a group of agents to take action with respect to their exposure to risks, prompting a move that reaches the tipping point. If agents believe other agents will not invest in security, their incentive to do so is reduced. The end result may be that no one invests in protection, although all would have been better off were they to incur the cost of protective measures. On the other hand, if agents believe that other agents will also undertake to mitigate risks, their optimal course of action will be to do the same.

A variety of schemes exist that are designed to mitigate networking risks, but the majority of these schemes depend on centralized control and full knowledge of the system. Furthermore, centralized designs are frequently more susceptible to limited situational awareness, making them inadequate and resulting in increased vulnerability and disastrous consequences. Understanding the mechanisms and determinants of risk can only help to describe what is going on, and these mechanisms have limited predictive power. Bidirectional causal relationships are an essential component in the study of networking risk. Understanding the relationship between the different levels at which macroscopic phenomena can be observed as networking risk is possible with the tools and insights generated by combining agent and network-based models. Agent-based modeling combined with network theory can explain certain networking risks. Instead of looking at the details of particular failures of nodes or agents, this approach involves investigating a series of failures caused by the dependencies among agents. Allen and Gale (2000), for instance, introduced network theory as a means for enriching our understanding of networking risks. They explored critical issues in their study of networking risk by answering fundamental questions, such as how resilient financial networks are to contagion, and how financial institutions form connections when exposed to the risk of contagion. They showed that by increasing the connections between financial institutions, the overall risk of contagion can be reduced because the risk is shared.

9.2 Systemic Risks

The concept of systemic risk pertains to something undesirable happening in a system that is significantly larger and worse than the failure of any one node or component. Global financial instability and economic crises are typical examples of systemic risks. The scope and speed of the diffusion of risk in recent financial crises have stimulated an analysis of the conditions under which financial contagion can actually arise. Systemic risk describes a situation in which financial institutions fail as a result of a common shock or a contagion process. A contagion process refers to the systemic risk that the failure of one financial institution will lead to defaults in other financial institutions through a domino effect to the interbank market.

Network interdependencies are apparent in financial networks, where the actions of a single agent (e.g., a financial institution) in an interconnected network can impact other agents in the network. Increased globalization and financial innovation have prompted a sudden increase in the creation of financial linkages and trade relationships between agents. In financial systems, therefore, there is a tendency for crises to spread from one agent to another, and this tendency can lead to systemic failures on a significantly large scale.

In this section, we analyze the mechanisms to systemic risk using agent-based modeling and network theory. In an interbank market, banks facing liquidity shortages may borrow liquidity from other banks that have liquidity surpluses. This system of liquidity swapping provides the interbank market with enhanced liquidity sharing, and it decreases the risk of contagion among agents when unexpected problems arise. However, solvency and liquidity problems faced by a single agent can also travel through the interbank market to other agents and cause systemic failures.

May and Arinaminpathy (2010) provided the models to analyze financial instability and the contagion process based on mean-field analysis. The assumption in their approach is that each agent in the network is identical. External or internal shocks may lead to the collapse of the entire system. If a single agent is disrupted and this causes a failure, this initial failure can lead to a cascade of failures. Several critical constellations determine whether this initial failure remains local or grows to a point where it affects the system, leading to systemic risk.

One of the main objectives in research on systemic risk is to identify agents that are systemically important to the contagion process. When these agents default, they influence other agents through the interconnectivity of the networked system. Systemic risk is not a risk of failure caused by the fundamental weakness of a particular agent. Because failed agents are not able to honor their commitments in the interbank market, other agents are likely to be influenced to default as well, which can affect more agents and cause further contagious defaults. For an agent, maintaining interconnections with other agents always implies a trade-off between risk sharing and the risk of contagion. Indeed, the more interconnected a balance sheet is, the more easily a negative shock, say a liquidity shock, can be dissipated and absorbed when an agent has multiple counterparties with whom to discharge the negative hit. Therefore, studying the role of the level and form of connectivity in the interbank network is crucial to understanding how direct contagion works, i.e., how an idiosyncratic shock may travel through the network of agents.

We can identify the critical agents for systemic risk with the threshold-based cascade model presented in Chapter 6. In the cascade model, each agent has a threshold ϕ, which determines its capacity to sustain a shock. If agents fail, they redistribute their debt to neighboring agents in a network. Once we can specify the threshold distribution of the agents, we can analytically derive the size of the default cascade. Systemic risk depends much more on factors such as the network's topology.

Gai and Kapadia (2010) were the first to analyze the mechanism of systemic risk with a threshold-based cascade model. They developed an analytical model to study the potential impact of contagion influenced by idiosyncratic shocks. Their model also explains the *robust-yet-fragile* tendency exhibited by financial systems. This property explains a *phase transition* in contagion occurring when connectivity and other properties of the network vary. In their model, every agent in the network is identical, i.e., all agents have the same number of debtors and creditors. They investigated how the system responds when a single agent defaults, and, in particular, when this results in contagion events in which a finite number of agents default as a result. The agent network was developed using a random graph where each agent had an identical threshold. The main results were (i) as a function of the average connectivity, the network displayed a window of connectivity for which the probability of contagion was finite; (ii) increasing the net worth of agents reduced the probability of contagion; (iii) when the network was well connected (i.e., when the average degree was high), the network was robust-yet-fragile (i.e., the probability of contagion was very low, but in those instances where contagion occurred, the entire network was shut down).

Nier et al. (2007) modeled systemic risk in financial networks with heterogeneous agents and discovered that the agents with the most capital were more resilient to contagious defaults. They also modeled part of the tiered structure by classifying the agents in the network into large and small agents. They found that tiered structures are not necessarily more prone to systemic risk, and that whether they are depends on the degree centrality, that is, the number of connections to the central agent. As the degree centrality increases, contagious defaults increase at first, but then they begin to decrease as the number of connections to the central agent leads the dissipation of the shock.

We portray N agents (viz. banks), randomly linked together in a weighted directed network where the weighted links represent interbank liabilities. The financial state of each agent is described by the balance sheet. The balance sheets of agents are modeled according to their assets and liabilities. Agent i's assets (denoted by A_i) include interbank loans (denoted by I_i) and external assets (denoted by E_i). Liabilities (denoted by L_i) consist of interbank borrowings (denoted by B_i), deposits (denoted by D_i), and the net worth (denoted by C_i), as shown in Figure 9.1. Agent vulnerability depends on the net worth, which is defined as follows:

$$C_i \equiv I_i + E_i - B_i - D_i. \tag{9.1}$$

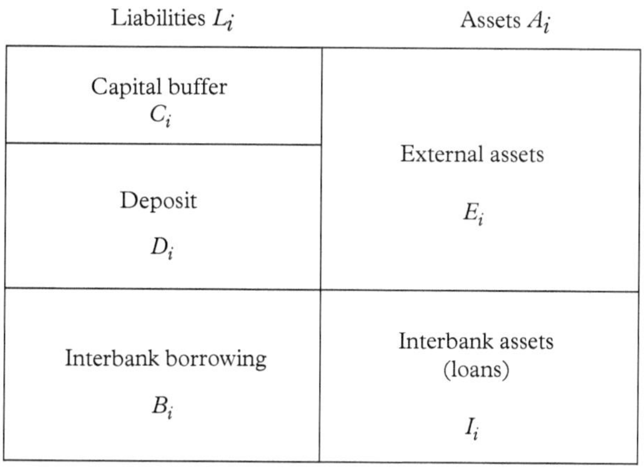

Liabilities L_i	Assets A_i
Capital buffer C_i	External assets E_i
Deposit D_i	
Interbank borrowing B_i	Interbank assets (loans) I_i

Figure 9.1 *Balance sheet for a bank i*

As an additional assumption, the total interbank asset positions are assumed to be evenly distributed among all incoming links, which represent loans to the other agents. The defaulting condition is as follows:

$$\text{Agent } i \text{ defaults if}: \ (1-p)A_i - qE_i - B_i - D_i < 0, \tag{9.2}$$

where p is the number of agents with obligations to agent i that have defaulted, and q is the resale price of the illiquid external assets, which takes a value between 0 and 1. When an agent fails, *zero recovery* is assumed and all of that agent's assets are lost.

The contagion process begins by selecting one defaulting agent randomly. Then, we observe whether there is a chain of defaults in the interbank network. Initially, all agents are solvent, and defaults can spread only if the neighboring agents of a defaulted agent are vulnerable. By definition, an agent is vulnerable whenever the default of one of neighboring agents with a credit relation causes a loss to the balance sheet such that it meets the defaulting condition from eqn. (9.2). The defaulting condition from eqn. (9.2) can thus be rewritten as follows:

$$p > \frac{C_i - (1-q)E_i}{L_i}. \tag{9.3}$$

By setting $q = 1$, we can derive the following solvency condition:

$$\text{Agent } i \text{ defaults if}: \ p > \frac{C_i}{L_i} \equiv \phi. \tag{9.4}$$

Agent vulnerability depends on the threshold ϕ, which is the ratio of the net worth (C_i) to liability (L_i). The defaulting condition is the same as the threshold rule from the cascade model in eqn. (6.7).

Leverage generally refers to using credit to buy assets. It is commonly used in the context of financial markets. However, the concept of leverage covers a range of techniques from personal investments to the activities in financial markets on a national scale. One widely used measure of financial leverage is the debt-to-net-worth ratio. This is the ratio of the liabilities of an agent with respect to the net worth. The inverse of the threshold ϕ in eqn. (9.4) represents the debt-to-net-worth ratio of an agent.

The process of contagion gains momentum and suddenly spreads to a large number of agents after the failure of some critical agents. Therefore, finding these critical agents is crucial to preventing a cascade of defaults. The criticality of an agent does not directly depend on the size of liabilities. Rather, it is determined by the debt-to-net-worth ratio. Another crucial factor is the location of an agent within the interbank network. It is on this latter point that the agent-based model may yield information useful for identifying critical agents.

Real-world financial networks often have fat-tailed degree distributions (Boss et al., 2003). Many authors also note that there is some form of community structure to the network. For example, Soramaki et al. (2007) showed that the so-called core-periphery networks include a tightly connected core of money-center banks to which all other banks connect, and that the interbank market is tiered to a certain extent. In their study of tiered banking systems, Freixas et al. (2000) demonstrated that a tiered system of money-center banks—where banks on the periphery are linked to the center but not to each other—may also be susceptible to contagion.

Fricke and Lux (2012) conducted an empirical study as the starting point for investigating interbank networks. They studied the network derived from the credit extended via the electronic market for interbank deposits (e-MID) trading platform for overnight loans between 1999 and 2010. e-MID is a privately owned Italian company and currently the only electronic brokerage market for interbank deposits. They showed that the Italian interbank market has a hierarchical core-periphery structure. This set of highly connected core banks tends to lend money to other core banks and a large number of loosely connected peripheral banks. These banks in turn tend to lend money to a small number of selected core banks, but they appear to have relatively little trade among themselves.

A core-periphery network structure is a division of the nodes into a densely connected core and a sparsely connected periphery. The nodes in the core should also be reasonably well connected to the nodes in the periphery, but the nodes in the periphery are not well connected to the core. Hence, a node belongs to the core if and only if it is well connected both to other nodes in the core and to nodes assigned to the periphery. Thus, a core structure to a network is not merely densely connected, it also tends to be central to the network in terms of short paths through the network. The latter feature also helps to distinguish a core-periphery structure from a community structure. However, many networks can have a perfect core-periphery structure as well as a community structure, so it is desirable to develop measurements that allow us to examine the various types of core-periphery structures.

Understanding the structure of a financial network is the key to understanding its function. Structural features exist at both the microscopic level, resulting from differences in the properties of single nodes, and the mesoscopic level, resulting from properties shared by groups of nodes. In general, financial networks contain unique structures, such as a core-periphery structure, by which a densely connected subset of core nodes and a subset of sparsely connected peripheral nodes coexist, or a modular structure with highly clustered subgraphs. With a core-periphery structure, the cores consist of the important financial institutions playing a critical role in the interactions within the financial network. In a modular network, the connecting nodes (i.e., bridge nodes) of multiple modules play a key role in network contagion. Therefore, identifying the intermediate-scale structure at the mesoscopic level allows us to discover features that cannot be found by analyzing the network at the global or local scale.

We use the different characteristics of network organizations to identify the influential nodes in some typical networks as a benchmark model. We use particular local measures, based either on the network's connectivity at a microscopic scale or on its community structure at a mesoscopic scale. We then use a mesoscopic approach to contagion models for describing risk propagation in a network, and we investigate the manner by which the network's topology impacts risk propagation. An agent fails depending on the states of its neighboring agents, and systemic risk is measured according to the proportion of failed agents. Failures are usually absorbed by neighboring agents provided that each agent is connected to other agents, even if failures occur relatively frequently with some agents. Such failures may, however, propagate to other agents under certain conditions. When that happens, things worsen throughout the chain and the network can experience widespread failure.

We distinguish between amplification, whereby specific agents cause failures in other agents by contagion, and vulnerability, determined by the number of failures brought down by the failure of other agents. Thus, we use two indices to quantify systemic risk:

1. *The vulnerability index* to quantify the ratio of agent defaults when each agent is selected in turn as an initial defaulting agent.
2. *The amplification index* to quantify the ratio of defaulting agents caused by an agent.

We consider two mock networks, each network consisting of the $N = 100$ agents, as seen in Figure 9.2, and we identify the influential agents in these networks. Figure 9.2(a) shows four symmetric star networks connected via one bridge agent in the center. Figure 9.2(b) shows four asymmetrical star networks of different sizes with the hub agents connected in the center of each subnetwork. These four hub agents play the role of the bridge agent.

The parameter θ denotes the ratio of total interbank loans $I = \sum_i I_i$ to total interbank assets $A = \sum_i A_i$, assuming that the ratio θ for every agent is the same across the interbank market. Given this assumption, the interbank loans I_i for agent i can be calculated from the assets A_i or the external assets E_i as follows:

$$L_i = \theta A_i = (\theta/1-\theta)E_i. \tag{9.5}$$

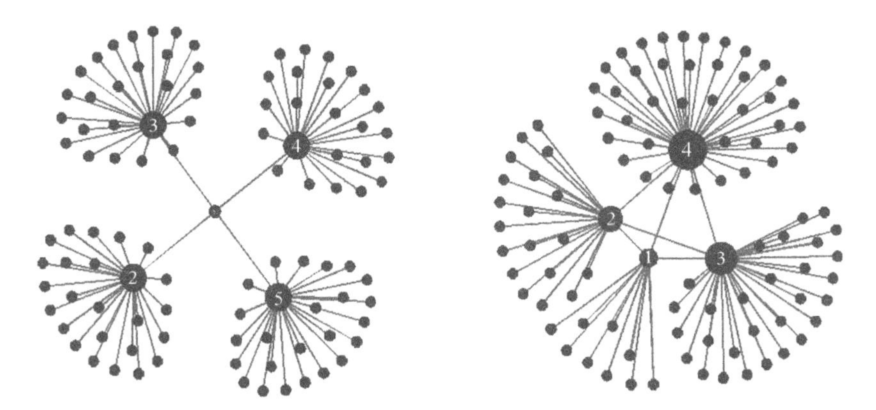

Figure 9.2 *Two mock networks: (a) Four symmetrical star networks are connected via one bridge agent (#1) at the center. (b) Four asymmetrical star networks of different sizes are connected through four hub agents (#1–#4). Each mock network consists of 100 agents*

Assuming that the unit amount of interbank loans (denoted by ω) between any two agents is the same (i.e., the amount of interbank loans represented by one link is fixed to a certain value ω), the interbank loans I_i can be calculated by $k_{out,i}$, denoting the out-degree of agent i (i.e., the number of agents who borrow from agent i). The assets A_i and the external assets E_i of agent i can be computed using the relation in eqn. (9.5). The interbank borrowings B_i is $\omega k_{in,i}$, where $k_{in,i}$ denotes the in-degree for agent i (i.e., the number of agents who lend to agent i). If we assume an undirected network to represent interbank transactions, we have $k_{out,i} = k_{in,i} = k_i$. In the simulation study, we set these parameters at $\omega = 1$, $\theta = 0.45$, and $\phi = 0.03$.

We select each agent in turn as an initial defaulting agent. We set the initial shock to a selected agent about to default, by stipulating a loss of 50% of the external assets. The propagation of defaulting contagion begins with the loss of a half of the external assets E_i owned by the agent i that failed initially. If the loss of the external assets E_i cannot be absorbed by the net worth (i.e., by the agent's capital buffer) C_i, the remaining the interbank borrowings B_i will be absorbed by each neighboring agent j, and agent j loses a portion of the interbank loans I_j. If the loss of assets A_j from agent j cannot be absorbed by the net worth C_j, the interbank borrowings B_j will be used. This series of debt propagation continues until all losses are finally absorbed in the interbank network.

We conducted 100 simulations by selecting each agent in turn as an initial defaulting agent. We derived the vulnerability index and the amplification index for each of these. Figure 9.3 shows the vulnerability and amplification indices for each agent in the network seen in Figure 9.2(a). The vulnerability index of the bridge agent at the center (#1) is the highest. Its amplification index is the lowest, however, and the same as those of the other peripheral agents (#6–#100). The respective vulnerability indices for the four hub agents (#2–#5) are very low but their respective amplification indices are the highest,

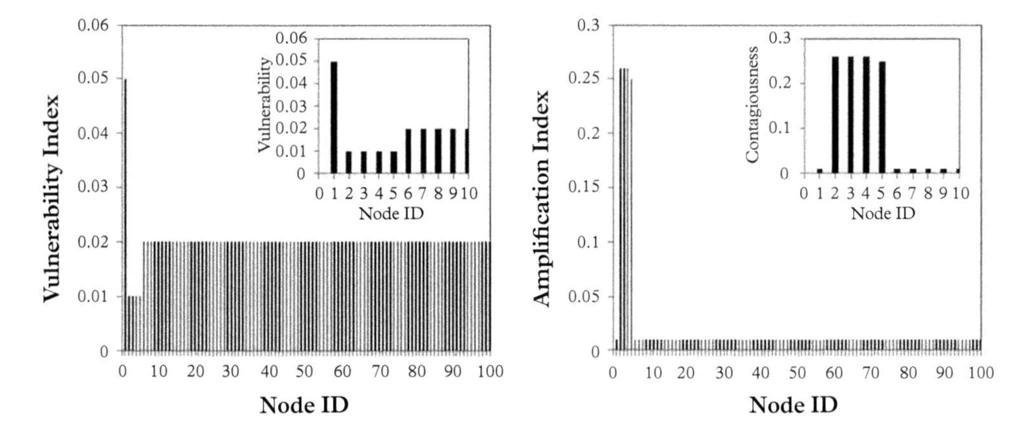

Figure 9.3 *The ratio of defaults when the cascade is initiated, by selecting each agent from Figure 9.3(a) to default in turn: (a) vulnerability index for each agent; (b) amplification index for each agent. The inserts in the figures highlight these respective indices for the hub agents*

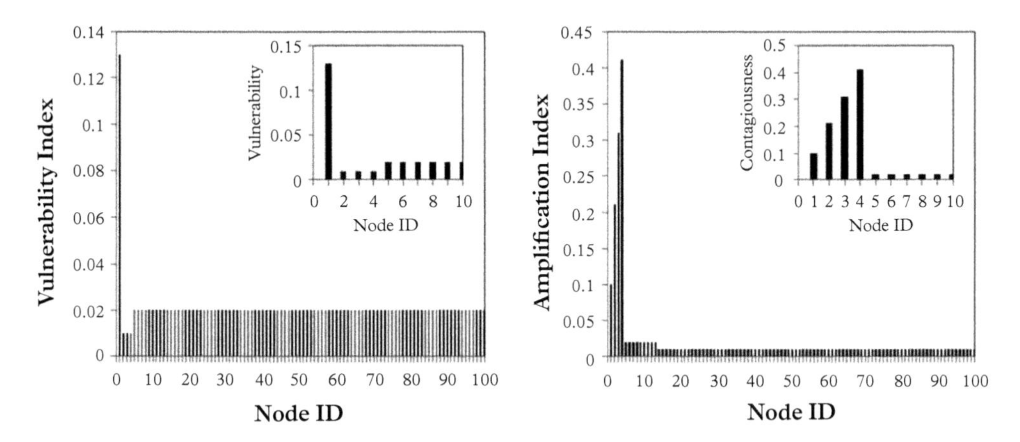

Figure 9.4 *The ratio of defaults when a cascade is initiated by selecting each agent in Figure 9.3(b) to default in turn: (a) vulnerability index for each agent; (b) amplification index for each agent. The inserts in the figures highlight these respective indices for the hub agents*

implying that when one of these hub agents is selected to default, it causes a large-scale cascade of defaults.

Figure 9.4 shows the vulnerability and amplification indices for each agent in the network depicted in Figure 9.3(b). In this case, the vulnerability of the hub agent (#1) in the smallest star network is extremely high, whereas those of the other three hub agents (#2, #3, #4) are lower than those of the peripheral agents (#5–#100). The amplification index for each hub agent (#1–#4) in each star network increases in proportion with the degree, implying that the amplification for a cascade of defaults is proportional to the degree of the hub agent. In summary, we found strong correlations between the position of an agent in the interbank network and the likelihood that it either causes contagion or will be affected by contagion.

It is known that interbank networks in general evolve over time without a fixed topology. On the other hand, most studies on systemic risk have been conducted on networks with fixed topologies. In recent studies, the underlying network topology is important for the occurrence of systemic risk. However, little is known about what type of network topology will minimize the degree of systemic risk. We turn now to an evaluation of the systemic risk in larger networks evolved from the two mock networks above. In Chapter 6, we considered the network design model, and Figure 9.5 shows two networks evolved from the two mock networks in Figure 9.3, by setting the parameters from Figure 6.7 in Chapter 6 as $p = 0.1$ and $\beta = 3$. Both networks in Figure 9.5 consist of 500 agents, and the average degree is $z = 5$.

Figure 9.6 shows the results from the simulation of default contagion on the evolved core-periphery network in Figure 9.5(a). Agents with a higher degree in the mock network (#2–#5) gain more connections during the network-evolving process and comprise the core of the core-periphery network. Hub agents in the mock network remain part of the core of the larger network, and they remain important agents, owing to their tremendous size. Figure 9.6(a) shows the relationship between the vulnerability index and the agent's degree. The vulnerability index for each core agent (#2–#5) is especially minimal. Figure 9.6(b) shows the relationship between the amplification index and each agent's degree. This result indicates that the amplification index for each agent has the tendency to grow in proportion to the degree. Core agents (#2–#5) have a higher amplification index, owing to the prominence of their degree.

Figure 9.7 shows the results from a simulated default contagion for the network in Figure 9.5(b). Both the vulnerability index and amplification index increase in

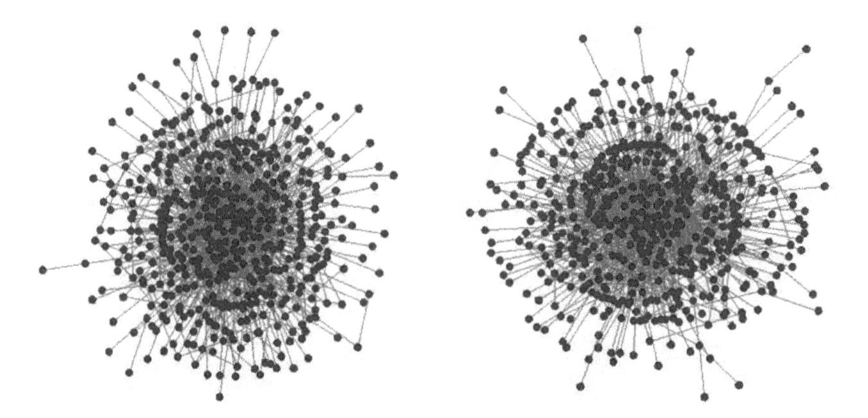

Figure 9.5 *Evolved networks from the mock networks in Figure 9.3. Both networks consist of 500 agents, and the average degree is $z = 5$. (a) Evolved network from Figure 9.3(a). (b) Evolved network from Figure 9.3(b)*

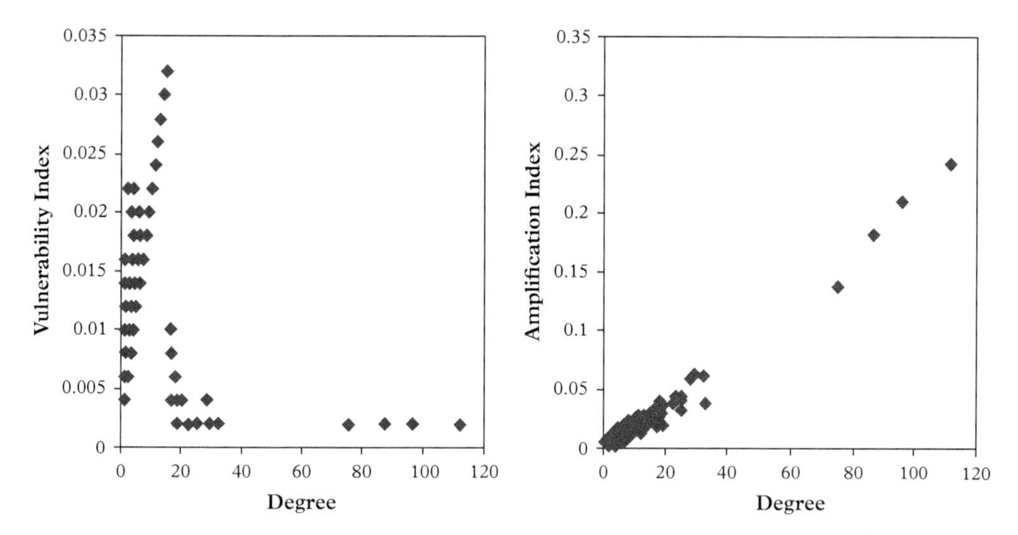

Figure 9.6 *Results from the simulation of default contagion on the evolved core-periphery network in Figure 9.5(a): (a) relationship between the vulnerability index and the degree; (b) relationship between the amplification index and the degree*

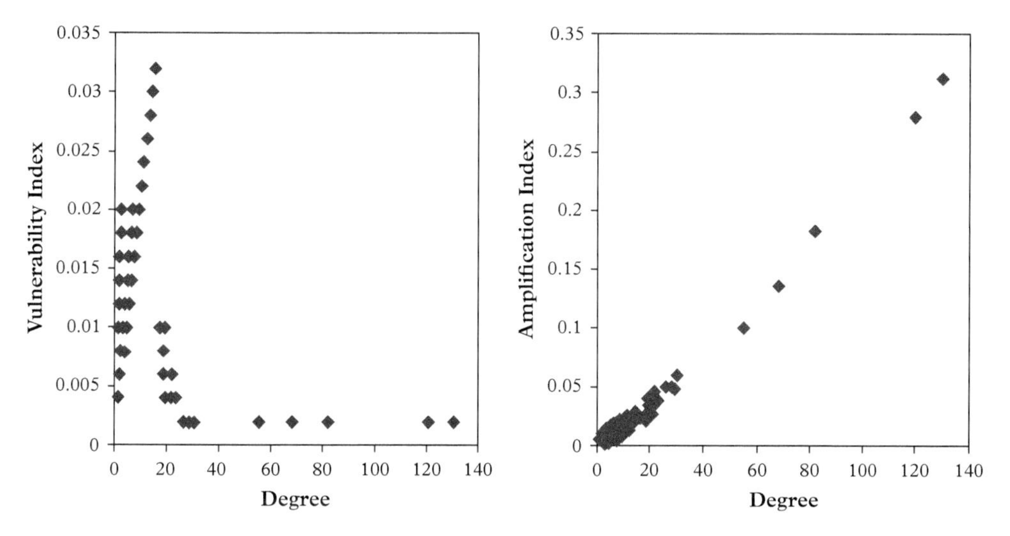

Figure 9.7 *Results from the simulation of default contagion on the evolved core-periphery network in Figure 9.5(b): (a) relationship between the vulnerability index and the degree; (b) relationship between the amplification index and the degree*

proportion with the agent's degree. The hub agents in the mock network (#1–#4) received more connections during the evolution of the network and they comprise the core of the evolved network in Figure 9.5(b). Figure 9.7(a) shows the vulnerability index, and Figure 9.7(b) shows the amplification index.

These results highlight the fact that it is not only the macroscopic features of the network, such as the degree distribution, that are relevant to predicting vulnerabilities. The mesoscopic features, such as the core-periphery structure and the network position (i.e., whether an agent is located close to the core or the periphery) are essential for predicting vulnerable agents and amplifiers of default contagion.

9.3 Cascading Failures and Defense Strategies

The ongoing progress in networking essential utilities has provided significant benefits to the quality of our lives. However, networks of this sort pose certain dangers insofar as a failure of a single node in the network can wipe out all the other nodes. In the past, people have repeatedly experienced large-scale blackouts. As mentioned above, a single power line failure in Oregon led to a massive cascade of power outages over six weeks in 1996, leaving tens of millions of people without electricity (Dobson et al., 2007). A large blackout may begin as a fairly routine disruption in a particular area that is prone to failure. Such failures are usually absorbed by neighboring nodes. However, when this happens under certain conditions, things worsen all along the network and the network can experience widespread failures.

Networked systems are routinely subjected to external shocks, and if one node fails as a result of an external shock or an excessive load, this failure is propagated to other nodes, often with disastrous consequences. A cascading failure is a failure in a network of interconnected nodes in which the failure of one node triggers the failure of successive nodes. Such failures can occur in many types of networked systems. A cascading failure usually begins when one node in the network fails. When this happens, nearby nodes must then take up the slack for the failed node. This, in turn, overloads the nodes causing them to fail and prompting additional nodes to fail. This basic mechanism to cascading failures implies that something detrimental can happen in a system that is a much bigger and worse than the initial failure.

The threshold-based cascade model introduced in Chapter 6 can also be applied to an understanding of the mechanism behind cascading failures, as discussed in Section 9.2. A cascading failure can be modeled using the threshold rule, which dictates that the state of each node is determined by the states of the nodes adjacent to it. Nodes fail when the number of failed neighboring nodes exceeds a certain threshold. Under specific conditions, the failure of a single node can change the state of the adjacent nodes sequentially and its negative impact can spread throughout the network. The failure of a single node is sufficient for triggering a cascading failure, and it can result in significant damage to the whole network.

A cascading failure can also be triggered by the initial failure of a single node from overloading. In this section, we investigate cascading failures from overloading critical nodes. In electric and traffic networks, high loads on particular nodes can cause failures, such as electrical line failures or traffic jams, with the potential to sever links or remove nodes from the network. If the load increase exceeds the capacity of any node, that node will fail, triggering a redistribution and possible subsequent failures in other nodes.

When all nodes are operational, the network operates steadily. However, the removal of a node will cause a redistribution to the load with the shortest path routing. This will generally increase the load at some nodes. If the load increase exceeds the capacity of any node, that node will fail, triggering a new redistribution with the possibility of subsequent failures in other nodes. Eventually the cascade of failures will cease, when all remaining nodes can handle the collective load. However, there is a critical load at which risk sharply increases toward a threshold for cascading failures.

The size of a cascading failure depends on the heterogeneously distributed load and the topological structure of the network. In many networks, each node is designed with some extra capacity and resources to respond to accidents, such as traffic bursts, from the failures in other nodes. The removal or failure of some nodes by accident or for other reasons usually causes a redistribution of the network flow. The impact of those initial failures can appear not only in adjacent nodes as a direct consequence of the failure, but also in distant nodes as an indirect consequence. The overflow model sets the node-specific capacity for each node, and a load for each node and connection is determined by traffic-routing schemes, such as the shortest path or other routing strategies. If the shortest path routing strategy is used, most traffic is directed through hub nodes. In such a case, the failure of the hub node will require a major redistribution of the flow. Preventing cascading failures resulting from amplified loads remains a challenging research issue.

Motter and Lai (2002) studied the overflow model, showing that a redistribution of loads can cause a cascading failure. They addressed the issue of a cascading failure, focusing on the traffic flow in networks. They showed that load turbulence triggered by the failure of a single node can cause catastrophic damage to the network. Their model is simple enough to support a tractable analysis, and it is generally applicable to actual networks such as the Internet and power grids. Their model consists of several key elements:

1. Traffic is simulated by the exchange of one unit of the relevant quantity (information, energy, etc.) between every pair of nodes along the shortest path connecting each pair of nodes. The load placed on a node is equivalent to the total number of shortest paths passing through the node.
2. The capacity of each node is defined as the maximum load that a node can handle. The capacity C_i of node i is assumed to be linearly proportional to its initial load, $L_i(0)$:

$$C_i = \alpha L_i(0), \quad i = 1, 2, \ldots, N, \tag{9.6}$$

where $\alpha \geq 1$ is the tolerance parameter, and N is the initial number of nodes in a network.

The tolerance parameter a in eqn. (9.6) also represents the loading level for each node. For instance, if $\alpha = 4$, each node is loaded with 25% of its capacity and the network load is low, but when $\alpha = 1$, the loading level is 100%, a full load. If a node has a relatively small load, its removal will not cause major changes in the load balance, and subsequent overload failures are unlikely. However, when the load at a node is relatively large, its removal is likely to affect loads at other nodes significantly, and this can lead to a sequence of overload failures.

We quantify the damage caused by a cascading failure. One commonly used measure is the ratio of the surviving nodes that are connected to each other. The largest connected component (LCC) is the largest component for which there is a path between any pair of nodes in a network. We quantify the network robustness using R defined in eqn. (9.7), which represents the ratio of surviving nodes in the LCC after a cascading failure caused by the failure of a single node with the highest initial load:

$$R = N'/N, \tag{9.7}$$

where N and N' denote the size of the LCC in the network before and after a cascading failure, respectively. A network has high integrity if $R \approx 1$, i.e., almost all nodes have survived and they are almost fully connected, meaning that the impact from a cascading failure was negligible. On the other hand, if $R \approx 0$, the network is almost completely disconnected and the cascading failure has significantly affected the network.

The traffic is simulated by sending one unit of flow from node i to node j, for every ordered pair of nodes (i, j). If there is more than one shortest path that connects node i and node j, then the traffic is divided evenly at each branching point. The load on node k is then given by:

$$L_k = \sum L_k^{(i,j)}, \tag{9.8}$$

where $L_k^{(i,j)}$ is the contribution of the ordered pair (i, j) to the load on node k.

The simulated cascade can be divided into two events:

1. The initial failure, where a single node with the highest load (and *betweenness*) is removed by a breakdown as a result of some external shock.
2. The propagation of the cascade, where a number of nodes are removed owing to subsequent failures from overloaded nodes.

We begin with the network $N = N(0)$ at time 0 with no overload in any node. The traffic is simulated by sending one unit of flow from node i to node j with the shortest path connecting the two nodes, i and j. The initial failure is performed at time 1, by removing the node with the highest load from $N(0)$ to form the network $N(1)$. The redistribution of the load is calculated and the load for each node k, $L_k(1)$, is updated. Nodes in which the load exceeds their capacities are overloaded, and their loads are simultaneously redistributed. After the load is redistributed, any node in which the load exceeds its capacity is removed to form the resulting network, $N(2)$. The redistribution of the load and the formation of a new network proceed in this manner until the cascading failure ends.

A cascading failure depends on the underlying network topology because the topology of a network crucially determines the traffic flow in it. There are interesting questions to explore related to what kind of network will be the *most vulnerable* to a cascading failure, and which network topology will be the *least susceptible* to one. The failure of a node can drastically change the balance of the load. Thus, we must take into account not only the performance of each node, but also the networks topology, in designing networks that are robust to cascading failures.

We can classify the nodes in a network into two categories: hubs and peripheral nodes. Hub nodes play an important role because they connect other nodes and guarantee the connectivity of the network. Hub nodes are usually selected as a pathway to reduce the average shortest path lengths between each pair of nodes. A hub node usually increases the efficiency of the traffic flow in a network because it often carries a large amount of flow as a relay point. Therefore, the failure of a hub node can lead to major changes in the network connectivity as well as load balance. Hence, the location of hubs are a weak point in the network, and the manner by which connections are made between hub nodes plays a key role in preserving the integrity of a network when hub nodes fail.

Based on some of the logical principles mentioned here, we observe the extent of a cascading failure in a specific network topology consisting of a clique with a peripheral network (*CPN*: *CPA or CRA*), as described in Chapter 6. We can observe that hub nodes are densely connected to form a clique, and many peripheral nodes are connected only to one of the hub nodes. Actually, the *CPN* consists of many tree-structured modules, and they are unified via a cluster of hub nodes. We evaluate and compare the robustness of the *CPN* with other network topologies: a scale-free network (*SF*) and a random network (*RND*). The size of the *CPN* is set at $n = 33$, and all networks have the same number of nodes $N = 500$, where the average degree is $z = 4$.

The robustness of each network is shown in Figure 9.8, when one node with the highest load was chosen as an initial failure node. The network robustness R in the *SF* and *RND*, as derived with eqn. (9.7), decreases drastically when the tolerance parameter α is low. For instance, even when every node has a capacity that is twice that of the initial load, the size of the largest component is reduced by more than 20%. When α is lower, and when every node in the network is running under a high load, the damage is considerably greater than when each node in the network is running under a lower load. The network robustness increases in proportion to the tolerance parameter α. Therefore, the most effective method for preventing a cascading failure in the *SF* or *RND* is to increase the tolerance parameter α, meaning that all nodes should have a sufficiently high capacity.

On the other hand, the network robustness R for the *CPN* remains close to one, even when the tolerance parameter α is very low. In fact, the robustness of the CPN does not depend on the tolerance parameter α, and cascading failures do not occur in the *CPN*. In general, the *CPN* has a wider margin between the capacity and the load compared with the other two networks. By forming a complete network (i.e., a clique) of hub nodes, each hub node prevents fragmentation from forming. At the same time, failures in the peripheral nodes decrease the total load of the system. Therefore, a *CPN*

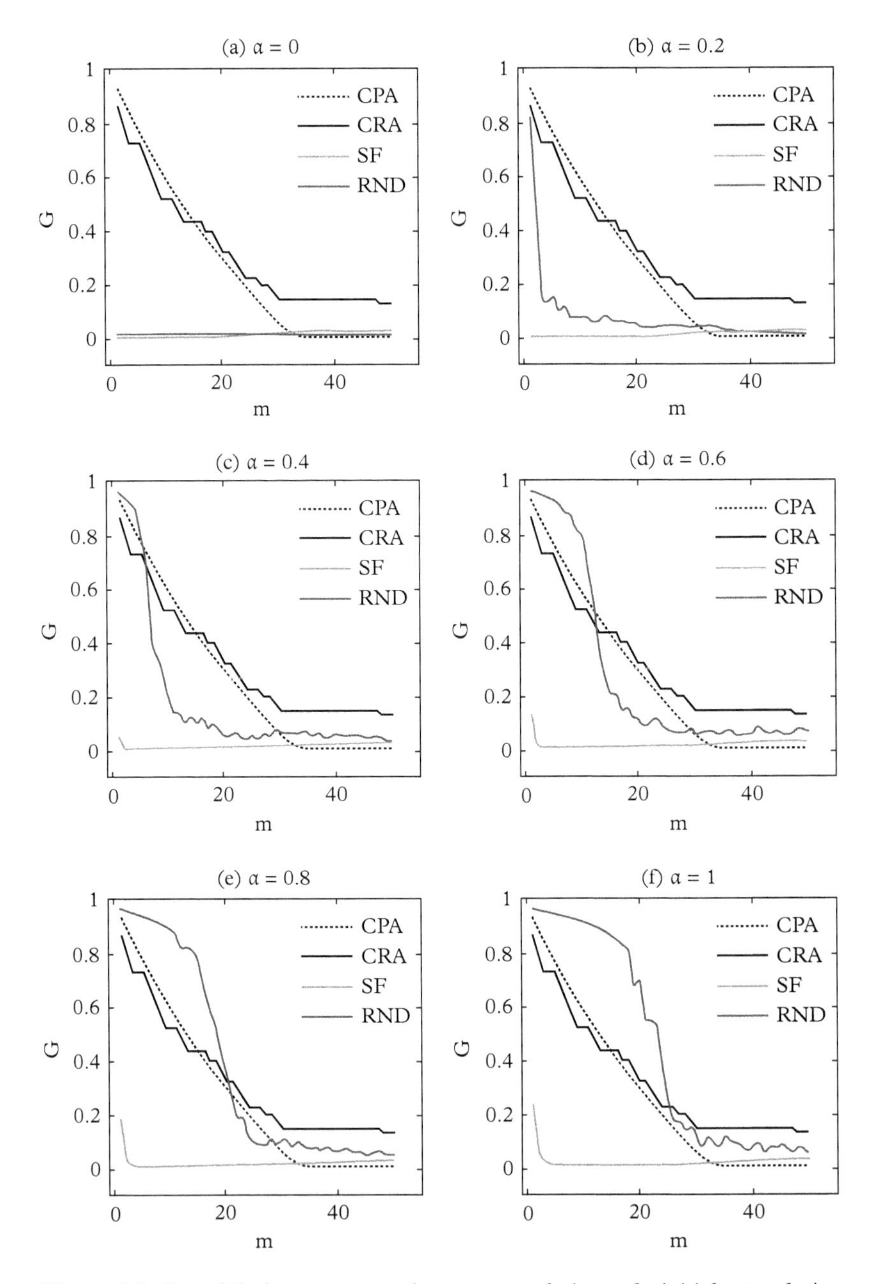

Figure 9.8 *Size of the largest connected component relative to the initial network size vs. the network capacity,* α*,* $N = 500$*,* $z = 4$*, (a) random network (RND), (b) scale-free network (SF), (c) a clique with a peripheral network (CPN)*

with a modular architecture is the most effective topology in suppressing load turbulence from node failures.

Motter et al. (2005) investigated strategies of defense in mitigating cascading failures. He introduced a relatively costless defense method based on selectively removing some nodes and edges immediately following an initial failure. Noticing that a cascading failure can be divided into two events—(i) the initial failure, in which a small number of nodes are removed either intentionally or from random breakdowns, and (ii) the propagation of the cascade, where more nodes are removed from subsequent failures resulting from overload—Motter's method of defense rests on the strategic removal of nodes after (i) but before (ii). The main idea is that insignificant nodes should be removed. The intentional removal of insignificant nodes reduces the overall load in the network, and the extent of the cascade can be drastically reduced by doing so. Although the intentional removal of nodes can result in an even smaller final-connected component, it is possible to reduce the magnitude of the cascade by carefully choosing which nodes to remove. The resulting robustness R of the network is greater than it is by not adopting this defense strategy.

However, Motter's proposed defense strategy is not particularly effective, especially when the tolerance capacity of each node is limited, or when the network is operated under a high load with a low α. This defense strategy also comes with significant disadvantages: it is difficult to detect cascading failures early, and the strategy requires a comprehensive understanding of the network's topology to find the correct nodes to remove. The speed of a cascading failure is typically much faster than the speed at which the network grows. Intentionally removing nodes with a shut-down procedure is feasible, but removing nodes during a cascading event may be unfeasible. Moreover, in general we do not have a precise understanding of actual networks, and this limitation makes it difficult to discover which nodes should be removed.

We now turn to an exploration of alternative defense strategies for mitigating cascading failures. There are essentially two types of strategies for doing so.

1. *Hard strategies*: A way to minimize damage from cascading failures by changing the connections in the network (i.e., the network topology), and by increasing the capacity (or the tolerance parameter) of the load for each node.
2. *Soft strategies*: A way to minimize damage from cascading failures without any change to the connections in the network or to the tolerance parameter of the load.

A high load is an incentive for installing additional connections to alleviate the strain of the load. However, because most actual networks have a specific existing structure, hard strategies of changing network topologies are not applicable in many cases. Another effective and simple method is to increase the tolerance parameter of the load α, as derived in eqn. (9.6), and by doing so giving all the nodes sufficient resources to prevent failures as a result of overloading. However, increasing the tolerance parameter α is often limited by design costs.

To overcome the limitations to hard strategies, soft strategies have been proposed to counter cascading failures without impacting the connections in a given network (Chen and Gostoli, 2015). One effective soft strategy is to design an efficient routing strategy that improves the traffic efficiency by adjusting load flows in the underlying network. By properly choosing the respective weight for connections, a maximum level of robustness against cascading failures can be achieved. The weight of a link that connects node i with node j is assigned proportionally to the connectedness of the two nodes as follows:

$$w_{ij} = (k_i k_j)^{\beta}, \tag{9.9}$$

where k_i and k_j are the degrees (i.e., the number of links) in node i and node j, respectively, and β is the weight control parameter. By adjusting the parameter β, the flow of networks can be controlled.

The weight of a link that connects node i with node j can be also referred to as the resistance of the connection against the flow. If the control parameter β in eqn. (9.9) is zero, all links are equally weighted, and the flow is transmitted along the shortest path. When the control parameter β is set to a positive number, the weight between hub nodes is high and hub nodes will be avoided during traffic transmissions. This situation reflects real world networks, because connections that are highly resistant will obstruct the flow. By contrast, if the parameter β is negative, lower weight is assigned to links connecting hub nodes, and the links connecting hubs with lower resistance are frequently used to transmit information.

The number of shortest paths flowing through each node is used to measure the load in an unweighted network. We extended the definition of load to weighted networks. The weight of a path from node m to node n that is passing through a set of intermediate nodes $S = 1, 2, \ldots, l$ is the sum of the link weights included along the path:

$$w_{m \to n} = \sum_{i=1}^{l-1} w_{ij}, \quad j = i + 1. \tag{9.10}$$

In particular, the load for node i is defined as the total number of shortest weighted paths that pass through that node.

We considered three specific cases: (i) $\beta = 0$, (ii) $\beta = -1$, and (iii) $\beta = 1$. In the first case, $\beta = 0$, traffic proceeds through the shortest path. Thus, a scale-free network (*SF*), for instance, has a heterogeneous load distribution, and a higher-degree node carries a higher load, because all nodes tend to use hubs as shortcuts to transmit information through the network. For (ii), the parameter $\beta = -1$, giving it the most heterogeneous load distribution of the three cases. In this case, lower weight is assigned to links connecting hub nodes and they are more frequently used to transmit information. We refer to this weight scheme as the hub-oriented strategy. With (iii) the parameter $\beta = 1$, a homogeneous load distribution is realized such that links connecting hub nodes will be mostly avoided. We refer to this weight scheme as the hub-avoidance strategy.

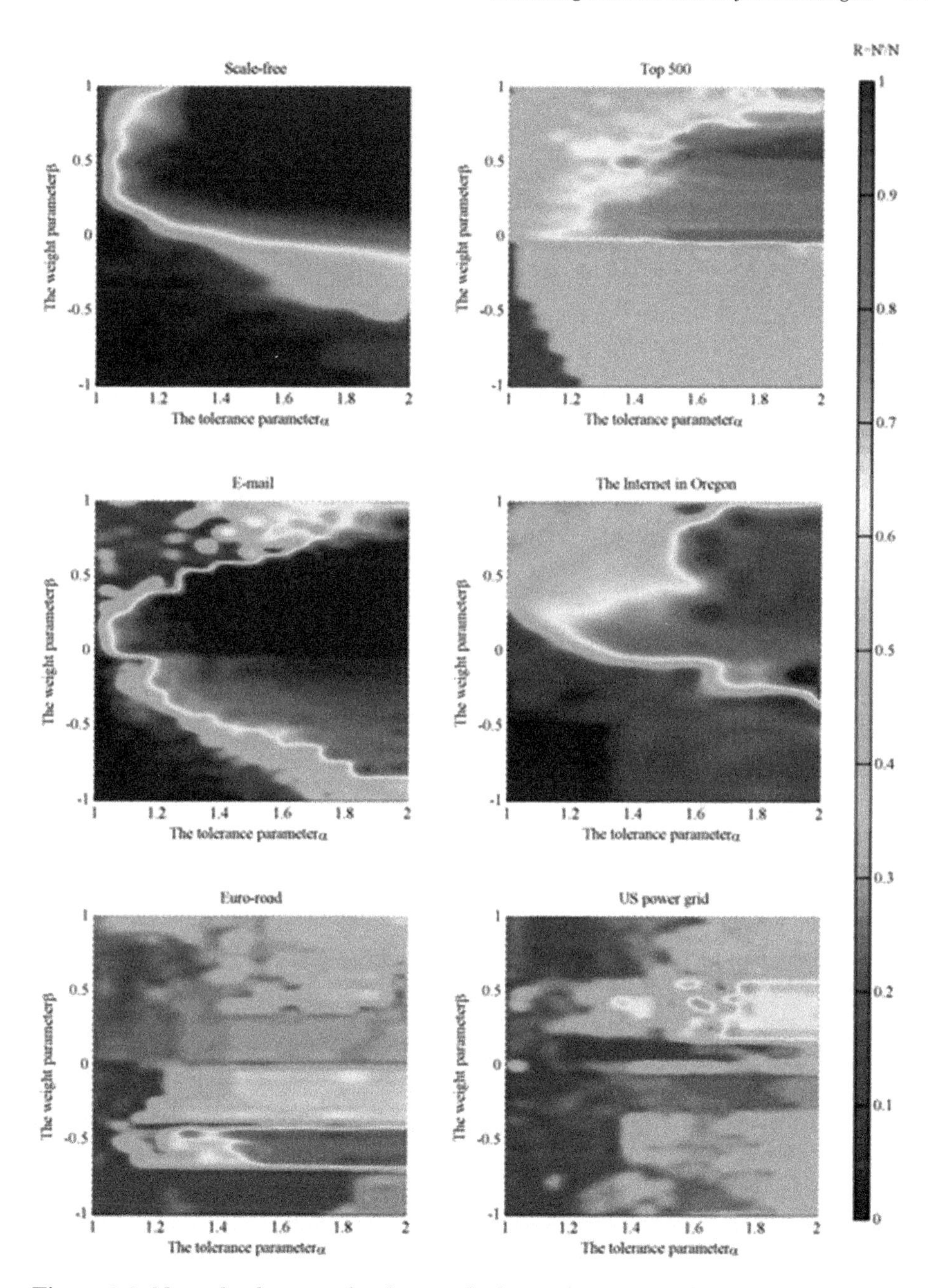

Figure 9.9 *Network robustness of each network: the x-axis represents the tolerance parameter* α*, the y-axis represents the control parameter* β

We evaluated the effectiveness of the load redistribution strategy with the assumption that only a single node with the highest load will initially fail. We considered the following typical networks: (i) a scale-free network (*SF*) that follows a power-law distribution in terms of the node degree $P(k) \sim k^3$, (ii) the u.s. power grid network, (ii) Euro road, (iv) the e-mail network, and (v) the Top 500 Airport network. Among them, the degree and load distributions in the Euro road and US power grid network are more likely homogeneous, and those of the *SF*, e-mail, and the Top 500 Airport are heterogeneous.

Figure 9.9 shows the network robustness for each network, according to the ratio of the surviving nodes (R). The x-axis represents the tolerance parameter α, and the y-axis represents the weight control parameter β. The network robustness for each network can be enhanced if the tolerance parameter α is increased. The effectiveness of the load redistribution strategy was evaluated by changing the weight control parameter β. In order to enhance network robustness, we adjusted both the tolerance parameter α and the weight control parameter β. When the control parameter β is positive, a heterogeneous load distribution tends to be more homogeneous. Figure 9.9 shows that network robustness was enhanced with a hub avoidance strategy (i.e., with $\beta > 0$) in the *SF* and Top 500 Airport networks. On the other hand, the Euro road network behaved in a different manner, such that the network was more robust with the hub oriented strategy (i.e., with $\beta < 0$). In this case, the load is redistributed such that hub nodes handle more traffic to realize more efficient network performance (Ann and Namatame, 2015).

In summary, the simulation results shown in Figure 9.9 indicate that the network robustness against cascading failures can be enhanced with a shortest path strategy for the weighted networks specified in eqn. (9.9) by choosing the proper control parameter β. Moreover, we can classify strategies that enhance network robustness into the following two groups:

1. The hub avoidance strategy ($\beta > 0$) is effective for scale-free networks and the Top 500 Airports network.
2. The hub oriented strategy ($\beta < 0$) is effective for the Euroroad network.

However, these two soft strategies are not as effective for the U.S. power grid network, and a hard strategy is consequently needed to make it more robust.

9.4 Gain from Risk Sharing

In this section, we discuss the particular challenges to enhancing the resilience of the networked agents in coping with amplified risks. In particular, the mechanism of risk sharing is proposed, in which the shocks or excess load of some agents are properly transferred to other agents in the network so that allocated shocks or extra loads are within the capacity of each agent. The mechanism of risk sharing is specified by the local shock transfer rule to achieve a globally balanced shock. We investigate how external shocks tend to be shared in the agent network and how this reallocation mechanism minimizes the probability of the agents failing.

Having studied the dynamics of cascading failures and systemic risk in networked systems, we may suspect that the most damaging failures are impossible to anticipate with any confidence. This type of risk is invisible because it results from the unpredictable interplay of a myriad of agents in the network. In power grids, for instance, we can reliably assess the risk that any single power generator in the network will fail under some conditions. However, once a cascade begins, we can no longer recognize what those conditions will be for each generator, because these conditions could change dramatically depending on what else happens in the network. In the financial system, risk managers are only able to assess their own exposure given that conditions in the rest of the financial system remain predictable, but in a crisis these conditions change unpredictably.

When networked agents are subjected to an external shock, the disruption will be amplified by the internal shock transfer process within the network, and therefore there is a growing need to focus on how to manage risk collectively. One solution is to make the system less complex, in order to reduce the possibility that any one agent will trigger a catastrophic chain of events. With respect to protection and risk reducing measures taken by agents, we need to assess the risk of each agent as well as how their actions affect one another. Because risks arise within the agent network, effective solutions usually require looking beyond individual agents. The fact that the risk is often determined in part by the behavior of other agents imparts a complex structure to the incentives that agents face in reducing risk by investing in risk mitigation measures.

Complex adaptive systems found in ecologies are often capable of showing remarkable resilience. Despite major changes in their surrounding environment, these systems are somehow able to adapt and survive by overcoming negative effects. Risk sharing in complex adaptive systems is a key mechanism for resilience in recovering from the disruption. The capacity of a system to absorb shock is just as important as its performance. Rather than concealing the actual risk, it is important for agents to be open if they are to share the risk according to some sharing mechanism. Unfortunately, agents tend to avoid openness. Recently developed tools in network analysis provide the possibility for understanding the deep connective structure of networked systems and for identifying critical nodes in the network that, when removed, can lead to catastrophic systemic collapse. These tools also help us to design large-scale networked systems that are resilient to various sorts of disruptions in different parts of the system.

Risk management is the process of identifying the measures needed to control and mitigate the damage an event may cause. External shocks to some agents are propagated in other agents, and amplification in the network often results in disastrous consequences. One promising method for mitigating a series of failures is to design a risk-sharing mechanism. The solution to the management problem of networked risk involves coordinating efforts to decrease risk collectively. Coordinating efforts and collective incentives can push agents toward the tipping point, which then reinforces behavior that benefits the entire system. With the right incentives, risk sharing can encourage actions by individual agents that reduce networked risks with full coordination. Coordination is the fundamental underpinning of the agent system, allowing agents to manage the associated risk with interacting agents before they fail. We focus on how to mitigate the impact from external shocks by risk sharing as coordinated risk management.

There is a large body of research on risk sharing, especially in finance. Gallegati et al. (2008) examined the trade-off between the mutual insurance of financial institutions and systemic risk. Risk sharing in networks generally considers voluntary connections and focuses on the effect of individual actions on the configuration and stability of the network. In particular, gain from risk sharing is generally acknowledged to require complete connectivity and full insurance among all members in the network. Cainelli et al. (2012) extended the work of Gallegati to analyze how production and financial networks can affect the probability that agents will default. They considered an equal shock transfer model in equilibrium and investigated how idiosyncratic shocks tend to be allocated in the agent network and how this allocation changes the probability of agent failure. They showed that risk sharing is beneficial only when the overall economic environment is favorable, whereas risk sharing might be detrimental when it is not. They also analyzed efficient and stable configurations in agent networks for risk sharing, that is, the connections that guarantee the agents' bilateral mutual insurance.

Cabrales et al. (2011) investigated how the capacity of an economic system to absorb shocks depends on the specific pattern of interconnections established among agents. They showed how the agent network's topology affects the probability that any given agent will fail. The key trade-off is between the advantages gained from risk sharing and the costs resulting from increased exposure to risk. They sought the optimal network topology that depends on the structure of the shocks, and discovered that risk is maximally shared when all the agents form a cohesive and fully connected network. However, this configuration also yields the highest exposure to large-scale shocks, and this can lead to widespread defaults across the entire agent network. Therefore, there are two alternative methods to reduce exposure to risk. One method involves segmentation, which isolates the agents in each component from any shock that might hit the other agents. The second method is to reduce the density of agent connections, which buffers the network and mediates the propagation of shocks throughout it.

In the analysis that follows, we refer to danger as the risk of being hit by a negative shock that is generated somewhere in the agent network. There are two basic ways by which shocks are diffused through the agent network. They can be transferred from an agent to connected agents based on a specified rule, or they can be transmitted and spread to other agents. The first process implies preserving the original quantity of a shock, whereas the latter implies its multiplication, as with epidemic diffusion. Here, we consider the first case, where the original quantity of a shock is preserved.

A shock migration strategy with risk sharing involves transferring the negative effects of a shock among agents so that it can be balanced. The goal of risk sharing is to mitigate a shock that targets specific agents. A surplus shock can be diffused through the network to reach a steady balanced state. If transferring risk does not result in a balanced solution immediately, the risk transfer process is reiterated until the shock difference between any two agents is smaller than a specified value. In order to achieve this balance, a portion of the shock excess for the agents is transferred to other agents. One agent is placed at each node in a network, where each agent applies a risk-sharing protocol to mitigate a shock or overload. Coordination underpins agent networks in a fundamental manner, allowing the agent to mitigate the associated risk with the other agents before that agent fails as a result of a shock beyond its capacity. A "rule" in risk sharing specifies how to propagate and share the incoming shock among agents. Each agent considers not only the risk to itself but also the risk to nearby agents. Thus, the agents act with reciprocal altruism.

A dynamic description of risk sharing among agents requires a preliminary definition of the topology of an agent network expressed as a graph. Risk sharing in the agent network is formalized using a weighted network, where a direct connection between two agents reflects the agreement of two agents to undertake a direct exchange of their risk (in the event of a shock). We allow these risk swaps to occur repeatedly. Thus, agents with only an indirect connection through intermediaries will be exposed to some other agent's risk reciprocally, if the exchanges are repeated.

We investigate the network effect on the gain from risk sharing by assuming that each agent i in a network with N agents will be subject to an external shock. We represent the transfer mechanisms for these idiosyncratic shocks in a weighted directed network through a matrix $W = (w_{ij})$, where the valued-directed edge w_{ij} from agent j to agent i measures the portion of the shock that agent j can transfer to agent i. We model the initial exogenous shock $x(0) = \{x_i(0), 1 \leq i \leq N\}$ for all agents as a random vector in which a generic element $x_i(0)$ is an identical and independent random variable with the same probability distribution. The shared risk for each agent i after one iteration is equal to:

$$x_i(1) = \sum_{j=1}^{N} w_{ij} x_j(0). \tag{9.11}$$

We consider the effects of the shocks after repeated iterations, where agents are assumed to be more resilient to the shock in the sense that they have a relatively high failure threshold. The long-term behavior of the shock transfer mechanism depends crucially on the network's structure as specified in the weight matrix W. In general, the shared risk of all agents at round $t + 1$ is calculated by:

$$x(t+1) = Wx(t) = W^t x(0). \tag{9.12}$$

Our goal is to derive the conditions for the channels between each pair of agents that guarantee that the risk-sharing mechanism will eventually converge upon globally sustainable risk in a totally distributed manner. We can measure the gain from risk sharing in terms of the ratio of safe agents when they are subjected to large-scale external shocks. The gain from risk sharing depends on the capacity of each agent to absorb shock, and on the interconnected patterns among agents. If too much time is needed to absorb the shock, owing to a slow convergence, the risk-sharing mechanism will be unsuitable. Thus, it is also important to reduce the time needed to achieve a risk-sharing equilibrium. Risk transfer has some disadvantages, resulting from the local nature of the risk information that is used. First, the shared shocks may not be balanced, and some agents may need to bear a larger share of the shock. Second, the number of iterations required may be high.

In particular, we consider the following two risk-sharing rules: an average-based sharing rule (*ASR*) and a diffusion-based sharing rule (*DSR*). For the *ASR*, each agent transfers an equal share of the shock, and the transfer mechanism works to equalize the shock to all agents in equilibrium. For the *DSR*, each agent transfers an adjusted portion of the shock according that agent's degree. The choice between these risk-sharing rules depends on the specific application of interest.

9.4.1 Average-based sharing rule (*ASR*)

Essentially, the *ASR* operates as follows: Each agent i compares the current shock x_i with the shocks to each of his/her neighbors in turn, and transfers the difference in order to balance the shock at a local level. Therefore, agents must know the current shock levels of their neighbors. The shared risk for agent i at round $t+1$ is defined as the proportion to the sum of the differences to the other connected agents as follows:

$$x_i(t+1) = x_i(t) + \alpha \sum_{j \in N(i)} (x_j(t) - x_i(t)) = (1 - \alpha k_i) x_i(t) + \alpha \sum_{j \in N(i)} x_j(t), \tag{9.13}$$

where $N(i)$ is the group of agents connected to agent i, and k_i is the cardinality of $N(i)$ (i.e., the degree) of agent i. The level of risk transfer as well as the risk receipt for each agent under the *ASR* is illustrated in Figure 9.10.

In general, it is not possible to reach an equilibrium in a single step because agents are restricted to using local risk information. Therefore, the risk-transfer process should

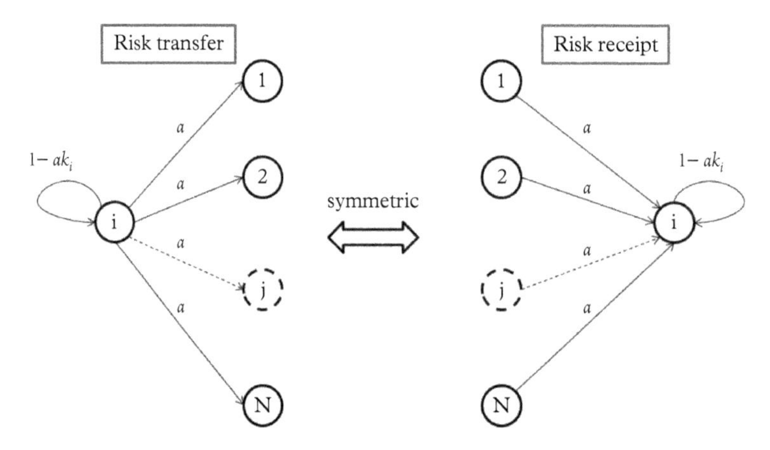

Figure 9.10 *Risk transfer and risk receipt for each agent under the ASR*

be repeated until all agents detect the shock to be locally balanced. From eqn. (9.10), we know that the *ASR* has a risk preserving property:

$$\sum_{j=1}^{N} x_j(t) = \sum_{j=1}^{N} x_j(0), \quad t = 1, 2, \ldots. \tag{9.14}$$

In matrix form, we can describe the risk transfer dynamics with the *ASR* as:

$$x(t+1) = (I - \alpha(D - A))x(t), \tag{9.15}$$

where $D = (k_i)$ is a diagonal matrix with a diagonal element k_i, and A is the adjacency matrix $A = (a_{ij})$, which is symmetric with elements $a_{ij} = a_{ji}$.

The weight matrix W from eqn. (9.12) under the *ASR* is given as:

$$W = I - \alpha(D - A). \tag{9.16}$$

Because this weight matrix is symmetric, i.e., $W = W^T$, it is a doubly stochastic matrix. Therefore, the risk distribution in equilibrium under the *ASR* is equalitarian, i.e., each agent in equilibrium receives a common shock amounting to the average of the initial shocks to all agents:

$$x_1(t) = x_2(t) = \ldots = x_i(t) = \ldots = x_N(t) = \sum_{i=1}^{N} x_i(0)/N, \quad (t \to \infty). \tag{9.17}$$

With the *ASR*, each agent tries to balance its shock with the shocks to the other agents with whom he/she is connected. In such a situation, the *ASR* provides for shock balancing within the network, although the number of steps required depends both on the initial distribution of the shock among the agents and on the networks topology.

9.4.2 Diffusion-based sharing rule (*DSR*)

The diffusion-type protocol (*DSR*) operates such that each agent i transfers the adjusted shock according to its degree with respect to other connected agents. This process is repeated until all agents can detect the shock to be locally balanced. The shared risk of each agent i at round $t + 1$ is defined as:

$$x_i(t+1) = (1 - \alpha)x_i(t) + \alpha \sum_{j \in N_i} x_j(t)/k_j. \tag{9.18}$$

The sharing weight w_{ij} for agent i to the risk of agent j is:

$$w_{ii} = 1 - \alpha_i, \quad w_{ij} = \alpha_{ij}/k_i, \quad j \neq i. \tag{9.19}$$

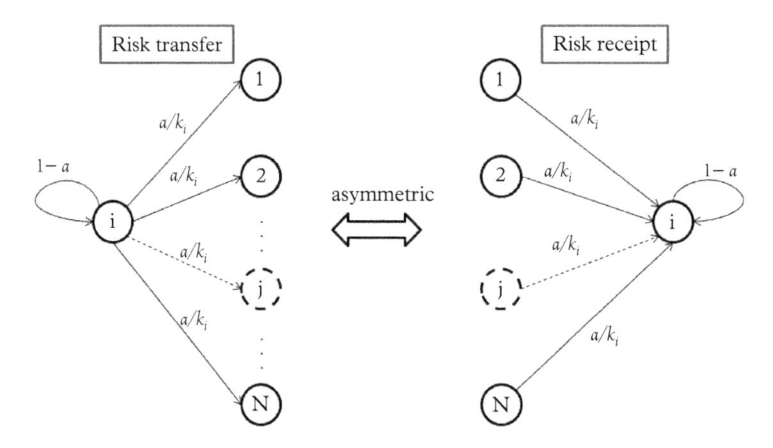

Figure 9.11 *Risk transfer and risk receipt for each agent under the DSR*

The level of risk transfer, as well as risk receipt for each agent under the *DSR*, is illustrated in Figure 9.11.

The *DSR* also has the risk preserving property, exhibited by satisfying the condition:

$$\sum_{j=1}^{N} x_j(t) = \sum_{j=1}^{N} x_j(0), \quad t = 1, 2, \dots . \tag{9.20}$$

In matrix form, we can describe the risk transfer dynamics with the *DSR* as:

$$x(t+1) = ((1-\alpha)I + \alpha AD^{-1})x(t). \tag{9.21}$$

The weight matrix W from eqn. (9.12) under the *DSR* is given as:

$$W = (1-\alpha)I + \alpha AD^{-1}. \tag{9.22}$$

The shared risk of each agent in equilibrium under the *DSR* is obtained by the weight proportional to the sum of the shocks to all agents, which is given as:

$$x_i(t) = \left(k_i / \left(\sum_{i=1}^{N} k_i \right) \right) \sum_{j=1}^{N} x_j(0), \quad i = 1, 2, \dots, N \ (t \to \infty). \tag{9.23}$$

A risk pool is a risk management strategy employed by insurance companies to ensure that losses from claims do not become catastrophic. Risk pools are seen as a way to provide protection to the individual, while also protecting those who are providing the insurance. In the risk pool, everyone shares equal risk, meaning that it is unlikely that even a large-scale event would affect everyone in the same way. Therefore, the safest risk pool is one that has a large number of people involved. As with the law of large numbers in statistics, the more people added to the risk pool, the less risk for the entity as a whole.

We can apply the risk sharing rules to the above situation in which agents issue debt to finance risky projects. If projects are too risky, agents can default if subjected to a shock, defined as occurring when the actual return after subtracting the investment and costs is greater than a certain amount. If the risk of the projects is independently distributed, agents have an incentive to diversify, in order to lower the probability that an individual project will fail, by exchanging part of the investment in their own risky project with other agents. Each agent can reduce the probability of failing by diversifying risky projects with other agents through risk sharing.

Suppose each agent i has a different investment portfolio subject to a stochastic return X_i, which is drawn from some probability distribution with the average $E(X_i)$ and the variance $Var(X_i)$. If each agent operates in isolation, an agent expects to obtain $E(X_i)$ as an individual return. Alternatively, if they share their risk with all other agents, agents pool their risk and receive the average of the stochastic return:

$$Y = \sum_{i=1}^{N} X_i/N. \tag{9.24}$$

Risk sharing is often beneficial for risky projects, for instance, but only if the stochastic return is greater than the expected return $E(X_i)$. The probability of failure can be compared by the limiting distribution of the stochastic average returns R with an individual expected return $E(X_i)$ when each agent operates in isolation. We shall investigate two basic issues: (i) how the allocation of risky projects is affected by the rule of risk sharing, and (ii) how risk allocation affects the probability of agent failure. In particular, we shall evaluate the advantages to risk sharing in terms of lowering the probability of agent failure.

To do so, we must take into account the distribution of the initial shocks and that of the final shock to each agent in equilibrium. We can assess the share of the overall shock the agent should sustain in equilibrium. The gains from risk sharing are evaluated in terms of lowering the probability of agent failure. We assume that the initial shock to each agent i, $X_i = x_i(0), (1 \leq i \leq N)$, is drawn from the identical probability distribution with the average $E(X_i)$ and the variance $Var(X_i)$. Each agent has a basic capacity to absorb external shocks, up to a certain threshold. We define the agent's threshold, which is proportional to the average shock, as $\theta E(X_i)$, where $\theta (> 0)$ is the tolerance parameter.

After obtaining the limiting distribution for the shocks in a risk-sharing network, we can investigate its impact on the probability of agent failure. If the agents receive a shock in equilibrium that is less than their resilience to failure, these agents are safe from failure. However, if the share in equilibrium exceeds an agents threshold, that agent will fail. We can investigate whether the risk shared in equilibrium actually causes the expected value of the shock to exceed the agent's threshold. We can then compare the ratio of safe agents when they share risk with the case in which all agents operate in isolation.

(Case 1). All N agents have access to an investment project with some stochastic shock (loss) $X_i(= x_i(0))$ with a uniform distribution between $(0,N)$.

(i) Each agent operates in isolation. If each agent operates in isolation, the ratio of safe agents is derived as follows:

$$P(x_i(t) \leq \theta E(X_i)) = \theta/2. \tag{9.25}$$

(ii) Risk sharing with the *ASR*. From eqn. (9.14), each agent shares the same shock in equilibrium, which is the average of the initial shocks to all agents:

$$x_i(t) = \sum_{i=1}^{N} x_i(0)/N = Y, \quad (t \to \infty). \tag{9.26}$$

From the central limit theorem, the average Y of N stochastic variables X_i with a uniform distribution converges to the normal distribution with the mean $E(Y) = N/2$ and variance $V(Y) = V(X_i)/N = N/12$.

With the *ASR*, which is an equal-risk sharing protocol, if the risk in equilibrium is less than or equal to the agent's threshold, i.e., $Y \leq \theta E(X)$, then all agents are safe; otherwise, all agents fail together. The probability that all agents will be safe is:

$$P(Y \leq \theta E(X)) = \int_0^{\theta E(X)} \frac{1}{\sqrt{2\pi}\sigma} e^{-\frac{(x-\mu)^2}{2\sigma^2}} dx. \tag{9.27}$$

(iii) Risk sharing with the *DSR*. Let N agents manage shares of risky projects using the *DSR*. The risk in equilibrium with the *DSR* is given in eqn. (9.23), and it depends on the agent network's topology. Thus, we must specify the topology. Here, we assume that N agents form a scale-free network (*SF*) with a degree distribution:

$$f(k) \propto k^{-(\gamma+1)}. \tag{9.28}$$

In this case, agent i shares the shock in proportion to the agent's degree, which is given:

$$x_i(t) = (k_i/k_{avg})Y, \quad 1 \leq i \leq N, \quad (t \to \infty), \tag{9.29}$$

where $k_i/\sum_{i=1}^{N} k_i \equiv k_i/Nk_{avg}$.

An agent with a lower than average degree k_{avg} shares a lower shock than the average agent. Therefore, agents with a lower degree are much safer than agents with a higher degree. The ratio of safe agents when they operate according to the *DSR* is obtained as:

$$P(x_i(t) \leq \theta E(X_i)) = P((k_i/k_{avg})/Y \leq \theta E(X_i)). \tag{9.30}$$

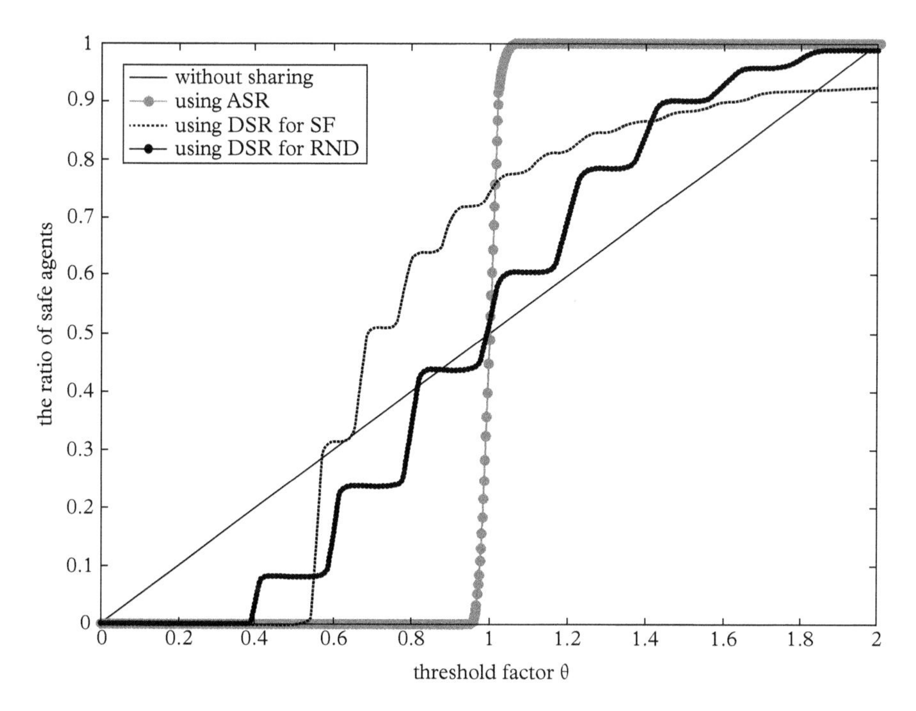

Figure 9.12 *Ratios of safe agents operating under different risk-sharing rules (viz. ARS and DRS), and all agents operating in isolation. The initial shock distribution for each agent is uniform between (0,N)*

In order for sharing risk to be advantageous, each agent should have a lower probability of failure. Figure 9.12 shows the ratios of safe agents as a function of the tolerance parameter α, comparing the ratio from operating in isolation with the ratio from agreeing to share risk. By not sharing risk and opting to operate in an isolated manner, some isolated agents may experience trouble when they receive a shock exceeding their threshold. If agents operate by sharing risk, however, the probability of failure depends on the ratio between the share of the overall shock in equilibrium and the agent's tolerance to the shock. If the agents that, in equilibrium, obtain higher shares of the overall shocks are also agents with a higher tolerance level, then the ratio of the safe agents will be relatively high when simultaneously experiencing mild shocks.

We compare risk sharing between the *ASR* and the *DSR*. The *ASR*, which is an equal risk-sharing protocol, reduces failure rates when the tolerance to the shock is high ($\theta > 1$). On the other hand, when the tolerance is low ($\theta < 1$), risk sharing becomes dangerous. For instance, if agents swap assets in order to diversify their individual risk, they will hold risky portfolios in common. If each agent's resilience to shocks is low, then the probability that many agents will fail together will be high. On the other hand, with the *DSR*, risk sharing in equilibrium differs depending on the agent's degree, and risk sharing is safer when there is breadth to the tolerance parameter. What matters is the correlation between the equilibrium of risk, as determined by the network topology, and the agents' tolerance to shocks. Some agents at higher degrees obtain more of the overall shocks, and most agents at lower degrees share low shocks in equilibrium. In that case, the ratio of safe agents is high. Risk sharing with the *DSR*, however, is dangerous when tolerance is low, and with extremely high tolerance levels.

In summary, if the agent's tolerance to shock is low ($\alpha < 0.5$), then any kind of risk sharing is dangerous, and sharing risks increases the chance of simultaneous failures. When the overall agent tolerance is very high, the agents that are especially tolerant to shocks receive a higher proportion of the overall shocks, and they can fail with the *DSR*. In general, with the *DSR*, more agents with lower tolerance to shocks can survive, but a very small proportion of the agents can fail by having a larger share of the total shocks to all agents.

We now consider the situation in which each agent faces a heavy-tailed shock, which is drawn from the power-law distribution function:

$$f(x) = \gamma k^{\gamma} x^{-(\gamma+1)} \quad (x \geq k). \tag{9.31}$$

This stochastic shock has an average $E(X) = k\gamma/(\gamma - 1)$ and variance $V(X) = k^2\gamma/(\gamma-1)^2(\gamma-2)$.

In order to determine the gain from sharing risk in a heavy-tailed shock environment, each agent should have a lower probability of defaulting. In particular, if the risk is shared equally, and the expected value of the shock exceeds the agent's threshold, which is proportional to the average shock, i.e., $x_i(t) > \theta E(X_i)$, then that agent will default.

(1) Agents operating in isolation
If each agent operates in isolation, then the ratio of agents with the initial shock within the threshold is:

$$P(X_i \leq \theta E(X_i)) = 1 - ((\gamma - 1)/\gamma\theta)^{\gamma}. \tag{9.32}$$

(2) Risk sharing with the *ASR*
The average N stochastic variables, each of which follows the power-law distribution in eqn. (9.31) also follows a power-law distribution with the same power index $\gamma + 1$. Therefore, when all agents operate together with the *ASR*, the ratio of safe agents with the shock in equilibrium falling within the threshold is obtained by:

$$P(Y \leq \theta E(Y)) = 1 - ((\gamma - 1)/\gamma\theta)^{\gamma}. \tag{9.33}$$

Therefore, the *ASR*, which involves sharing risk equally, does not necessarily reduce the failure rate of each agent, and there is a small chance that all agents will fail simultaneously.

(3) Risk sharing with the *DSR*
If each agent operates with the *DSR*, the probability that agent i will be safe is obtained by:

$$P(x_i(t) \leq \theta E(X_i)). \tag{9.34}$$

Each agent i receives the shock in equilibrium:

$$x_i(t) = (k_i/k_{avg})(Y/N), \quad 1 \leq i \leq N, \tag{9.35}$$

and the probability that agent i with the degree k_i will be safe is:

$$P(k_i \leq \theta k_{avg}) = 1 - ((\gamma - 1)/\gamma\theta)^{\gamma}. \tag{9.36}$$

Therefore, the ratio of safe agents with the *DSR* is also the same when each agent operates in isolation. Therefore, risk sharing does not reduce the probability of each agent failing if the shock distribution they encounter is a thick-tailed type, such as a power-law distribution.

We can classify risky situations in which risk sharing in networks is beneficial and works as a risk-pooling mechanism to enhance the agents' capacity to absorb shock. The key point is that the gain from risk sharing depends on the capacity of each agent to absorb shock and the network topology that specifies the interconnection patterns among agents. The gain from risk sharing can be classified into three groups, depending on the capacity of the agents to absorb shock: (i) low capacity, (ii) intermediate capacity, and (iii) high capacity. Thus, we can conclude that risk sharing is beneficial only if the agents' capacity to absorb shock is at intermediate level. At the low- or high-capacity level, agents do not gain from risk sharing, and operating in isolation is preferred. Sharing risk according to the *ASR* works especially well under any network topology and shock distribution. However, the *ASR* does not necessarily reduce the rate of failure, unless agent resilience to shock is high. On the other hand, the *DSR* has an advantage over isolated operations. We have revealed those risky environments where shocks are characterized by the shock-distribution pattern. Thus, models for sharing risk reveal that the network acts as a perfect risk-pooling mechanism, when the network is strongly connected. However, resorting to sharing risk does not necessarily reduce the rate of failure in an agent network, unless the shock they experience is distributed as a heavy-tailed function and the agents are willing to accept that there is a slight chance of facing a very large-scale shock.

10

Agent-Based Modeling of Economic Crises

This chapter presents an agent-based model for studying the interactions between the financial system and the real economy. The dynamics of credit depend on the demand for credit from firms in order to finance production, and on the supply of credit from the financial system. The model can reproduce a wide array of macro- and micro-empirical regularities and stylized facts regarding fundamental economic activities. This model also contributes to enriching our understanding of potential vulnerabilities and the paths through which economic crises can propagate across economic and financial networks. Combining an agent-based model with network analysis can help to shed light on how a failure in a firm propagates through dynamic credit networks and leads to a cascade of bankruptcies.

10.1 Agent-Based Economics

Over the last few decades, macroeconomies have been studied with established approaches, and several tools are available to help us understand their complexities. However, most existing models tend to be based on a top-down and partial-equilibrium view of markets, and they are unable to incorporate the complexity of behavior among heterogeneous agents whose behavior often changes (Delli Gatti et al. 2011). The debate on the methodological foundations of macroeconomic theory has gained renewed momentum during the recent economic crisis, and there is a need for alternative methodologies in order to understand the mechanism of economic crises. Agent-based economics (*ABE*) from the bottom-up are lately receiving more and more traction in financial analysis. Gallegati and Kirman (2012) argue that a view of the economy as an interactive heterogeneous agent system allows us to explain the mechanisms behind recent economic crises, crises that were inconsistent with the equilibrium models central to the standard approach of macroeconomics. *ABE* follows a different methodology, based on an analysis of systems with many heterogeneous interacting agents. *ABE* generates data that appears to be more consistent with observed empirical regularities emerging

Agent-Based Modeling and Network Dynamics. First Edition. Akira Namatame and Shu-Heng Chen.

from the system as a whole; these are the regularities that cannot be identified by looking at any single agent in isolation.

ABE divides into two categories. The first category attempts to mimic real-world economies in a highly detailed manner. The most complete agent-based model developed to date is the model known as EURACE (Cincotti et al. 2010). The second category consists of abstract models from real economies. They contain only a small number of different types of heterogeneous agents and the rules respecting their interaction. Delli Gatti et al. (2005, 2010) provided the framework and promising models from the second category in a series of articles. Their models focus on the relationship between the real economy and financial systems. Understanding the feedback loops between these two sectors leads to a deeper understanding of the stability, robustness, and efficiency of the economic system. In particular, they studied the linkage dependence among agents (viz. firms and banks) at the micro-level to estimate their impact on macro-activities. Their model is able to explain how some crucial macroeconomic phenomena can be derived from the interactions of heterogeneous agents where their linkages may evolve endogenously over time.

From the viewpoint of mainstream macroeconomics, *ABE* suffers from a few drawbacks. For instance, *ABE* makes it impossible to think in aggregate terms. Aggregate variables are only constructed from the bottom up by summing the individual quantities. As a consequence, the interpretation of the mechanism of shock propagation, for instance, is somehow arbitrary. In the context of models created to describe realistic economic phenomena, emphasis must be placed on the reproducibility of simulated experiments to validate the results as scientific. However, agent-based models are rarely replicated. One of the major problems in replication is the ambiguity in modeling. In most cases, agent-based models contain several parameters, and the degrees of freedom for these parameters must be correctly tuned. With enough degrees of freedom, any simulation output can be generated. These models are designed to describe or explain real-world phenomena. However, they contain non-linear and recursive interactions with multiple feedbacks, making it difficult to interpret the simulation results. The explanatory and predictive power of *ABE* is increasingly concerned with the development of models to account for the complexity of real economic dynamics. Therefore, the approach of *ABE* remains complex and challenging. From this perspective, a promising research path can be found by complementing very simple models that are particularly suited to illuminating the core dynamics of economic phenomena with increasingly complex and empirically grounded simulations.

Fagiolo et al. (2007) provided a critical guide to empirical validation developed in *ABE*. In practice, the approach of *ABE* tends to be hampered by a lack of high-quality datasets. When this is the case, we should use stylized facts—findings based on statistical regularities—to pin down values or value ranges for key parameters. By stressing the reproduction, explanation, or prediction of a set of stylized facts, one hopes to circumvent problems of data availability and reliability. For instance, economic growth processes have attracted much interest. Stylized facts typically concern macro-level regularities, such as the relationship between the volatility in the growth rate of a gross domestic product, the unemployment rate, properties of economic cycles, the shape of the distributions of the size of firms, and the growth rates of firms. Recent empirical studies suggest that the firm size distribution, for instance, follows a power-law distribution. Axtell (2001) and Stanley et al. (1996) discovered another empirical feature, one regarding growth in firms, showing that it follows a tent shape distribution in a logarithmic scale. Delli Gatti et al.'s model generates these empirically observed regularities or stylized facts.

The real economy is accelerated by the financial system, which itself has a profound effect on the real economy. The dynamics of credit are shaped as a result of the demand for credit by firms to finance productive activity, and the supply of credit from banks, which is constrained by financial conditions. Understanding the linkage dependency between firms and banks can help to uncover the mechanism that drives and reins economic activity while generating economic risk. Grilli et al. (2012) extended the model from Delli Gatti et al. (2010) by incorporating an interbank network of multiple interactive banks. Multiple banks operate not only in the credit market but also within the interbank market. They modeled the credit network between firms and banks and the interbank networks as random graphs and studied the network's resilience by changing the degree of connectivity of the interbank network. A firm may ask for a loan from a bank in order to increase its production and profit. If this bank faces a liquidity shortage such that it cannot cover the firm's requirements, the bank may borrow from other banks in the interbank market. Therefore, banks share the risk with the bank for the loan to the firms. The bank can default if the firm is unable to pay back its debt to the bank owing to economic fluctuations. In the interbank market, a defaulting borrower bank can lead to defaults in lender banks and other healthy firms. Therefore, a firm's bad debt can be the catalyst for a cascade of bankruptcies among banks and firms.

In the past, financial markets were driven by the real economy, which in turn had a reciprocal effect on it. In recent decades, a massive transfer of resources from the manufacturing sector to the financial sector has come to be a defining characteristic of the globalized economy. However, it is widely believed that this transfer process is mainly responsible for growing economic instability. The productive sector has witnessed a dramatic increase in output volatility and uncertainty. The growth in the number of connections in the financial sector plays an important role in accelerating economic instabilities. The global financial crisis in particular has shed light on the importance of the connections between the financial sector and the real economy. One source of financial risk is the dense interconnections of agents created through financial transactions. These transactions generate complex financial networks. Nevertheless, it is unclear exactly how financial networks function and how robust they are. Understanding systemic risk in financial networks is of critical importance in order to establish defense strategies and regulations to manage it effectively.

Recent financial crises have triggered numerous studies on systemic risk caused by the contagion effects from interconnections in the financial networks. Systemic risk can lead to continual large-scale defaults or systemic failures in financial institutions (i.e., agents). Allen and Gale (2000) introduced the use of network theory to enrich our understanding of the vulnerability of financial systems. They studied how such systems respond to contagion when agents are connected in different network topologies. An increasing body of recent research is focused on an investigation of the interbank market from the perspective of network theory. Indeed, by understanding single banks as nodes, and their credit relationships as links between nodes, one can represent such data as a network in a straightforward way, with credit volumes defining the adjacency matrix of links between agents. With this surge in network research, the data from interbanks can be investigated analytically from such a network perspective.

The agent-based approach presents the study of the relationship between credit and economic instability. This issue is of primary importance, because a lower variability in output and inflation has numerous economic benefits. The dynamics of credit are endogenous, and depend on the supply of credit from banks. This credit is constrained by the net worth base, and on the demand for credit from firms to finance projects. In particular, we need to investigate the effect of credit connections on firms' activities in order to explain some key elements that occur during an economic crisis. From this perspective, network theory is a natural candidate for the analysis of networked agents. The financial sector can be regarded as a set of agents who interact with each other through financial transactions. These interactions are governed by a set of rules and regulations that can be represented on an interaction graph that depicts each connection among agents. The network of mutual credit relations among agents plays a key role in the spread of contagious defaults.

10.2 Model Descriptions

In this section, we build an integrated model to provide a baseline that has the key elements of both the real economy and the financial system, making it possible to study the interactions between them. An agent-based economic system consists of heterogeneous agents who interact with each other through economic and financial transactions. These interactions are governed by a set of rules and regulations, and they take place within the network.

In our simulated economy, three markets coexist: the goods market, the credit market, and the interbank market. Given the structure of the model, we can analyze the interaction among agents, as well as their behavior in different markets. Understanding the linkage dependency between the financial sector and the real economy can help to design a regulatory paradigm that can control systemic risk while encouraging economic growth. We focus on the linkage dependence among agents at the micro-level, and on estimating the impact of such agents on macro-activities, such as the aggregate production and its growth rate and the agent size and its growth rate in terms of net worth. The dynamics of credit are endogenous, and they can be characterized by the demand for credit from firms and the supply of credit from banks, which is constrained by its net worth base.

The goods market is implemented following the model from Delli Gatti et al. (2010). In their model, output is supply determined. That is, firms sell all the output they optimally decide to produce. Firms use capital stock and labor as the input of production, and the production is determined by the investment. The investment depends on the interest rate and the firm's financial fragility, which is inversely related to its net worth. Each period, firms enter in the credit market asking for credit. The amount of credit requested by a firm is related to its investment expenditure and depends on the interest rate and the firm's financial condition.

The credit channel involves the balance sheet from both the banks and firms. The primary purpose of banks is to channel funds towards loans for firms. Banks can grant the requested loan only when they have enough available liquidity. The supply of credit is a percentage of banks' net worth, because they adopt a system of risk management based upon the debt ratio. Whereas a bank's balance sheet affects the potential supply of loans, owing to the capital-adequacy ratio, the net worth of a highly leveraged firm influences the banks willingness to lend it money. When banks do not have liquidity to lend, they can draw on the interbank market to avoid losing the opportunity to invest in a firm.

Credit relationships will constitute the main connections between the financial sector and the real economy. Here, we consider two cases: (i) a credit market with no interbank market, and (ii) multiple banks operating not only in the credit market but also in the interbank market. These two cases specify the clearing mechanism and the bailout procedure when a bank defaults, as explained below. A network of mutual credit relations between banks and firms plays a key role in the risk of contagious defaults, which implies the importance of the connectivity and the credit network's topology in the analysis of a cascade of defaults. In the second case, banks may have the stocks or security bonds of other banks through the interbank market if they have sufficient liquidity. Increasing the connectivity of banks in the interbank network may leave them less exposed to systemic risk, owing to risk sharing. However, if bank connectivity becomes too high and things go awry, the financial linkage of highly leveraged banks may become a propagation channel for contagion and the source for a cascade of defaults.

Firm behavior. Delli Gatti et al. (2010) introduced the model to study the properties of a credit network and the causes of the emergence of so-called *financial acceleration*. In their model, there are two types of firms: (i) downstream firms (D firms), producing goods for consumption using labor and intermediate goods, and (ii) upstream firms (U firms), producing intermediate goods demanded by D firms. These firms may ask for loans from the banks if their net worth is insufficient to finance the wages needed.

The core assumption of the model is that the level of production $Y_i(t)$ for the ith D firm at time t is an increasing concave function of the financial robustness in terms of its net worth $A_i(t)$, such that:

$$Y_i(t) = \varphi A^{\beta}{}_i(t), \tag{10.1}$$

where $\varphi > 1$, $0 < \beta < 1$ are parameters that are uniform across all D firms. Eqn. (10.1) represents the financially constrained output function. Goods for consumption are sold at a stochastic price that is a random variable extracted from a uniform distribution in the interval (0,1), and this is the only source of disturbance from the outside of the system.

The ith D firm demands both labor $N_i(t)$ and intermediate goods $Q_i(t)$, depending on its financial condition, as understood by its net worth $A_i(t)$. The demand for labor and intermediate goods from the ith D firm are respectively given as follows:

$$N_i(t) = c_1 A_i(t)^{\beta}, \quad Q_i(t) = c_2 A_i(t)^{\beta}. \tag{10.2}$$

The U firms produce intermediate goods by employing labor exclusively. We assume that many D firms can be connected to a single U firm, and each D firm has only one supplier of intermediate goods from the U firms. The production of the jth U firm is determined by the sum of the demands from the D firms, which is given as follows:

$$Q_j(t) = c_3 \sum_{i \in D_j} A_i(t)^{\beta}, \tag{10.3}$$

where D_j is the set of the D firms to be supplied from the jth U firm. The demand for labor from the jth U firm is determined by its production level $Q_j(t)$, derived as follows:

$$N_j(t) = c_4 \sum_{i \in D_j} A_i(t)^{\beta}. \tag{10.4}$$

The price of intermediate goods from the jth U firm is determined by:

$$p_j(t) = 1 + r_j(t) = \alpha A_j(t)^{-\alpha}, \tag{10.5}$$

where $r_j(t)$ is the margin rate of the jth U firm. The margin rate is determined by the interest on trade credit, which is assumed to be exclusively dependent on the financial condition of the jth U firm, as determined by its net worth A_j. In particular, if the jth U firm is not performing well, it will be offered credit at a less favorable margin rate, owing to discounting.

Whereas the scale of production for each D firm is financially constrained, and it is determined by the degree of financial robustness in terms of its net worth, the scale of production for each U firm is demand constrained, and it is determined by the demand for intermediate goods from the D firms. Therefore, the financial conditions of D firms are also the driving forces for the intermediate goods produced by U firms. If the financial conditions of D firms are healthy, the scale of productive activity is higher and the demand for intermediate goods will increase.

If the net worth of each D or U firm is insufficient for paying wages $W = c_5 N$, it will request credit from a bank, and the credit requested by each D or U firm is calculated as follows:

$$B(t) = c_5 N(t) - A(t). \tag{10.6}$$

Bank behavior. Firms may ask banks for loans to increase their production and profits. Many firms can be connected to a single bank, but each firm is assumed to have only one supplier of credit (i.e., only one bank per firm). The primary purpose of banks is to channel their funds in the form of loans to firms. A bank, in analyzing both its own credit risk and the firm's risk, may grant the requested loan when it has enough liquidity. The supply of credit depends on the bank's net worth, because banks adopt risk management based on the net worth debt ratio.

Leverage is generally referred to as the use of credit to finance something. One widely used measure for financial leverage is the debt-to-net worth ratio, which represents the financial soundness of firms and banks. The interest rate on a loan is largely determined by the perceived repayment ability of a firm; a higher risk of payment default almost always leads to higher interest rates for borrowing capital. We adopt the principle according to which the interest rate charged by banks incorporates an external finance premium that increases according to the leverage. It is, therefore, inversely related to the borrower's net worth (Delli Gatti et al. 2010).

The interest rate of a bank loan to a firm decreases with the financial soundness of the bank in terms its net worth A_z, and it increases with the firm's leverage ratio $l_i = B_i/A_i$. Then, bank z offers its interest rate to the borrower (D or U) firm i:

$$r_{z,i} = \alpha A_z^{-\beta} + \theta l_i^{\theta}, \tag{10.7}$$

where A_z is the net worth of the zth bank, $l_i = B_i/A_i$ is the leverage ratio of the ith D or U firm, and α, β, and θ are positive parameters.

Growth of net worth and bankruptcy condition. At the end of each period t, after transactions have taken place, the firms and banks update their net worth based on profits. The profits of the firms and banks are evaluated from the difference between their income and the costs they faced.

Goods for consumption are sold at a stochastic price $u_i(t)$, a random variable extracted from a uniform distribution in the interval (0,1). Therefore, the profit of the ith D firm at time t is:

$$\pi_i(t) = u_i(t)Y_i(t) - (1 + r^i{}_j(t))Q_i(t) - (1 + r^i{}_z(t))B_i(t), \tag{10.8}$$

where $r^i{}_j(t)$ is the price of the intermediate good supplied from the jth U firm, and $r^i{}_z(t)$ is the interest rate of bank z. Similarly, the profit from the jth U firm at time t is given by:

$$\pi_j(t) = S(1 + r^i{}_j(t))Q_i(t) - (1 + r^i{}_z(t))B_j(t) - BD^i{}_j(t), \tag{10.9}$$

where $BD^i{}_j(t)$ is the bad debt, taking into account the possibility that the ith D firm cannot pay back its loan when it goes bankrupt.

The profit of a bank depends on the interest generated from loans to D and U firms:

$$\pi_z(t) = \sum_i (1 + r^i{}_z(t))B_z(t) + \sum_j (1 + r^j{}_z(t))B_z(t) - \sum_i BD^i{}_z(t) - \sum_j BD^j{}_z(t), \tag{10.10}$$

$BD(t)$ is the bad debt, taking into account the possibility that firms cannot pay back the loan when they go bankrupt.

The net worth of a generic agent (i.e., a firm or bank) evolves according to the current net worth and profit according to:

$$A(t+1) = A(t) + \pi(t). \tag{10.11}$$

Bankruptcies occur when financially fragile agents fail. When there are more loans than assets, the net worth is negative. Therefore, firms or banks go bankrupt when the net worth at time $t+1$ becomes negative with $A(t+1) \leq 0$, and in this case they leave the market. We consider a closed economy, where the total number of banks, U firms, and D firms are kept constant over time. Therefore, bankrupt firms or banks are replaced at a one-to-one ratio, and the initial net worth of the new entry is set to be small and randomly taken from the interval (0,1).

Defaulting cascade. If one or more firms are unable to pay back their debt to the bank, the bank's balance sheet decreases, and the firms' bad debt affects the net worth of banks that might also fail as a result. Therefore, a firm's bad debt can bring about a cascade of bankruptcies among firms and banks. The cause of this cascade may be the result of an indirect interaction between bankrupt firms and their lending banks through the credit market, or from direct interaction between banks through its interbank connections.

The U firms are lenders to D firms through the supply of products. Suppose that a random price fluctuation in eqn. (10.8) causes the bankruptcy of some D firms. Consequently, the loans they have will not be fulfilled, and the net worth of the lenders (i.e., the banks and U firms) will decrease. Eventually, this will result in some of them going bankrupt, and their failures cause an increase in the interest rates charged to their borrowers. The resulting financial situation will increase the probability of bankruptcy in a firm (i.e., a D or U firm) or a bank. When a highly connected firm or bank defaults, it can provoke a cascade of defaults, especially if the credit network has a particular structure.

We shall illustrate how the dynamics of the credit network relate to a cascade of default. Figure 10.1 shows the conceptual framework for a defaulting cascade on the credit network among firms and banks and the interbank connections among banks. Suppose that a random price fluctuation causes bankruptcies in some D firms (viz. D_1, D_7, and D_8). As a result of these bankruptcies, they will no longer be able to make payments on any loans they have, and the net worth of the U firms (viz. U_1 and U_3) and the banks (viz. B_1 and B_2) will decrease. Eventually, this will result in the bankruptcy of some of them and, more importantly, in an increase in the interest rates charged to their past and new borrowers. This, in turn, will increase the probability that other firms will go bankrupt, repeating and escalating the chain. In this scenario, the interbank connections between banks also brings about the default of bank B_4.

Bailout procedure and clearing mechanism. In a closed economy, there are no extraneous sources of money, except from new entries in the network, where a small net

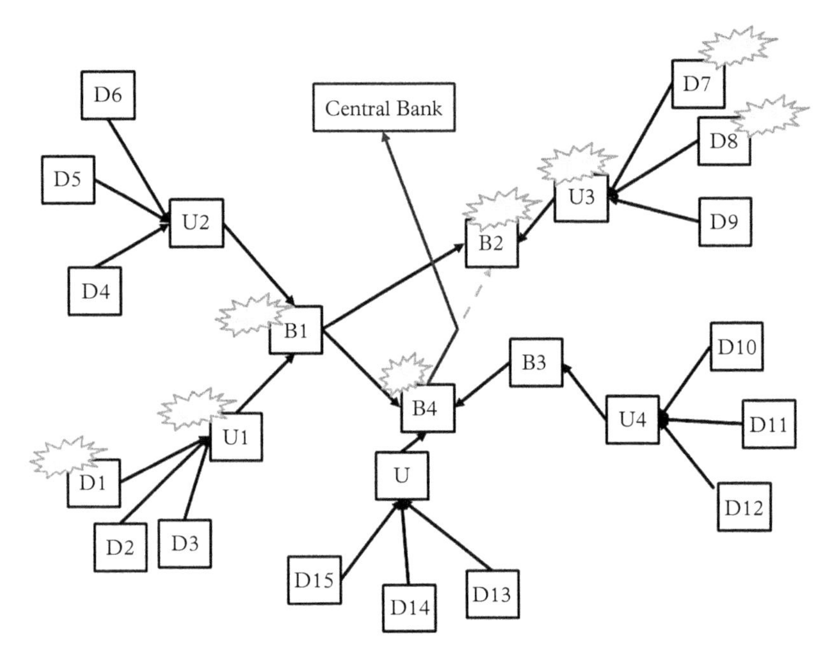

Figure 10.1 *Conceptual framework of a defaulting cascade*

worth is their only source of money. If a firm goes bankrupt, it enters a bailout procedure through loan consolidation, and credited banks absorb the bad debt. A bank goes bankrupt if its net worth becomes negative, which can be triggered by the bankruptcy of a firm. Liquidity problems can also occur when banks are unable to raise sufficient liquidity through the interbank market.

We consider the following bailout modes and the clearing mechanisms:

Bailout mode 1. A bailout is financed by the central bank. This bailout procedure is said to occur when where there are no interbank connections, and other banks do not pay for the debt of the defaulting bank. We refer to this case when banks operate outside the interbank network.

Bailout mode 2. This bailout procedure operates in the case when there are interbank connections, and other banks in the network absorb some of debt of the defaulting bank. The bank is bailed out, and the required funds are collected from other banks within the interbank network. If (and only if) other banks cannot absorb the debt, the bailouts are financed by the central bank. We refer to this case when banks operate under the interbank network.

The interbank market is particularly useful to the financial sector. When a bank cannot offer a loan to a firm, it can get the necessarily funding through the interbank market. The connections between banks are also effective for shock absorption: if some bank risks going bankrupt, other banks can partially absorb its bad debt. This mechanism of risk sharing and shock absorption brings positive effects to the real economy. In order to simplify the bad debt sharing mechanism, we assume that the bad debt is shared in proportion with the net worth of the banks to which the defaulting bank has the obligation to make payments. We are explicitly concerned with the potential role of the interbank network to act as a mitigating factor on contagion, as well as the risk sharing mechanisms for economic crises to determine the effect of the interbank network on economic activity.

Network formation. We define network formation dynamics by how D firms seek a partnership with a U firm, and by how D and U firms find banks to ask for loans. In order to establish the product supply partnership, D firms adopt a partner choice rule. That is, they search for the minimum prices charged by a randomly selected set of possible suppliers. Of course, the firm can change suppliers if a better price is found. When firms need loans, they contact a number of randomly chosen banks. In order to establish credit connections, the firms also adopt the partner choice rule: they seek out the minimum interest rate charged among the banks offering loans. Credit connections between firms and banks are defined by a bipartite graph. Banks offer an interest rate, as determined by eqn. (10.7).

Each borrowing agent must choose a lending agent to establish a credit relationship. Initially, the credit network is random—i.e., the links between the D and U firms and between firms and banks are established at random. Then, in every period, each borrowing agent observes the interest rates from randomly selected prospective lending agents. We assume that the borrowing agent sticks to the current lending agent if the previous partner's interest rate, r_{old}, is smaller or equal to the minimum interest rate set by the new prospective lending agent, r_{new}. If this is not the case, the probability p_s of switching to a new lending agent decreases according to the difference between r_{old} and r_{new}, such that:

$$p_s = 1 - exp^{\lambda(r^{new} - r^{old})/r^{new}}, \tag{10.12}$$

where λ is a positive parameter (Delli Gatti et al. 2010). Therefore, the number of links among agents changes over time, owing to changing prices or interest rates charged by the U firms or banks, respectively, and because the topology of the network is evolving over time. The procedure for choosing a partner is activated in every period, but firms change partners infrequently, owing to something akin to the law of inertia, and only when the interest rates observed by the borrowing agents is much lower than the rate at which they are currently charged. Therefore, the business relationships between firms and the credit relationships between firms and banks may last longer.

10.3 Simulation and Results

We consider a closed economy consisting of 500 firms (250 D firms and 250 U firms) and 100 banks. This economy is simulated over a time period T = 1000. We consider

Table 10.1 *Sets of parameter values*

Parameter	Case 1	Case 2	Case 3
$A_{i,0}$	1	5	1
$A_{j,0}$	1	1	1
$E_{z,0}$	1	10	1
α	0.01	0.01	0.03
σ	0.01	0.01	0.05

Parameter	Value	Parameter	Value
I	500	δ_d	0.5
$\mathcal{J}$	250	γ	0.5
Z	100	δ_u	1
T	500	w	1
φ	2.0	θ	0.01
β	0.9	λ	1

three cases with different parameter values, as listed in Table 10.1. In Case 1 (i.e., the baseline case), all firms and banks were homogeneous, with the same initial net worth. In Case 2, the D firms and the banks begin with a larger net worth than that of the U firms. In Case 3, the price and interest rate determinant parameters from eqn. (10.5) and eqn. (10.7), respectively, are set higher than those in Case 1. We repeat the simulation 100 times, with different random seeds, and obtain the averages for the values of interest.

10.3.1 Distributions of agent sizes and growth rates

Although the simulation begins with homogeneous firms and banks of the same size, the size of firms and banks in terms of their respective net worth became extremely heterogeneous. The size distribution of D firms basically obeys a power-law distribution in the region where the net worth is greater than one, and the size distributions for the U firms and the banks follow a log-normal distribution. Technically, a power law distribution is characterized by the complementary cumulative distribution function (*CCDF*). The probability distribution for observing a net worth A greater than or equal to a is proportional to a power of γ:

$$Pr(A \geq a) \propto a^{-\gamma}. \tag{10.13}$$

Figure 10.2 shows the log–log plot for the *CCDF* of the net worth size of the D firms in two cases: without an interbank network (left panel), and with an interbank network (right panel). We can observe the skewed right distribution of D firm sizes, even when all firms begin at the same size.

Figure 10.3 shows the corresponding growth rate (*GR*) distribution:

$$GR = log_{10}(A_{t+1}/A_t), \tag{10.14}$$

where A_t is the size of a D firm in terms of the net worth at t, and A_{t+1} is its size during the subsequent period. The *GR* distribution obeys a tent-shaped distribution.

Figure 10.4 shows the log–log plot of the *CCDF* for the net worth size of the U firms in two cases: without an interbank network (left panel), and with an interbank network (right panel). We can observe the skewed right distribution of U firm sizes, even when

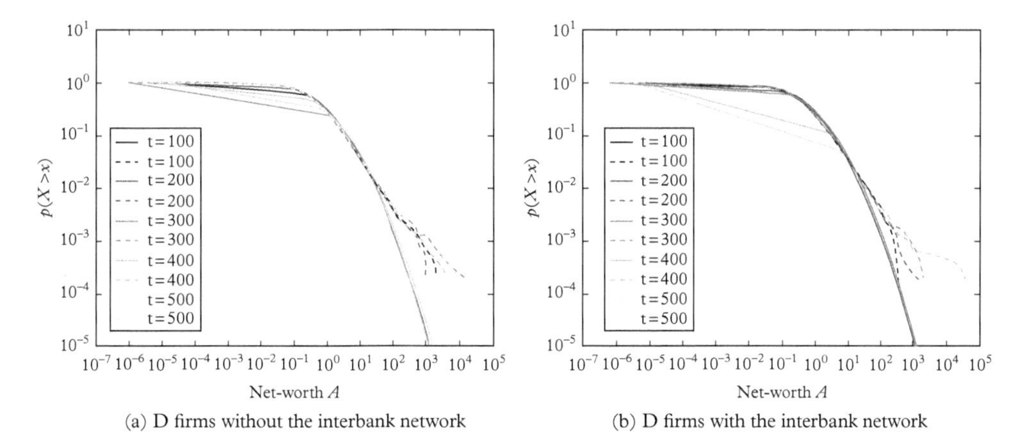

(a) D firms without the interbank network (b) D firms with the interbank network

Figure 10.2 *Complementary cumulative distribution of the D firms' size in terms of net worth at $t = 100, 200, 300, 400$, and 500. The parameter values are for Case 1. The plots are in log–log scale.*

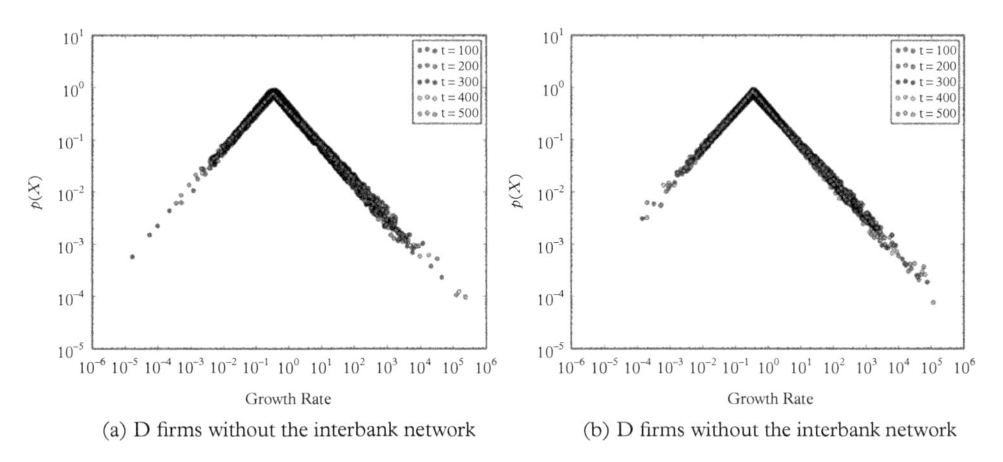

(a) D firms without the interbank network (b) D firms without the interbank network

Figure 10.3 *Distribution of the growth rate* (GR) *of the D firms' size: $GR = log_{10}(A_{t+1}/A_t)$. The plots are in log–log scale: (a) without interbank network, (b) with interbank network*

all firms begin at the same size. Figure 10.5 shows the corresponding *GR* distribution, which also obeys a tent-shaped distribution.

Figure 10.6 shows the log–log plot for the *CCDF* of the net worth size of the banks in two cases: without an interbank network (left panel), and with an interbank network (right panel). We can observe the skewed right distribution of the bank sizes, even when all banks begin at the same size. Figure 10.7 shows the corresponding *GR* distribution, which also obeys a tent-shaped distribution.

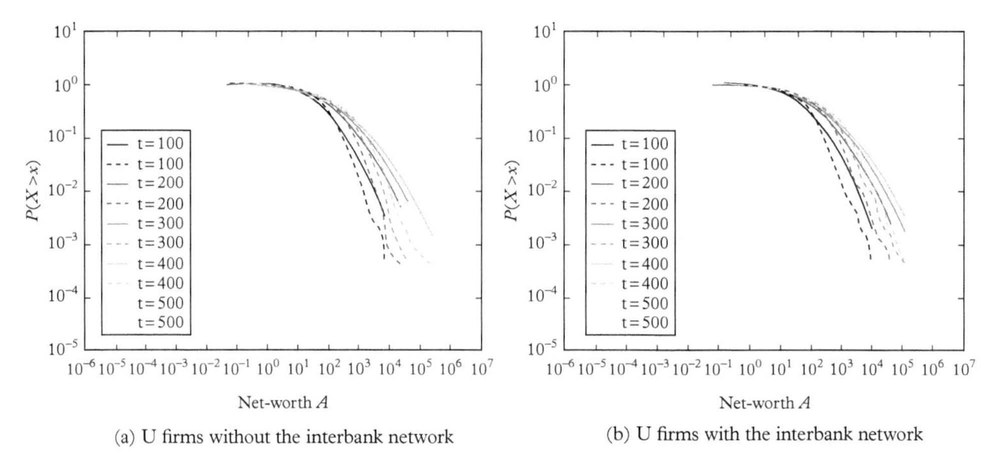

(a) U firms without the interbank network (b) U firms with the interbank network

Figure 10.4 *Complementary cumulative distribution of the U firms' size in terms of net worth at t = 100, 200, 300, 400, and 500. The plots are in the log–log scale. (a) without the interbank network, (b) with the interbank network*

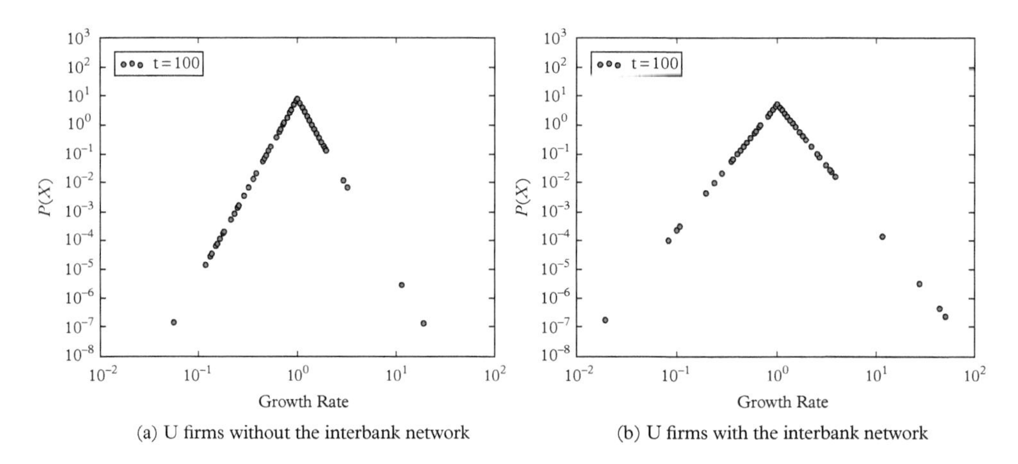

(a) U firms without the interbank network (b) U firms with the interbank network

Figure 10.5 *Distribution of the growth rate* (GR) *of the U firms' size: $GR = log_{10}(A_t/A_{t-1})$. The plots are in the log–log scale*

These simulation results show that a significant heterogeneity emerges in terms of the size of the agent. Some key stylized facts observed with real economic data are that the firms' size obeys a power-law distribution, and that the corresponding *GR* obeys a tent-shaped distribution (Axtell, 2001) (Stanley et al. 1996). The peculiarity of a power-law distribution is explained by the relative abundance of agent groups whose size significantly exceeds the average, and these agents (i.e., large firms and mega-banks) explain their large proportion of economic and financial transactions.

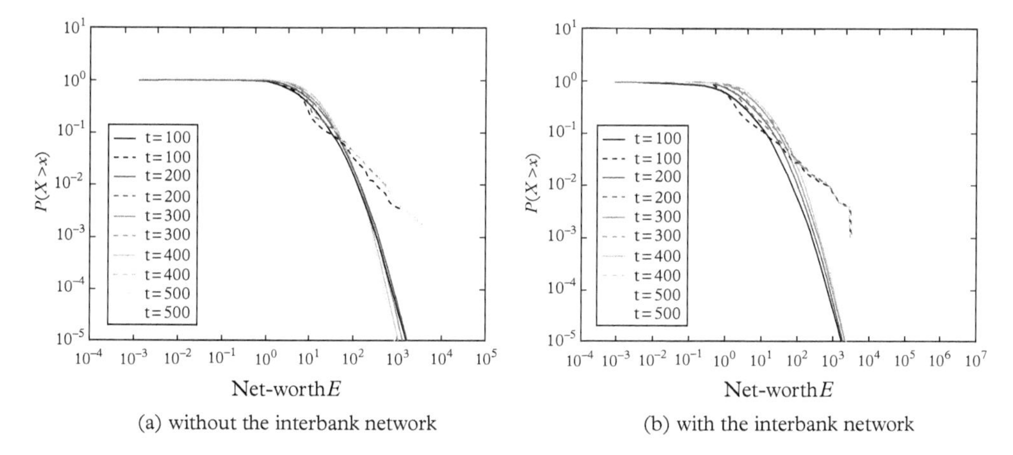

Figure 10.6 *Complementary cumulative distribution function of the banks' size in terms of net worth at t = 100, 200, 300, 400, and 500. The plots are in the log–log scale: (a) without the interbank network, (b) with the interbank network*

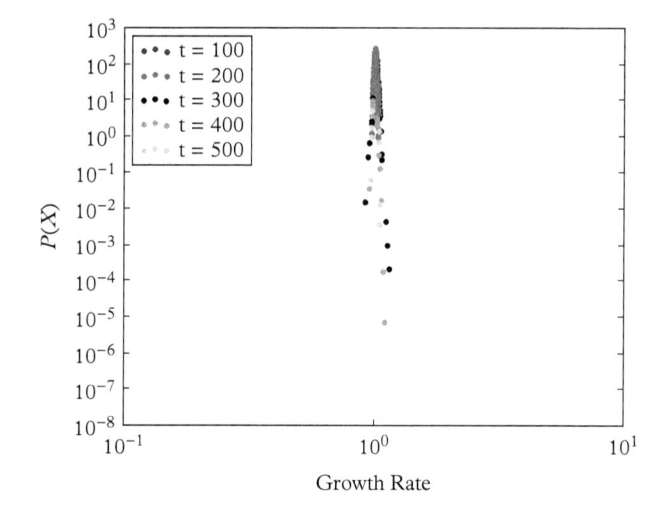

Figure 10.7 *Distribution of the growth rate* (GR) *of the banks' size:* $GR = log_{10}(A_t/A_{t-1})$. *The plots are in log–log scale.*

10.3.2 The distributions of aggregate profit and growth rates

We now turn to an investigation of the relationship between the credit market and economic growth. The *GDP* is an aggregate of all economic activity, and some stylized facts have been observed as macro-level regularities, such as the volatility of the *GDP*'s growth rate.

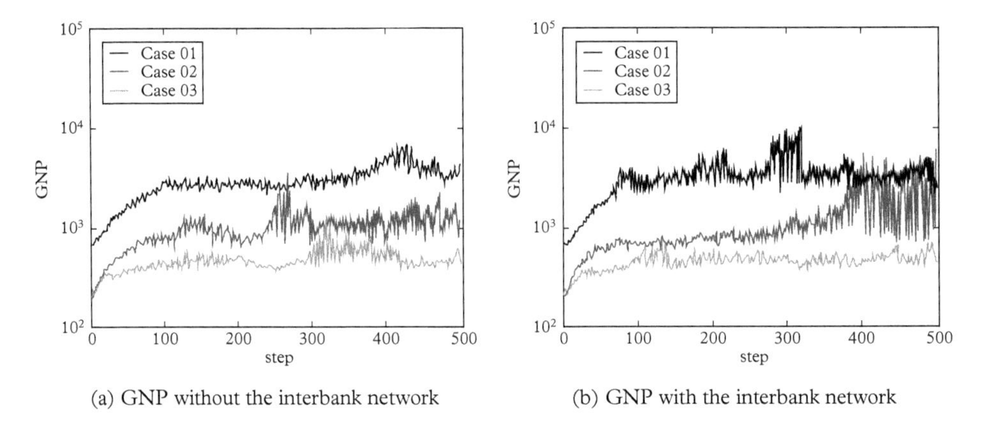

(a) GNP without the interbank network

(b) GNP with the interbank network

Figure 10.8 *Time series of logarithms for the aggregate profit* (GNP), *obtained by adding up the profits from all firms and banks. (a) without the interbank network, (b) with the interbank network*

Figure 10.8 shows the time series for the logarithm of the aggregate profit obtained by adding up the profits from all firms and banks. The aggregate profit time series shows an upward trend. Beginning from homogeneous agents of the same size, this time series behaves in complex ways with irregular fluctuations. In particular, we can observe the frequent expansion and contraction of the aggregate profit. Amplitude and periodicity vary wildly from sub-period to sub-period. With interactive structures among firms and banks, the model reveals the source of fluctuations in the matured economy (after the time period $T = 200$). This is because of the indirect interactions between bankrupt firms and their lending banks through the credit market, and because of the interactions among banks through the interbank network. Figure 10.9 shows the volatility of

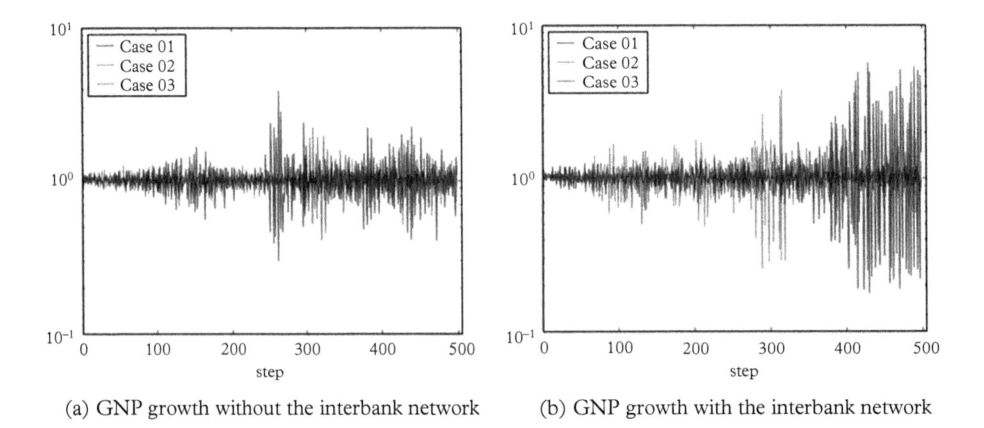

(a) GNP growth without the interbank network

(b) GNP growth with the interbank network

Figure 10.9 *Time evolution for the aggregate profit* (GNP) *growth rate: (a) without the interbank network, (b) with the interbank network*

aggregate profit expressed in terms of the *GR*. The model is able to exhibit endogenous economic cycles. The main source of observed economic cycles is the strict relation between real economic activity and its financing through the credit market.

We aim at an understanding of the relationship between firms and banks in the cycle of economic growth and decline. By developing simple interactive structures among agents, and by tracing the feedback effects, we can reproduce the source of economic fluctuations from indirect interactions between bankrupt firms and their lending banks through the credit market, and from direct interactions between lending and borrowing banks in the interbank network. Studying the interconnectivity enables us to emphasize the role of financial connections as the root cause of economic cycles and instability. Figures 10.8 and 10.9 compare the aggregate profit and the corresponding *GR* without the interbank network (left panel) and with the interbank network (right panel). In particular, we can observe the effect of the interbank network on the expansion and contraction of the aggregate profit. The results of the simulation indicate that the existence of the interbank network increases economic fluctuations. We considered the three cases described above, using the parameter values from Table 10.1. Among them, the largest aggregate profit was observed in Case 2, where the *D* firms and the banks begin with a larger net worth than the *U* firms.

The origin of the development and unfolding of economic crises is related to the model's parameters. In reality, it is never the case that all the relevant data are available such that we can identify all the crucial parameters. However, some parameters can be identified as relevant. Moreover, it is possible to follow the establishment of a new regime after a shock occurs, such as net worth distributions and recovery rates. This issue is of primary importance, because lower variability in terms of the output and interest rate has numerous economic benefits.

10.3.3 Agent bankruptcies

In this model, bankruptcy is defined as occurring when the net worth of a financially fragile firm or bank is negative. When one or more firms are unable to pay back debts to the bank, the bank's net worth decreases. Therefore, a firm's bad debt can lead to a bank failure by affecting the balance sheet of that bank. The source of a cascade of bankruptcies is the result of either the indirect interaction between bankrupt firms and their lending banks through the credit market, or from the direct interaction between banks through the interbank network.

The profit from each *D* firm in eqn. (10.8) is subject to exogenous disturbances (i.e., external disturbances) because goods for consumption are sold at a stochastic price—a random variable extracted from a uniform distribution in the interval (0,1). Therefore, the rate of bankruptcy for *D* firms remains nearly constant over time, owing to the fact that the disturbances fall outside of the system. Figure 10.10 shows the ratio of failed *D* firms over time without the interbank network (left panel) and the same ratio with the interbank network (right panel). In each case, a high rate of bankruptcy for *U* firms and banks is observed in the early stages of the simulation (before time period $T = 200$).

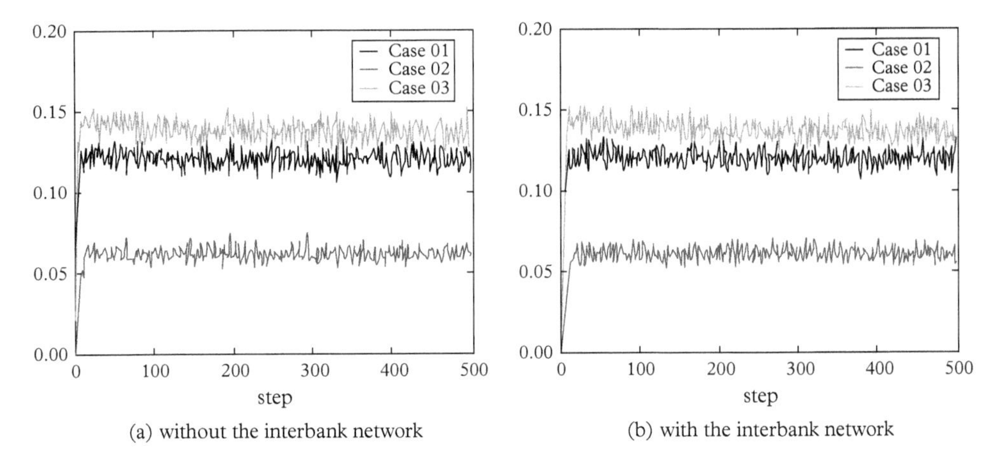

Figure 10.10 *Time evolution for the ratio of failed D firms: (a) without the interbank network, (b) with the interbank network*

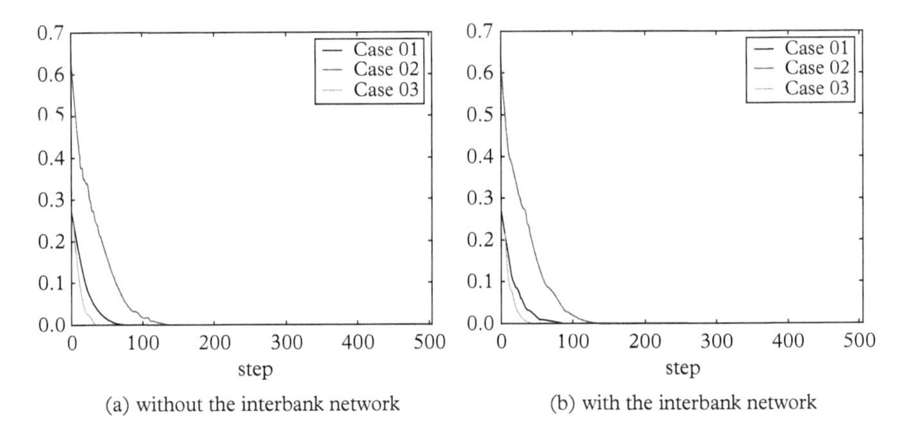

Figure 10.11 *Time evolution for the ratio of failed U firms: (a) without the interbank network, (b) with the interbank network*

Subsequently, only a few U firms and banks go bankrupt. Among the three cases described above, using parameter values from Table 10.1, the rate of bankruptcy is the lowest in Case 2, where D firms and banks begin with a larger net worth than U firms.

Figures 10.11 and 10.12 show the ratios for failed U firms and banks, respectively, over time without the interbank network (left panel) and with the interbank network (right panel). The simulation begins with homogeneous firms and banks, and some firms and banks grow extremely heterogeneous, with the agent size distribution in terms of net worth obeying a power law distribution. Only a few U firms and banks default in later stages. Among the three cases using the parameter values in Table 10.1, the rate of bankruptcy is the highest in Case 2. This is the opposite result from the case of D firms.

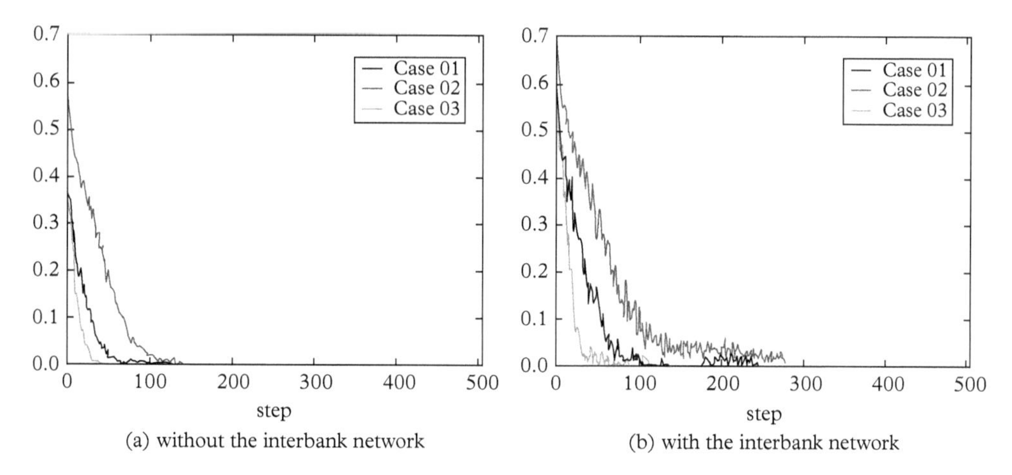

Figure 10.12 *Evolution of the ratio of failed banks: (a) without the interbank network, (b) with the interbank network*

These results suggest that the presence of multiple homogeneous agents in a credit network makes it more susceptible to a cascade of defaults. In this case, in fact, when failures occur, many agents with a relatively small net worth are vulnerable.

We now turn to an investigation of the effects of the interbank network on the contagion process. Figure 10.12 shows the ratio of failed banks over time, both without the interbank network (left panel) and with the interbank network (right panel). As expected, the interbank network results in a larger cascade of defaults, owing to the connections among banks. More generally, the interbank network implies a more severe trade-off between the stabilization effect from risk diversification and the higher systemic risk associated with bankruptcy cascades and frequent and durable recessions triggered by a stronger money supply. In particular, the existence of the connections among banks generates a larger cascade of bankruptcies.

10.3.4 Effects of the clearing mechanism

We are concerned with the potential role of an interbank network in serving to escalate contagion. Thus, we compared two debt clearing mechanisms. We modeled two clearing mechanisms from banks that have defaulted. Under Bailout Mode 1, when a bank fails, its bad debt is financed by the central bank. This corresponds to a case where there is no interbank network and no other banks pay for the bad debt of any other bank. Under Bailout Mode 2, when a defaulting bank is bailed out, the funds needed to clear bad debts are collected from other banks with which the defaulting bank has a connection in the interbank market.

Figure 10.13 shows the accumulated bad debt absorbed by the central bank over time under Bailout Mode 1, both without the interbank network (left panel), and under Bailout Mode 2, with the interbank network (right panel). The total amount of bad

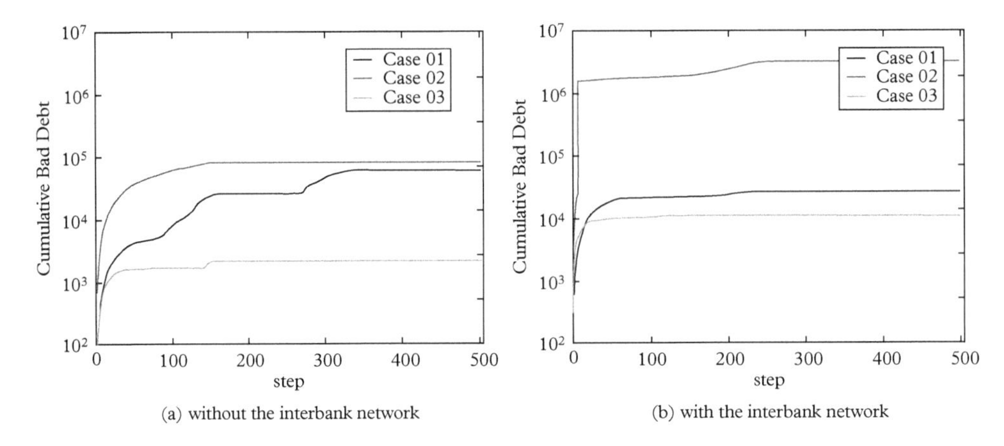

Figure 10.13 *Total debt absorbed by the central bank over time: (a) without the interbank network, (b) with the interbank network*

debt absorbed by the central bank is more under Bailout Mode 2, and when multiple banks fail, more bad debt must be financed by the central bank. Among the three cases, using the parameter values in Table 10.1, the largest aggregate profit is seen in Case 2. However, the central bank must absorb the most bad debt when there is an interbank network. These results also suggest that there is an issue with the role of the interbank network in the real economy in pursuing economic growth. Indeed, in our model, the central bank plays a significant role in clearing bad debt. With an interbank network, multiple banks can absorb a large amount of the bad debt from firms subject to market price shocks, and subsequently the economy can grow at a constant rate. At the same time, however, the financial fragility of firms and banks also increases, leading to an exacerbated economic crisis and distress contagion among firms and banks.

10.3.5 Net worth-based analysis

Agent-based modeling is able to show the possibility of explaining the emergence of an economic cycle based on the complex internal functioning of an economy without any exogenous shocks. The dynamics of credit money are endogenous, and they depend on the supply of credit from the banking system, which is constrained by the net-worth base, and the demand for credit from firms to finance productive activity. The simulation results show that the emergence of endogenous economic cycles are mainly the result of the interplay between real economic activity and its financing through the credit market. In particular, the amplitude of an economic cycle strongly increases when the net worth of firms is higher—that is, when firms are not constrained in borrowing money.

The agent model can reproduce a wide array of macro- and micro-empirical regularities in fundamental economic activities, as well stylized facts concerning economic dynamics and economic crises. Our results support the claim that the interaction among agents (viz. firms and banks) is a key element in reproducing important stylized facts concerning economic growth and fluctuations. The emergence of endogenous economic cycles is mainly caused by the interplay between real economic activity and the financial sector through the credit market. Firms are, in fact, constrained in borrowing money to fund economic activity. This can be explained by leverage. Leverage is defined as the ratio of debt-to-net-worth and can be considered a proxy for the likelihood of a firm's bankruptcy, which can trigger a cascade of bankruptcies. A cascade of bankruptcies arises from the complex nature of agent interactions. To better analyze this complexity, we must undertake a simulation-driven data analysis in terms of the net worth.

Basically, there are two regimes: (i) when the net worth of each firm is less than one, and (ii) when the net worth of each firm is more than one. The simulation begins with homogeneous firms and banks with a small net worth (Regime 1). At the early stages of the simulation and at the premature stage of the economy, most agents find themselves under Regime 1, and many U firms and banks go bankrupt. A growing firm will balance increasing net worth with increased debt exposure. As leverage increases, the economy is riskier, with higher volatility in terms of the aggregate production and an increase in the bankruptcies of firms and banks. The increased net worth boosts a firm's growth until the economic system reaches a critical point of financial fragility and the cycle is reversed through an increase in bankruptcies.

However, after the expansionary phase, profits allow firms to accumulate net worth, and there is a decrease in firm bankruptcies and an overall reduction in bad debt. This allows many firms with a large net worth to finance themselves because they are more capitalized. When the net worth of a firm increases, its leverage also increases (i.e., there is a positive correlation between the net worth and leverage).

10.4 Evolving Credit Networks

Financial credit connections play an important role in economic growth, as well as in the emergence of economic and financial instabilities. In particular, for debt propagation, the financial conditions as well as the positions in the credit network of firms and banks are pertinent. Addressing the question of why firms and banks form sparsely or densely connected networks that are prone to systemic risk is important for understanding how effectively the financial sector can perform economic functions. By combining the network approach with agent-based modeling, we can better understand the primary role of lending banks in a growing economy, and how banks are likely to interact with firms and other banks. Understanding such issues can also help in designing policy to enhance economic development, and for designing resilient economic and financial systems. The question of how firms and banks form networks is also important for understanding how effectively the financial sector performs economic functions.

Our model is based on a situation where the supply of loans from banks is matched with the demand for loans originating in financially constrained firms facing random market price shocks. In our model, firms can seek loans from banks to increase their production and profit. Banks also hold the securities of other banks in the interbank system. Therefore, credit networks and interbank connections evolve over time. We scrutinized

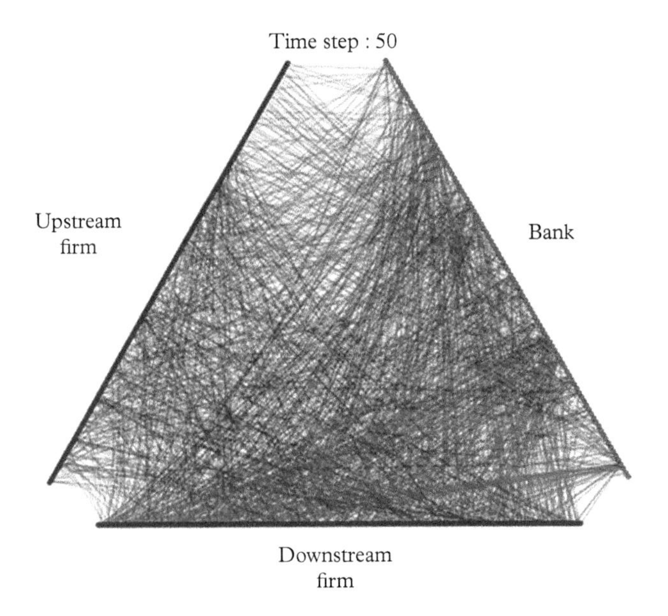

Figure 10.14 *Snapshot of evolving credit connections between firms and banks*

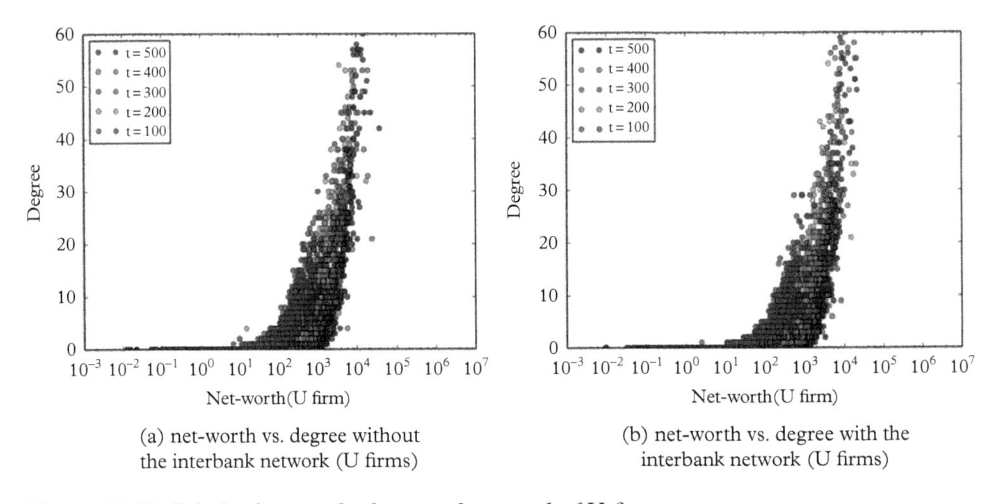

Figure 10.15 *Relation between the degree and net worth of U firms*

the characteristics of such evolving networks. Some snapshots of the credit connections among firms and banks are shown in Figure 10.14. These networks are continuously evolving with changes in terms of connectivity.

Figures 10.15 and 10.16 show the relations between the degree and the net worth of *U* firms and the banks that are not part of the interbank network (left panel) and those

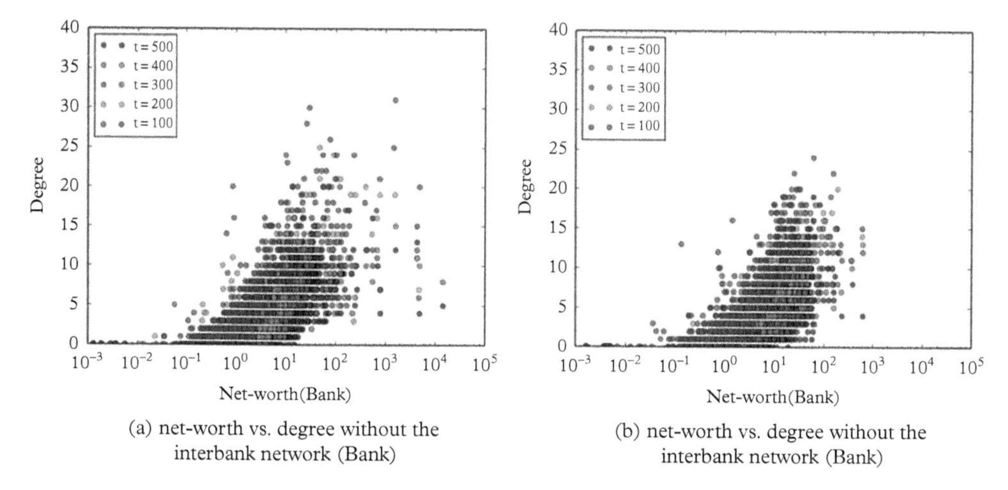

(a) net-worth vs. degree without the interbank network (Bank)

(b) net-worth vs. degree without the interbank network (Bank)

Figure 10.16 *Relation between the degree and net worth of banks*

that are (right panel). These figures show that U firms and banks with more net worth have higher degrees and are more densely connected to D firms and banks.

However, time-dependent resolution for the properties of economic and financial networks that evolve must move beyond a single snapshot approach. This allows the researcher to identify conditions for the dynamical or path-dependent evolution of networks by combining findings with complementary information, such as the correlations between economic network evolution and economic dynamics. This network evolution challenges both theoretical concepts and the use of real data for empirical validation.

An understanding of how networked systems work depends on our ability to obtain more and better data. Accumulating knowledge also fosters the transition from a qualitative study to data-driven studies and evidence-based science (Schweitzer et al. 2009). As computational power increases, it allows for large-scale network data to be gathered on different levels in the economy, as well as testing models that reflect the generation of large synthetic datasets. This includes detailed data on specific firm interactions and firm–bank credit market interactions. Interbank credit is also a crucial component in modern economies. Disruptions in the financial system can lead to severe problems for liquidity management, and a sudden rapture of established funding lines can trigger an avalanche of liquidity problems across the financial system. Understanding the structure of a network of credit relationships is essential for accurately assessing economic risk. However, until very recently little was known about the topology of credit connections and their salient features.

We must pay increased attention to the risks emerging from the various connections between banks in order to make the economic and financial systems safer against contagion effects. However, extracting the network structure from reported data is very difficult, in particular for financial transactions. The financial sector does not make all debt and credit relationships publicly available. Even central banks and regulators have only a dim view of the interconnections between banks at any given moment. Thus, the systemic risk to the financial system, and each bank's contribution to this risk, is poorly understood.

A natural starting point is to utilize a complementary approach based on agent-based modeling and simulations. Simulated transactions between firms and banks can generate real-time data, and mapping such data into feature types will lead to an improved understanding of the stability and resilience of economic and financial systems, and perhaps ultimately to pragmatic rules that are beneficial for agents. This line of inquiry also relates strongly to other areas of scientific research, from which we may draw additional ideas and strategies. For instance, to understand the most critical agent behavior and interactions, simulated data descriptions can be extended to wider jurisdictions that can record more detail. The ability to process large data streams will require new tools to squeeze out every last drop of available information reflecting agent interactions and network properties. Such databases, therefore, may complement both economic and financial network experiments and empirical studies by allowing large-scale observations in real-time.

Agent-based models allow us to estimate the probability of economic risk—i.e., the large-scale collective behavior based on the individual behavior of interactive agents. The primary advantage to doing so is that an agent-based model can generate data where there is none or little available in the financial system. The collective effects can be studied by running multiple simulations. Then, estimations can be made about the effect that changing the rules for interactions between agents will have on the collective outcome in the financial system. For example, in this work, the likelihood of a firm or bank defaulting was shown to depend on particular regulations, such as the bailout procedure. With agent models, we can demonstrate that such regulatory measures will have specific effects; these results would be difficult, if not impossible, to surmise otherwise. The origin, development, and unfolding of economic crises is tied to the model's parameters. In reality, the relevant data are never completely available. If it were, we could more easily identify the crucial parameters. However, salient parameters can indeed be identified through a modeling and simulation process.

References

Abramson, G. and Kuperman, M. (2001). Social games in a social network. *Physical Review E*, 63(3), 030901.

Acemoglu, D., Como, G., Fagnani, F., and Ozdaglar, A. (2013). Opinion fluctuations and disagreement in social networks. *Mathematics of Operations Research*, 38, 1–27.

Adamatzky, A. (ed.) (2010). *Game of Life Cellular Automata*. London: Springer.

Albin, P. (1975). The Analysis of Complex Socioeconomic Systems, Lexington, MA: Lexington Books.

Albin, P. (1992). Approximations of cooperative equilibria in multi-person prisoners' dilemma played by cellular automata. *Mathematical Social Sciences*, 24(2), 293–319.

Albin, P. (1998). *Barriers and Bounds to Rationality: Essays on Economic Complexity and Dynamics in Interactive Systems*. Edited by D. Foley. New Jersey: Princeton University Press.

Albin, P. and Foley, D. (1992). Decentralized, dispersed exchange without an auctioneer. *Journal of Economic Behavior and Organization*, 51, 18–27.

Allen, F. and Gale, D. (2000). Financial Contagion. *The Journal of Political Economy*, 108, 1–33.

Almog, A., Squartini, T., and Garlaschelli, D. (2015). A GDP-driven model for the binary and weighted structure of the International Trade Network. *New Journal of Physics*, 17(1), 013009.

Alpern, S. and Refiners, D. (2002). Spatial dispersion as a dynamic coordination problem. *Theory and Decision*, 53, 29–59.

Arthur, W. B. (1989). Competing technologies, increasing returns, and lock-in by historical events. *Economic Journal*, 99, 116–131.

Arthur, W. B. (1994). Complexity in economic theory: inductive reasoning and bounded rationality. *American Economic Review*, 82(2), 406–411.

Assenza, S., Gómez-Gardeñes, J., and Latora, V. (2008). Enhancement of cooperation in highly clustered scale-free networks. *Physical Review E*, 78(1), 017101.

Auyang, S. Y. (1998). Foundations of complex system theories in economics. *Evolutionary Biology and Statistical Physics*, Cambridge: Cambridge University Press.

Axelrod, R. (1984). *The Evolution of Cooperation*. New York: Basic Books.

Axelrod, R. (1997). The dissemination of culture: A model with local convergence and global polarization. *Journal of Conflict Resolution*, 41(2), 203–226.

Axtell, R. (2001). Zipf Distribution of U.S. Firm sizes. *Science*, October 2001.

Bala, V. and Goyal, S. (2000). A noncooperative model of network formation. *Econometrica*, 68(5), 1181–1229.

Ball, P. (2004). *Critical Mass*. New York Farrar, Straus and Giroux.

Bandura, A., Dorothea, R., and Sheila, A. (1963). Vicarious reinforcement and imitative learning. *The Journal of Abnormal and Social Psychology*, 67(6), 601–607.

Banerjee, A. (1992). A simple model of herd behavior. *Quarterly Journal of Economics*, 110, 797–817.

Banos, A. (2012). Network effects in Schelling's model of segregation: new evidences from agent-based simulation. *Environment and Planning B: Planning and Design*, 39(2), 393–405.

Barabasi, A. L. and Albert, R. (1999). Emergence of scaling in random networks. *Science*, 286, 509–512.

Bargigli, L. and Tedeschi, G. (2014). Interaction in agent-based economics: A survey on the network approach. *Physica A: Statistical Mechanics and its Applications*, 399, 1–15.

Barnes, J. (1954). Class and committees in a Norwegian Island Parish. *Human Relations*, 7, 39–58.

Barrat, A., Barthelemy, M., Pastor-Satorras, R., and Vespignani, A. (2004). The architecture of complex weighted networks. *Proceedings of the National Academy of Sciences of the United States of America*, 101(11), 3747–3752.

Bass, F. M. (1969). A new product growth model for consumer durables. *Management Science*, 13(5), 215–227.

Battu, H., Seaman, P., and Zenou, Y. (2011). Job contact networks and the ethnic minorities. *Labour Economics*, 18(1), 48–56.

Beckage, B., Kauffman, S., Gross, L. J., Zia, A., and Koliba, C. (2013). More complex complexity: Exploring the nature of computational irreducibility across physical, biological, and human social systems. In, *Irreducibility and Computational Equivalence*, Berlin Springer: Heidelberg, 79–88.

Bell, A. M. and Sethares, W. A. (1999). The El Farol problem and the internet: congestion and coordination failure. *Society for Computational Economics*, Working Paper No. 1999.

Bendor, J., Huberman, B. A., and Wu, F. (2009). Management fads, pedagogies, and other soft technologies. *Journal of Economic Behavior and Organization*, 24.

Benkler, Y. (2006). *The Wealth of Networks: How Social Production Transforms Markets and Freedom*. New Haven: Yale University Press.

Berg, J., Dickhaut, J. and McCabe, K. (1995). Trust, reciprocity, and social history. *Games and Economic Behavior*, 10, 122–142.

Berninghaus, S. K., Ehrhart, K. M., and Ott, M. (2006). A network experiment in continuous time: The influence of link costs. *Experimental Economics*, 9(3), 237–251.

Berninghaus, S. K., Ehrhart, K. M., Ott, M., and Vogt, B. (2007). Evolution of networks? An experimental analysis. *Journal of Evolutionary Economics*, 17(3), 317–347.

Bewley, T. F. (1999). *Why Wages Don't Fall During a Recession*. Cambridge, MA: Harvard University Press.

Bhattacharya, K., Mukherjee, G., Saramäki, J., Kaski, K., and Manna, S. S. (2008). The international trade network: weighted network analysis and modelling. *Journal of Statistical Mechanics: Theory and Experiment*, 2008(02), P02002.

Biggs, N., Lloyd, E. K., and Wilson, R. J. (1986). *Graph Theory*, New York: Clarendon Press, 1736–1936.

Billen, J., Wilson, M., and Baljon, A. (2009). Eigenvalue spectra of spatial-dependent networks. *Physical Review E*, 80, 046116.

Blume, L. E. (1993). The statistical mechanics of strategic interaction. *Games and Economic Behavior*, 5, 387–424.

Bonabeau, E., Dorigo, M., and Theraulaz, G. (1999). *Swarm Intelligence: from Natural to Artificial Systems*. New York: Oxford University Press.

Boorman, S. A. (1975). A combinatiorial optimization model for transmission of job information through contact networks. *The Bell journal of economics*, 6(1): 216–249.

Boss, M., Elsinger, H., Summe, M. and Thurner, S. (2003). The network topology of the interbank market. *arXiv:cond-mat/0309582v1*.

Boyd, R., Richerson, P., and Henrich, J. (2011). The cultural niche: why social learning is essential for human adaptation. *Proceedings of the National Academy of Sciences*, 108(Supplement 2), 10918–10925.

Boyd, S., Ghosh, A., Prabhakar, B., and Shah, D. (2006). Randomized gossip algorithms. *IEEE ACM Transactions on Networking*, 52, 2508–2530.

Bravo, G., Squazzoni, F., and Boero, R. (2012). Trust and partner selection in social networks: An experimentally grounded model. *Social Networks*, 34(4), 481–492.

Brynjolfsson, E. and McAfee, A. (2014). The second machine age: Work, progress, and prosperity in a time of brilliant technologies. New York *WW Norton & Company*.

Buchanam, M. (2007). *The Social Atom*. New York: Bloomsbury.

Cabrales, A., Calvo-Armengol, A., and Zenou, Y. (2011). Social interactions and spillovers. *Games and Economic Behavior*, 72(2), 339–360.

Cabrales, A., Gottardi, P., and Vega-Redondo, F. (2012). Risk-sharing and contagion in networks. Working paper, European University Institute.

Cahuc, P. and Fontaine, F. (2009). On the efficiency of job search with social networks. *Journal of Public Economic Theory*, 11(3), 411–439.

Cainelli G., Montresor S., and Marzetti G., (2012). Production and financial linkages in inter-firm networks: structural variety, risk-sharing and resilience. *Journal of Evolutionary Economics*, 22(4):711–734.

Caldarelli, G., Capocci, A., De Los Rios, P., and Muñoz, M. A. (2002). Scale-free networks from varying vertex intrinsic fitness. *Physical Review Letters*, 89(25), 258702.

Callander, S. and Plott, C. R. (2005). Principles of network development and evolution: An experimental study. *Journal of Public Economics*, 89(8), 1469–1495.

Calvo-Armengol, A. (2004). Job contact networks. *Journal of Economic Theory*, 115(1), 191–206.

Calvo-Armengol, A. and Jackson, M., (2004). The effects of social networks on employment and inequality. *American Economic Review*, 94(3), 426–454.

Calvo-Armengol, A. and Zenou, Y. (2005). Job matching, social network and word-of-mouth communication. *Journal of Urban Economics*, 57(3), 500–522.

Calvo-Armengol, A. and Jackson, M. O. (2007). Networks in labor markets: Wage and employment dynamics and inequality. *Journal of Economic Theory*, 132(1), 27–46.

Carlson, M. and Doyle, J. (2002). Complexity and robustness. *Proceedings of the National Academy of Sciences*, 99, 2538–2545.

Centola, D. (2010). The spread of behavior in an online social network experiment. *Science*, 329, 1194–1197.

Chakravorti, B. (2003). *The Slow Pace of Fast Change: Bringing Innovations to Market in a Connected World*. Boston: Harvard Business School Press.

Chatterjee, K. and Susan, H. (2004). Technology diffusion by learning from neighbors. *Advances in Applied Probability*, 36, 355–376.

Charness, G., Feri, F., Melendez-Jimenez, M., and Sutter, M. (2014). Experimental games on networks: Underpinnings of behavior and equilibrium selection. *IZA Discussion Paper*, No. 8104.

Chen, S. H. (1997). Would and should government lie about economic statistics: Understanding opinion formation processes through evolutionary cellular automata. In: Conte R., Hegselmann R., and Terna P. (eds.), Simulating Social Phenomena. *Lecture Notes in Economics and Mathematical Systems*, 456, 471–490.

Chen, S. H. (2006). Graphs, networks and ACE. *New Mathematics and Natural Computation*, 2(03), 299–314.

Chen, S. H. (2008). Computational intelligence in agent-based computational economics. In J. Fulcher and L. C. Jain (eds.), *Computational Intelligence: A Compendium*, Berlin: Springer, 517–594.

Chen, S. H. (2012). Varieties of agents in agent-based computational economics: A historical and an interdisciplinary perspective. *Journal of Economic Dynamics and Control*, 36(1), 1–25.

Chen, S. H. (2013). Reasoning-based artificial agents in agent-based computational economics. In Nakamatsu K, Jain L (eds.), *Handbook on Reasoning-based Intelligent Systems*, World Scientific, 2013, 575–602.

Chen, S. H., and Du, Y. R. (2014). Granularity in economic decision making: An interdisciplinary review. In: Pedrycz W., and Chen S-M. (eds.), *Granular Computing and Decision-Making: Interactive and Iterative Approaches*. Switzerland: Springer.

Chen, S. H., and Gostoli, U. (2015). Coordination in the El Farol Bar problem: The role of social preferences and social networks. *Journal of Economic Interaction and Coordination*, Forthcoming.

Chen, S. H. and Ni, C. C. (2000). Simulating the ecology of oligopoly games with genetic algorithms. *Knowledge and Information Systems: An International Journal*, 2, 310–339.

Chen, S. H., Chang, C. L., and Wen, M. C. (2014). Social networks and macroeconomic stability. Economics: The Open-Access, *Open-Assessment E-Journal*, 8(2014–16), 1–40.

Chen, S. H., Chie, B. T., and Zhang, T. (2015). Network-based trust game: An agent-based model. *Journal of Artificial Societies and Social Simulation*. Forthcoming.

Chen, S. H., Sun, L. C. and Wang, C. C. (2006). Network topologies and consumption externalities. In: *New Frontiers in Artificial Intelligence*, Berlin Springer: Heidelberg, 314–329.

Chung, F. K. (1997). *Spectral Graph Theory*. AMS.

Cincotti, S., Raberto, M., and Teglio, A. (2010). Credit money and macroeconomic instability in the agent-based model and simulator Eurace. *Economics: The Open-Access, Open-Assessment E-Journal*, 4(26).

Cochard, F., Van Nguyen, P., and Willinger, M. (2004). Trusting behavior in a repeated investment game. *Journal of Economic Behavior and Organization*, 55, 31–44.

Codd, E. (1968). *Cellular Automata*. New York: Academic Press.

Cohen, R., Erez, K., Avraham, D., and Havlin, S. (2001). Breakdown of the internet under intentional attack. *Phys. Rev. Lett.*, 86, 3682–3685.

Colizza, B., Barrat, A., Barthelemy, M., and Vespignani, D. (2006). The role of the airline transportation network in the prediction and predictability of global epidemics. *PNAS*, 103, 2015–2020.

Coleman, A. M. (2014). Game theory and experimental games: The study of strategic interaction. Elsevier.

Corbae, D. and Duffy, J. (2008). Experiments with network formation. *Games and Economic Behavior*, 64(1), 81–120.

Croson, R. T. and Marks, M. B. (2000). Step returns in threshold public goods: A meta-and experimental analysis. *Experimental Economics*, 2(3), 239–259.

Dasgupta, P. (1988). Trust as a commodity. In: Gambetta D (ed.). *Trust: Making and Breaking Cooperative Relations*. Oxford: Blackwell, 49–72.

David, P. A. (1985). Clio and economics of QWERTY. *American Economics Review*, 75, 332–337.

De Grauwe P. (2010). The scientific foundation of dynamic stochastic general equilibrium (DSGE) models. *Public Choice*, 144, 413–443.

De Grauwe P. (2011). Animal spirits and monetary policy. *Economic Theory*, 47, 423–457.

Deck, C. and Johnson, C. (2004). Link bidding in laboratory networks. *Review of Economic Design*, 8(4), 359–372.

De Montrs A., Barthélemy, M., Chessa, A., and Vespignani, A. (2005). The structure of inter-urban traffic: A weighted network analysis. *arXiv preprint physics/0507106*.

Deck, C. and Johnson, C. (2004). Link bidding in laboratory networks. *Review of Economic Design*, 8(4), 359–372.

DeGroot, M. (1974). Reaching a consensus. *Journal of the American Statistical Association*, 69, 345, 118–121.

Delli Gatti, D., Di Guilmi, C., Gallegati, M., and Palestrini, A. (2005). A new approach to business fluctuations: heterogeneous interacting agents, scaling laws and financial fragility. *Journal of Economic Behavior and Organization*, 56(4), 489–512.

Delli Gatti, D., Gallegati, M. Greenwal, B., and Stiglitz, J. (2010). The financial accelerator in an evolving credit network, *Journal of Economic Dynamic and Control*, 34, 1627–1650.

Delli Gatti, D., Desiderio, S., Gaffeo, E., Cirillo, P., and Gallegati. M. (2011). *Macroeconomics from the Bottom-Up*. Milan: Springer.

Delre, S. A., Jager, W., and Janssen, M. A. (2007). Diffusion dynamics in small-world networks with heterogeneous consumers. *Computational and Mathematical Organization Theory*, 13(2), 185–202.

Di, Cagno D. and Sciubba, E. (2010). Trust, trustworthiness and social networks: Playing a trust game when networks are formed in the lab. *Journal of Economic Behavior and Organization*, 75(2), 156–167.

Disdier, A. C. and Head, K. (2008). The puzzling persistence of the distance effect on bilateral trade. *The Review of Economics and Statistics*, 90(1), 37–48.

Dixit, A. (2006). Thomas Schelling's contributions to game theory. *The Scandinavian Journal of Economics*, 108(2), 213–229.

Dobson, L., Carreras, S., Lynch, D., and Newman, D. (2007). Complex systems analysis of series of blackouts: cascading failure, critical points, and self-organization. *CHAOS*, 17.

Doreian, P. (2006). Actor network utilities and network evolution. *Social Networks*, 28(2), 137–164.

Dueñas, M. and Fagiolo, G. (2013). Modeling the International-Trade Network: a gravity approach. *Journal of Economic Interaction and Coordination*, 8(1), 155–178.

Dulsrud, A. and Gronhaug, K. (2007). Is friendship consistent with competitive market exchange? A microsociological analysis of the fish export, Vimport business, *Acta Sociologica*, 50(1), 7–19.

Dunbar, R. (1998). The Social Brain Hypothesis. *Evolutionary Anthropology*, 6, 178–190.

Dunbar, R. and Shultz, S. (2007). Evolution in the social brain. *Science*, 317, 1344–1347.

Durlauf, S. and Young, P. (2004). (eds.), *Social Dynamics*, Cambridge, MA and London, MIT Press.

Durrett, R. (2007). *Random Graph Dynamics*. England: Cambridge University Press.

Eguiluz, V. M., Zimmermann, M. G., Cela-Conde, C. J., and San Miguel, M. (2005). Cooperation and the Emergence of Role Differentiation in the Dynamics of Social Networks. *American Journal of Sociology*, 110(4), 977–1008.

Elbittar, A., Harrison, R., and Muñoz, R. (2014). Network structure in a link-formation game: An experimental study. *Economic Inquiry* 52(4), 1341–1363.

Ellison, G. (1993). Learning, local interaction, and coordination. *Econometrica*, 61(5), 1047–1071.

Fagiolo, G., Moneta, A., and Windrum, P. (2007). A critical guide to empirical validation of agent-based models in economics: Methodologies, procedures, and open problems. *Computational Economics* October 2007, 3, 195–226.

Fagiolo, G., Reyes, J., and Schiavo, S. (2008). On the topological properties of the world trade web: A weighted network analysis. *Physica A: Statistical Mechanics and its Applications*, 387(15), 3868–3873.

Fagiolo, G., Reyes, J., and Schiavo, S. (2009). World-trade web: Topological properties, dynamics, and evolution. *Physical Review E*, 79(3), 036115.

Fagiolo, G., Reyes, J., and Schiavo, S. (2010). The evolution of the world trade web: a weighted-network analysis. *Journal of Evolutionary Economics*, 20(4), 479–514.

Fagiolo, G., Squartini, T., and Garlaschelli, D. (2013). Null models of economic networks: the case of the world trade web. *Journal of Economic Interaction and Coordination*, 8(1), 75–107.
Fagnani, F. (2014). Consensus dynamics over networks. *Technical Paper*. Politecnico di Torino.
Falk, A. and Kosfeld, M. (2012). It's all about connections: Evidence on network formation. *Review of Network Economics*, 11(3). DOI: 10.1515/1446-9022.1402
Farley, R. (1998). Blacks, hispanics, and white ethnic groups: are blacks uniquely disadvantaged? *American Economic Review*, 80, 237–241.
Fehr, E. and Gachter, S. (1998). Reciprocity and economics: The economic implications of homo reciprocans. *European Economic Review*, 42, 845–859.
Fehr, E. and Schmidt, K. (1999). A theory of fairness, competition, and cooperation. *The Quarterly Journal of Economics*, 114(3), 817–868.
Fehr, E., Kirchsteiger, G., and Riedl, A. (1993). Does fairness prevent market clearing? An experimental investigation. *The Quarterly Journal of Economics*, 108(2): 437–459.
Fogli, A. and Veldkamp, L. (2012). Germs, social networks and growth (No. w18470). National Bureau of Economic Research.
Fontaine, F. (2008). Why are similar workers paid differently? The role of social networks. *Journal of Economic Dynamics and Control*, 32(12), 3960–3977.
Fossett, M. and Dietrich, D. R. (2009). Effects of city size, shape, and form, and neighborhood size and shape in agent-based models of residential segregation: are Schelling-style preference effects robust? Environment and planning. B, *Planning and Design*, 36(1), 149–169.
Freixas, X., Parigi, B. M., and Rochet, J. C. (2000). Systemic risk, interbank relations and liquidity provision by the central bank. *Journal of Money, Credit and Banking*, 32, 3, 611–638.
Fricke, D. and Lux, T. (2012). Core-periphery structure in the overnight money market: Evidence from the e-Mid trading platform. *Kiel Working Paper*, 1759.
Friedkin, N. (2012). A formal theory of reflected appraisals in the evolution of power. *Administrative Science Quarterly*, 56, 501–529.
Friedkin, N. and Johnsen, E. C. (1999). Social influence networks and opinion change. *in Advances in Group Processes*, 16, 1–29.
Fronczak, A. and Fronczak, P. (2012). Statistical mechanics of the international trade network. *Physical Review E*, 85(5), 056113.
Gai, P. and Kapadia, S. (2010). Contagion in financial networks. *Proceedings of the Royal Society A: Mathematical, Physical and Engineering Science*, 466(2120), 2401–2423.
Gallegati, M., Greenwald, B., Richiardi, M. G., and Stiglitz, J. E. (2008). The asymmetric effect of diffusion processes: Risk sharing and contagion. *Global Economy Journal* 8(3), 1–20.
Gallegati, M. and Kirman, A. (2012). Reconstructing economics, *Complexity Economics*, 1(1), 24–35.
Galenianos, M. (2014). Hiring through referrals. *Journal of Economic Theory*. 152, 304–323.
Galeotti, A. and Goyal, S. (2010). The law of the few. *American Economic Review*, 100(4), 1468–1492.
Galeotti, A. and Merlino, L. P. (2010). Endogenous job contact networks (No. 2010-14). *ISER Working Paper Series*.
Garcia R. and Jager, W. (2011). Special issue: agent-based modeling of innovation diffusion. *Journal of Product Innovation Management*, 28, 148–151.
Garlaschelli, D. and Loffredo, M. I. (2004). Fitness-dependent topological properties of the world trade web. *Physical Review Letters*, 93(18), 188701.
Garlaschelli, D. and Loffredo, M. I. (2005). Structure and evolution of the world trade network. *Physica A: Statistical Mechanics and its Applications*, 355(1), 138–144.
Garlaschelli, D. and Loffredo, M. I. (2009). Generalized Bose-Fermi statistics and structural correlations in weighted networks. *Physical Review Letters*, 102(3), 038701.

Gemkow, S. and Neugart, M. (2011). Referral hiring, endogenous social networks, and inequality: an agent-based analysis. *Journal of Evolutionary Economics*, 21(4), 703–719.
Gigerenzer, G. and Gaissmaier, W. (2011). Heuristic decision making. *Annual Review of Psychology*, 62, 451–482.
Gil, S. and Zanette, D. H. (2006). Coevolution of agents and networks: Opinion spreading and community disconnection. *Physics Letters A*, 356, 2, 89–94.
Gilder, G. (1993). Metcalfe's law and legacy. *Forbes ASAP*, 13.
Gladwell, M. (2002). *The Tipping Point: How Little Things Can Make a Big Difference*. Back Bay Books.
Goeree, J. K., Riedl, A., and Ule, A. (2009). In search of stars: Network formation among heterogeneous agents. *Games and Economic Behavior*, 67(2), 445–466
Goldenberg, J., Han, S., Lehmann, D. R., and Hong, J. W. (2009). The role of hubs in the adoption process. *Journal of Marketing*, 73(2), 1–13.
Golub, B., and Jackson, M. (2010). Naïve Learning in Social Networks and the Wisdom of Crowds. *American Economic Journal: Microeconomics*, 2(1): 112–149.
Gómez-Gardeñes, J., Campillo, M., Floría, L. M., & Moreno, Y. (2007). Dynamical organization of cooperation in complex topologies. *Physical Review Letters*, 98(10), 108103.
Goyal, S. (2012). Connections: an introduction to the economics of networks. NewJersey: Princeton University Press.
Granovetter, M. (1973). The strength of weak ties. *English American Journal of Sociology*, 78, 6, 1360–1380.
Granovetter, M. (1978). Threshold models of collective behavior. *The American Journal of Sociology*, 83, 1420–1443.
Granovetter, M. (1993). The nature of economic relationships. In: R. Swedberg (ed.) *Explorations in Economic Sociology*, New York: Russel Sage Foundation, 3–41.
Granovetter, M. (1995). Getting a Job: A Study of Contacts and Careers, 2nd ed. Cambridge: Harvard University Press.
Granovetter, M. (2005). The impact of social structure on economic outcomes. *Journal of Economic Perspectives*, 19(1): 33–50.
Grilli R., Tedeschi, G. and Gallegati, M. (2012). Markets connectivity and financial contagion. Working paper 375, Universita Politecnica delle Marche.
Grimm, V. and Railsback, S. (2005). *Individual-based Modeling and Ecology*. New Jersey Princeton University Press.
Gross (2008). Adaptive coevolutionary networks: a review, Vol.5, pp.259–271
Gross, T. and Blasius, B. (2008). Adaptive coevolutionary networks: a review. *Journal of the Royal Society, Interface*, 5(20): 259–271.
Hauert, C. and Doebeli, M. (2004). Spatial structure often inhibits the evolution of cooperation in the snowdrift game. *Nature*, 428(6983), 643–646.
Hayashida, T., Nishizaki, I., and Kambara, R. (2014). Simulation Analysis for Network Formulation. *Computational Economics*, 43(3), 371–394.
Hegselmann, R. and Flache, A. (1998). Understanding complex social dynamics: A plea for cellular automata based modelling. *Journal of Artificial Societies and Social Simulation*, 1(3). http://jasss.soc.surrey.ac.uk/1/3/1.html
Helbing, D. (2010). Systemic risk in society and economics. *Technical Report*, IRGC-Emerging Risks.
Helbing, D. (2010). Systemic Risks in Society and Economics, International Risk Governance Council (IRGC) pp.1–25
Hines, P. and Talukdar, S. (2007). Controlling Cascading Failures with Cooperative Autonomous Agents. *International Journal of Critical Infrastructures,* 3, 192–220.

Hummon, N. P. (2000). Utility and dynamic social networks. *Social Networks*, 22(3), 221–249.
Ilachinski, A. (2001). *Cellular Automata: A Discrete Universe*. Singapore: World Scientific.
Innes, J., Thebaud, O., Norman-Lopez, A., and Little, L. R. (2014). Does size matter? An assessment of quota market evolution and performance in the Great Barrier Reef fin-fish fishery. *Ecology and Society*, 19(3), 1–14.
Ioannides, Y. M. and Loury, L. D. (2004). Job information networks, neighborhood effects, and inequality. *Journal of Economic Literature*, 1056–1093.
Isaac, R. M., Walker, J. M., and Thomas, S. H. (1984). Divergent evidence on free riding: An experimental examination of possible explanations. *Public Choice*, 43(2), 113–149.
Israeli, N. and Goldenfeld, N. (2004). Computational irreducibility and the predictability of complex physical systems. *Physical Review Letters*, 92(7), 074105.
Israeli, N. and Goldenfeld, N. (2006). Coarse-graining of cellular automata, emergence, and the predictability of complex systems. *Physical Review E*, 73(2), 026203.
Jackson, M. O. (2005). A survey of network formation models: stability and efficiency. In: Demange G., Wooders M. (eds.): Group Formation in Economics: Networks, Clubs, and Coalitions, Cambridge, 11–49.
Jackson, M. (2008). *Social and Economic Networks*. New Jersey: Princeton University Press.
Jackson, M. O. and Wolinsky, A. (1996). A strategic model of social and economic networks. *Journal of Economic Theory*, 71(1), 44–74.
Jackson, M. O. and van den Nouweland, A. (2005). Strongly stable networks. *Games and Economic Behavior*, 51(2), 420–444.
Kahneman, D. and Tversky, A. (1979). Prospect theory: An analysis of decision under risk. Econometrica. *Journal of the Econometric Society*, 47(2), 263–291.
Kakade, S. M., Kearns, M., Ortiz, L. E., Pemantle, R., and Suri, S. (2004). Economic properties of social networks. *Advances in Neural Information Processing Systems*. Cambridge, 633–640.
Kali, R. and Reyes, J. (2007). The architecture of globalization: a network approach to international economic integration. *Journal of International Business Studies*, 38(4), 595–620.
Kali, R. and Reyes, J. (2010). Financial contagion on the international trade network. *Economic Inquiry*, 48(4), 1072–1101.
Kandori, M. and Mailath, G. (1993). Learning, mutation and long-run equilibria in games. *Econometrica*, 61, 29–56.
Kar, S., Aldosari, S., and Moura, J. M. F. (2008). Topology for distributed inference on graphs. *IEEE Transactions on Signal Processing*, 56, 6, 2609–2613.
Katz, M. L. and Shapiro, C. (1985). Network externalities, competition, and compatibility. *The American Economic Review*, 424–440.
Katz, M. and Heere, B. (2013). Leaders and followers: An exploration of the notion of scale-free networks within a new brand community. *Journal of Sport Management*, 27(4), 271–287.
Keenan, D. C. and O'Brien, M. J. (1993). Competition, collusion, and chaos. *Journal of Economic Dynamics and Control*, 17(3), 327–353.
Kephart, J. and White, S. (1991). Direct-graph epidemiological models of computer virus prevalence. *Proceedings of the 1991 IEEE Computer Society Symposium on Research in Security and Privacy*, 343–359.
Kermack, W. O. and McKendrick, A. (1927). A contribution to the mathematical theory of epidemics. *Proceedings of the Royal Society of London*, A, 700–721.
Klemm, K. (2013). Searchability of central nodes in networks. *Journal of Statistical Physics*, 151, 707–719.
Kim, S. and Shin, E. H. (2002). A longitudinal analysis of globalization and regionalization in international trade: A social network approach. *Social Forces*, 81(2), 445–468.

Kirman, A. (2001). Market organization and individual behavior: Evidence from fish markets. *Networks and markets*, 155–196.
Kiss, C. and Bichler, M. (2008). Identification of influencers? Measuring influence in customer networks. *Decision Support Systems*, 46(1), 233–253.
Kitamura, Y. and Namatame, A. (2015). *The influence of stubborn agents in a multi-agent network for inter-team cooperation/negotiation.* International Journal of Advanced Research in Artificial Intelligence, 4(5), 1–9.
Kranton, R. E. and Minehart, D. F. (2001). A Theory of Buyer-Seller Networks. *The American Economic Review*, 91(3), 485–508.
Krauth, B. V. (2004). A dynamic model of job networking and social influences on employment. *Journal of Economic Dynamics and Control*, 28(6), 1185–1204.
Krebs, V. (2005). *It's the Conversations, Stupid!* Commonweal Institute.
Kunreuther, H. and Heal, G. (2003). Interdependent security. *Journal of Risk and Uncertainty*, 26, 231–249.
Lamm, E. and Unger, R. (2011). *Biological Computation.* Florida: CRC Press.
Laschever, R. (2005). The doughboys network: social interactions and labor market outcomes of World War I veterans. *Unpublished manuscript*, Northwestern University.
Laurie, A. J. and Jaggi, N. K. (2003). Role of 'vision' in neighbourhood racial segregation: A variant of the Schelling segregation model. *Urban Studies*, 40(13), 2687–2704.
Lazerson, M. (1993). Factory or putting-out? Knitting networks in Modena. In: Grabber G. (ed.) *The Embedded Firm: On the socioeconomics of industrial networks*, New York: Routledge. 203–226.
Le Bon, G. (1895). *The Crowd: A Study of the Popular Mind.* Dover Publications.
Leamer, E. E. (2007). A flat world, a level playing field, a small world after all, or none of the above? A review of Thomas L. Friedman's "The world is flat". *Journal of Economic Literature*, 83–126.
Liebowitz, S. J. and Margolis, S. (2002). *The Economics of QWERTY: History, Theory, and Policy.* New York: NYU Press.
Liedtke, G. (2006). *An Actor-Based Approach to Commodity Transport Modelling.* Baden-Baden: Nomos.
Lindner, L. and Strulik, H. (2014). The great divergence: A network approach. Discussion Papers, Center for European Governance and Economic Development Research, No. 193.
Lonkila, M. (2010). *Networks in the Russian Market Economy.* Palgrave Macmillan.
Lopetz-Pintado, D. (2006). Contagion and coordination in random networks. *International Journal of Game Theory*, 34, 371–382.
Lorentz, J., Rauhut, H, Schweitzer, F, and Helbing, D. (2011). How social influence can undermine the wisdom of crowd effect. *PNAS*, 108(22), 9020–9025.
Louzada, V. N., Araujo, A. M., Andrade, J. S., and Herrmann, H. J. (2012). How to suppress undesired synchronization. *Scientific Reports 2*, 658.
Macy, M. and Willer, R. (2002). From factors to actors: Computational sociology and agent-based modeling. *Annual Review of Sociology*, 28, 143–66.
Marsden, P. and Friedkin, N. (1993). Network studies of social influence. *Social Methods & Research*, 22(1): 127–151.
Mastrandrea, R., Squartini, T., Fagiolo, G., and Garlaschelli, D. (2014). Enhanced reconstruction of weighted networks from strengths and degrees. *New Journal of Physics*, 16(4), 043022.
Matsuo, K. (1985). Ecological characteristics of strategic groups in 'dilemmatic world', *Proceedings of the IEEE International Conference on Systems and Cybernetics*, 1071–1075.
May, R. M., Levin, S. A., and Sugihara. G. (2008). Complex systems: ecology for bankers. *Nature*, 451, 893–895.

May, R. M. and Arinaminpathy, N. (2010). Systemic risk: the dynamics of model banking systems. *Journals of the Royal Society*, 7, 823–838.

McCulloch, W. S. and Pitts, W. (1943). A logical calculus of the ideas immanent in nervous activity. *The Bulletin of Mathematical Biophysics*, 5(4), 115–133.

McIntosh, H. (2009). *One Dimensional Cellular Automata.* Frome: Luniver Press.

McPherson, M., Smith-Lovin, L. and Cook, J. (2001). Birds of a feather: Homophily in social networks. *Annual Review of Sociology*, 27, 415–444.

Meyers, L., Pourbohloul D., and Newman, M. E. J. (2005). Network theory and SARS: predicting outbreak diversity. *Journal of Theoretical Biology*, 232, 71–81.

Mieghem, P. V., Omic, J., and Kooij, R. (2009). Virus spread in networks. *IEEE/ACM Transactions on Networking*, 17, 1.

Milgram, S. (1967). The small world problem. *Psychology Today*, 2, 60–67.

Mills, K. (2014). Effects of Internet use on the adolescent brain: despite popular claims, experimental evidence remains scarce. *Trends in Cognitive Sciences*, 18(8), 385–387.

Montgomery, J. (1991). Social networks and labor market outcomes: Toward an economic analysis. *American Economic Review*, 81, 1408–1418.

Moore, T., Finley, P., Linebarger, J. (2011). Extending opinion dynamics to model public health problems and analyze public policy interventions. *SAND 2011-3189 C.*

Morris, S. (2000). Contagion. *Review of Economic Studies*, 67, 57–78.

Motter, A. E. and Lai Y. C. (2002). Cascade-based attacks on complex networks. *Physical Review E*, 66(6), 065102.

Motter, A. E., Zhou, C., and Kutth, J. (2005). Network synchronization, diffusion, and the paradox of heterogeneity. *Physical Review E*, 71, 016116.

Moussald, M., Kammer, P., Hansjorg, A., and Neth, H. (2013). Social influence and the collective dynamics of opinion formation. PLoS ONE 8(11): e78433. doi:10.1371/journal.pone.0078433.

Moussald et al. (2013). Social Influence and the Collective Dynamics of Opinion Formation PLoS ONE, Vol.8. Issue 11, e78433

Namatame, A. and Iwanaga, S. (2002). The complexity of collective decision, *Psychology, and Life Sciences*, 6, 137–158.

Newman, M. E. J. (2002). Assortative mixing in networks. *Physical Review Letters*, 89(20), 208701.

Newman, M. E. J. (2004). Fast algorithm for detecting community structure in networks. *Physical Review E*, 69, 066133.

Newman, M. E. J. and Park, J. (2003). Why social networks are different from other types of networks. *Physical Review E*, 68(3), 036122.

Nier, E., Yang, J., Yorulmazer, T., and Alentorn, A. (2007). Network models and financial stability. *Journal of Economic Dynamics and Control*, 31, 6, 2033–2060.

Nishiguchi, T. (1994). Strategic industrial sourcing: The Japanese advantage. Oxford: Oxford University Press.

Nowak, M. A. (2006). *Evolutionary Dynamics.* Cambridge, MA: Harvard University Press.

Nowak, M. A. and May, R. M. (1992). Evolutionary games and spatial chaos. *Nature*, 359(6398), 826–829.

Nowak, M. A. and May, R. M. (1993). The spatial dilemmas of evolution. *International Journal of Bifurcation and Chaos*, 3, 35–78.

Nowak, M. A. and Sigmund, K. (2000). Games on Grids. In: U. Dieckmann, R. Law and J. A. J. Metz (eds.) *The Geometry of Ecological Interactions*, Cambridge: Cambridge University Press, 135–150.

Oakley, B. (2007). *Evil genes: Why Rome Fell, Hitler Rose, Enron Failed, and My Sister Stole My Mother's Boyfriend.* New York Prometheus Books.

Olfati-Saber, R. and Murray. R. (2007). Algebraic connectivity ratio of ramanujan graphs. *Proceeding of American Control Conference.*

Olfati-Saber, R., Fax, J. A., and Murray, R . M. (2007). Consensus and cooperation in networked multi-agent systems. *Proceedings of the IEEE*, 95, 1, 215–233.

Olshevsky, A. and Tsitsiklis, J. N. (2011). Convergence speed in distributed consensus and averaging, *SIAM Review*, 53(4): 747–772.

Pancs, R. and Vriend, N. (2007). Schelling's spatial proximity model of segregation revisited. *Journal of Public Economics*, 91, 1–24.

Park, J. and Newman, M. E. J. (2004). Statistical mechanics of networks. *Physical Review E*, 70(6), 066117.

Pastor-Satorras, R. and Vespignani, A. (2001). Epidemic spreading in scale-free networks. *Physical Review Letters*, 86, 3200.

Patterson, O. (1998). *Rituals of Blood: Consequences of Slavery in Two American Centuries.* Civitas Counterpoint, Washington.

Pemantle, R. and Skyrms, B. (2004). Network formation by reinforcement learning: the long and medium run. *Mathematical Social Sciences*, 48(3), 315–327.

Perc, M., Gómez-Gardeñes, J., Szolnoki, A., Floría, L. M., and Moreno, Y. (2013). Evolutionary dynamics of group interactions on structured populations: A review. *Journal of the Royal Society Interface*, 10(80), 20120997.

Pinheiro, F. L., Santos, M. D., Santos, F. C., and Pacheco, J. M. (2014). Origin of peer influence in social networks. *Physical Review Letters*, 112(9), 098702.

Pusch, A., Weber, S., and Porto, M. (2008). Impact of topology on the dynamical organization of cooperation in the prisoners dilemma game. *Physical Review E*, 77(3), 036120.

Rahwan, I., Krasnoshtan, D., Shariff, A.. and Bonnefon, J. (2014). Analytical reasoning task reveals limits of social learning in networks. *Journal of The Royal Society Interface*, 11(93), 20131211.

Rand, W. and Rust, R. T. (2011). Agent-based modeling in marketing: Guidelines for rigor. *International Journal of Research in Marketing*, 28(3), 181–193.

Raub, W., Buskens, V., and Van Assen, M. A. (2011). Micro-macro links and microfoundations in sociology. *The Journal of Mathematical Sociology*, 35(1–3), 1–25.

Rees, A. (1966). Information networks in labor markets. *The American Economic Review*, 56, 1/2, 559–566.

Rendell, L., Boyd, R., Cownden, D., Enquist, M., Eriksson, K., Feldman, M., Fogarty, L., Ghirlanda, S., Lillicrap, T., and Laland, K. (2010) Why copy others? Insights from the social learning strategies tournament. *Science*, 328(5975), 208–213.

Rogers, E. M. (2003). *Diffusion of Innovations* (5th edn). New York: Free Press.

Rogers, E. M., Rivera, U. and Wiley, C. (2005). Complex adaptive systems and the diffusion of innovations. *The Innovation Journal*, 10(3), 1–26.

Rogers, T. and McKane, A. J. (2011). A unified framework for Schelling's model of segregation. *Journal of Statistical Mechanics, Theory and Experiment*, 2011(07), P07006.

Rong, R. and Houser, D. (2012). Emergent star networks with ex ante homogeneous agents. George Mason University Interdisciplinary Center for Economic Science Paper No. 12, 33.

Rong, Z., Yang, H. X., and Wang, W. X. (2010). Feedback reciprocity mechanism promotes the cooperation of highly clustered scale-free networks. *Physical Review E*, 82(4), 047101.

Rosenberg, N. (1972). Factors affecting the diffusion of technology. *Explorations in Economic History*, 10(1), 3–33.

Rosenthal, R. W. (1981). Games of perfect information, predatory pricing and the chain-store paradox. *Journal of Economic theory*, 25(1), 92–100.
Rossbach, S. (1983). *Feng shui: The Chinese Art of Placement*. New York: Dutton.
Rousseau, J. J. (1984). *A Discourse on Inequality*. London: Penguin.
Sakoda, J. (1971). The checkerboard model of social interaction. *Journal of Mathematical Sociology*, 1, 119–132.
Sander, R., Schreiber, D., and Doherty, J. (2000). Empirically testing a computational model: The example of housing segregation. *Proceedings of the workshop on simulation of social agents: Architectures and institutions*, 108–115.
Santos, F. C. and Pacheco, J. M. (2005). Scale-free networks provide a unifying framework for the emergence of cooperation. *Physical Review Letters*, 95(9), 098104.
Saramäki, J., Kivelä, M., Onnela, J. P., Kaski, K., and Kertesz, J. (2007). Generalizations of the clustering coefficient to weighted complex networks. *Physical Review E*, 75(2), 027105.
Schelling, T. C. (1971). Dynamic models of segregation. *Journal of Mathematical Sociology*, 1, 143–186.
Schelling, T. C. (1972). A process of residential segregation: Neighborhood tipping. In Pascal A (ed.), *Racial Discrimination in Economic Life*, D. C. Heath, Lexington, MA, 157–84.
Schelling, T. C. (1978). *Micromotives and Macrobehavior*. New York: Norton.
Schiff, J. (2008). *Cellular Automata: A Discrete View of the World*. Chichester: Wiley.
Schram, A. J., Van Leeuwen, B., and Offerman, T. (2013). Superstars Need Social Benefits: An Experiment on Network Formation. Available at SSRN 2547388.
Schweitzer, F., Fagiolo, G., Sornette, D., Vega-Redondo, F., and White, D. (2009). Economic networks: What do we know and what do we need to know? *Advances in Complex Systems*, 12, 4.
Seinen, I. and Schram, A. (2006). Social status and group norms: Indirect reciprocity in a repeated helping experiment. *European Economic Review*, 50(3), 581–602.
Serrano, M. Á. and Boguñá, M. (2003). Topology of the world trade web. *Physical Review E*, 68(1), 015101.
Simon, H. A. (1955). A behavioral model of rational choice. *The Quarterly Journal of Economics*, 99–118.
Skyrms, B. (2004). *The Stag Hunt and the Evolution of Social Structure*. Cambridge: Cambridge University Press.
Skyrms, B. and R. Pemantle. (2000). A Dynamic Model of Social Network Formation. *Proceedings of the National Academy of Sciences*, 97, 9340–46.
Skyrms, B. and Pemantle, R. (2010). Learning to network. In: Eells E., Fetzer J. (eds.) *The Place of Probability in Science: In Honor of Ellery Eells* (1953–2006). Springer, 277–287.
Smith, A. (1776). *The Wealth of Nations*. Random House, Inc.
Song, Y. and van der Schaar, M. (2013). Dynamic Network Formation with Incomplete Information. arXiv preprint arXiv:1311.1264.
Soramaki, K., Bech, M. L., Arnold, J., Glass, R. J., and Beyeler, W. E. (2007). The topology of interbank payment flows. *Physica A*, 379, 317–333.
Sorenson, O. and Waguespack, D. M. (2006). Social structure and exchange: Self-confirming dynamics in Hollywood. *Administrative Science Quarterly*, 51(4), 560–589.
Spielman, D. J. (2005). Systems of innovation: Models, methods, and future directions. *Innovation Strategy Today*, 2(1), 55–6.
Squartini, T., Fagiolo, G., and Garlaschelli, D. (2011a). Randomizing world trade. I. A binary network analysis. *Physical Review E*, 84(4), 046117.

Squartini, T., Fagiolo, G., and Garlaschelli, D. (2011b). Randomizing world trade. II. A weighted network analysis. *Physical Review E*, 84(4), 046118.

Stanley, M., Amaral, L., Sergey, V., Havlin, L., Leschhorn, H., Maass, P., Stanley, E. (1996). Scaling behavior in the growth of companies, *Nature*, 379, 804–806.

Sunstein A, and Kuran, T, (2007) "Availability cascade and risk regulation, The working paper, The University of Chicago Law School.

Surowiecki, J. (2004). *The Wisdom of Crowds: Why the Many are Smarter than the Few and how Collective Wisdom Shapes Business, Economies, and Nations*. New York: Random House.

Tadesse, B. and White, R. (2010). Cultural distance as a determinant of bilateral trade flows: do immigrants counter the effect of cultural differences? *Applied Economics Letters*, 17(2), 147–152.

Tam, C. M., Tso, T. Y., and Lam, K. C. (1999). Feng Shui and its impacts on land and property developments. *Journal of Urban Planning and Development*, 125(4), 152–163.

Tassier, T. (2006). Labor market implications of weak ties. *Southern Economic Journal*, 72, 704–719.

Tassier, T. and Menczer, F. (2001). Emerging small-world referral networks in evolutionary labor markets. *Evolutionary Computation*, IEEE Transactions on, 5(5), 482–492.

Tassier, T. and Menczer, F. (2008). Social network structure, segregation, and equality in a labor market with referral hiring. *Journal of Economic Behavior & Organization*, 66(3), 514–528.

Thiriot, S., Lewkovicz, Z., Caillou, P., and Kant, J. D. (2011). Referral hiring and labor markets: A computational Study. In: Osinga S, Hofstede G, Verwaart T (eds.). Emergent Results of Artificial Economics. *Lecture Notes in Economics and Mathematical Systems*, 652, 15–25.

Tinbergen, J. (1962). Shaping the World Economy: Suggestions for an International Economic Policy. New York: The Twelveth Century Find.

Topa, G. (2001). Social interactions, local spillovers and unemployment. *The Review of Economic Studies*, 68(2), 261–295.

Topa, G., Bayer, P., and Ross, S. (2009). Place of work and place of residence: informal hiring networks and labor market. *Journal of Political Economy*, 116, 1150–1196.

Tran, H. and Namatame, A. (2015). *Mitigating Cascading Failure with Adaptive Networking.* New Mathematics and Natural Computation, 11(2), 1–17. International Journal of Advanced Research in Artificial Intelligence.

Travers, J. and Milgram, S. (1969). An experimental study of the small world problem. *Sociometry*, 32, 425–443.

Tversky, A. and Kahneman, D. (1992). Advances in prospect theory. *Journal of Risk and Uncertainty*, 5, 4, 297–323.

Uzzi, B. (1996). The sources and consequences of embeddedness for the economic performance of organizations: The network effect. *American Sociological Review*, 674–698.

Valente, W. T. (2012). Network interventions. *Science*, 337, 49–53.

Van Dolder, D. and Buskens, V. (2014). Individual choices in dynamic networks: An experiment on social preferences. *PloS one*, 9(4), e92276.

Van Huyck, J. B., Battalio, R. C., and Walters, M. F. (1995). Commitment versus discretion in the peasant-dictator game. *Games and Economic Behavior*, 10(1), 143–170.

Van Leeuwen, B., Offerman, T., and Schram, A. (2013). Superstars need social benefits: an experiment on network formation. Preprints, Tinbergen Institute Discussion Paper, no T1 2013–112/1, Tinbergen Institute.

Van Putten, I., Hamon, K. G., and Gardner, C. (2011). Network analysis of a rock lobster. Quota lease market. *Fisheries Research*, 107(1), 122–130.

Vanin, P. (2002). *Network formation in the lab: a pilot experiment.* Universitat Pompeu Fabra.

Von Neumann, J. completed by Burks, A. (1966). *Theory of Self Reproducing Automata.* University of Illinois Press. More materials on cellular automata can be found in Schiff (2008).
Voorhees, B. (1996). *Computational Analysis of One-Dimensional Cellular Automata.* Singapore: World Scientific.
Wang, Y. and Chakrabarti, D. (2003). Epidemic spreading in real networks: An eigenvalue viewpoint. *Proceedings of the 22nd Symposium on Reliable Distributed Computing*, 242–262.
Wang, P. and Watts, A. (2006). Formation of buyer-seller trade networks in a quality-differentiated product market. *Canadian Journal of Economics*, 39(3), 971–1004.
Wasserman, H. and Yohe, G. (2001). Segregation and the provision of spatially defined local public goods. *The American Economist*, 13–24.
Watts, D. J. (2002). A simple model of global cascades on random networks. *Proceedings of the National Academy of Sciences*, 99, 9, 5766–5771.
Watts, D. J. and Strogatz, S. H. (1998). Collective dynamics of small world networks. *Nature*, 393, 440–442.
Watts, D. J. and Dodds, P. (2007). Influentials, networks, and public opinion formation. *Journal of Consumer Research*, 34, 441–458.
Webster, R. (2012). *Feng Shui for Beginners.* Llewellyn Worldwide.
Weisbuch, G., Kirman, A., and Herreiner, D., 2000. Market organization. *Economica*, 110, 411–436.
Whitehead, D. (2008). The El Farol Bar problem revisited: reinforcement learning in a potential game. *ESE Discussion Papers*, 186. University of Edinburgh.
White, R. (2007). Immigrant-trade links, transplanted home bias and network effects. *Applied Economics*, 39(7), 839–852.
Wilhite, A. (2001). Bilateral trade and 'small-world' networks. *Computational Economics*, 18, 49–64.
Wolfram, S. (1983). Statistical mechanism of cellular automata. *Review of Modern Physics*, 55, 501–644.
Wolfram, S. (1994). *Cellular Automata and Complexity: Collected Papers.* Boulder, Colorado: Westview Press.
Wolfram, S. (2002). *A New Kind of Science*, Linois: Wolfram Media.
Young, H. P. (1998). *Individual Strategy and Social Structures.* New Jersey: Princeton University Press.
Young, H. P. (2009). Innovation diffusion in heterogeneous populations: Contagion, social influence, and social learning. *American Economic Review*, 99(5), 1899–1924.
Young, H. P. (2011). The dynamics of social innovation. *Proceedings of the National Academy of Sciences*, 108(9), 21285–21291.
Yuan, H. (1988). A bound on the spectral radius of graphs. *Linear Algebra and its Applications*, 108, 135–139.
Zelmer, J. (2003). Linear public goods experiments: A meta-analysis. *Experimental Economics*, 6(3), 299–310.
Zimmermann, M. G. and Eguiluz, V. M. (2005). Cooperation, social networks, and the emergence of leadership in a prisoner's dilemma with adaptive local interactions. *Physical Review E*, 72(5), 056118.
Zimmermann, M., Eguiluz, V., and San Miguel, M. (2004). Coevolution of dynamical states and interactions in dynamic networks. *Physical Review E*, 69(6), 065102.
Zschache, J. (2012). Producing public goods in networks: Some effects of social comparison and endogenous network change. *Social Networks*, 34(4), 539–548.

Index

T